T0258287

Advanced Control Systems Handbook

Advanced Control Systems Handbook

Edited by **Chester Mann**

New York

Published by NY Research Press,
23 West, 55th Street, Suite 816,
New York, NY 10019, USA
www.nyresearchpress.com

Advanced Control Systems Handbook
Edited by Chester Mann

© 2015 NY Research Press

International Standard Book Number: 978-1-63238-013-5 (Hardback)

Printed in the United States of America.

Contents

Permissions

List of Contributors

Preface

The world is advancing at a fast pace like never before. Therefore, the need is to keep up with the latest developments. This book was an idea that came to fruition when the specialists in the area realized the need to coordinate together and document essential themes in the subject. That's when I was requested to be the editor. Editing this book has been an honour as it brings together diverse authors researching on different streams of the field. The book collates essential materials contributed by veterans in the area which can be utilized by students and researchers alike.

A descriptive study of advanced control systems has been provided to the readers in this book. It comprehensively presents current developments in advanced control from both theoretical and practical outlook. The book highlights basic and advanced research outcomes and also talks about evolution in technical aspects of control theory. The book is an excellent source of interface between the theorists and the researchers of control therapy to remain updated about developments in each other's domain. It also aids growth of new control schemes and their usage. Furthermore, the book presents new pathways for students and practitioners to learn about the frontiers of control technology.

Each chapter is a sole-standing publication that reflects each author's interpretation. Thus, the book displays a multi-facetted picture of our current understanding of application, resources and aspects of the field. I would like to thank the contributors of this book and my family for their endless support.

Editor

Highlighted Aspects from Black Box Fuzzy Modeling for Advanced Control Systems Design

Ginalber Luiz de Oliveira Serra
Federal Institute of Education, Science and Technology
Laboratory of Computational Intelligence Applied to Technology, São Luis, Maranhão
Brazil

1. Introduction

This chapter presents an overview of a specific application of computational intelligence techniques, specifically, fuzzy systems: **fuzzy model based advanced control systems design**. In the last two decades, fuzzy systems have been useful for identification and control of complex nonlinear dynamical systems. This rapid growth, and the interest in this discussion is motivated by the fact that the practical control design, due to the presence of nonlinearity and uncertainty in the dynamical system, fuzzy models are capable of representing the dynamic behavior well enough so that the real controllers designed based on such models can garantee, mathematically, stability and robustness of the control system (Åström et al., 2001; Castillo-Toledo & Meda-Campaña, 2004; Kadmiry & Driankov, 2004; Ren & Chen, 2004; Tong & Li, 2002; Wang & Luoh, 2004; Yoneyama, 2004).

Automatic control systems have become an essential part of our daily life. They are applied in an electroelectronic equipment and up to even at most complex problem as aircraft and rockets. There are different control systems schemes, but in common, all of them have the function to handle a dynamic system to meet certain performance specifications. An intermediate and important control systems design step, is to obtain some knowledge of the plant to be controlled, this is, the dynamic behavior of the plant under different operating conditions. If such knowledge is not available, it becomes difficult to create an efficient control law so that the control system presents the desired performance. A practical approach for controllers design is from the mathematical model of the plant to be controlled.

Mathematical modeling is a set of heuristic and/or computational procedures properly established on a real plant in order to obtain a mathematical equation (models) to represent accurately its dynamic behavior in operation. There are three basic approaches for mathematical modeling:

- White box modeling. In this case, such models can be satisfactorily obtained from the physical laws governing the dynamic behavior of the plant. However, this may be a limiting factor in practice, considering plants with uncertainties, nonlinearities, time delay, parametric variations, among other dynamic complexity characteristics. The poor understanding of physical phenomena that govern the plant behavior and the resulting model complexity, makes the white box approach a difficult and time consuming task.

In addition, a complete understanding of the physical behavior of a real plant is almost impossible in many practical applications.

- `Black box modeling`. In this case, if such models, from the physical laws, are difficult or even impossible to obtain, is necessary the task of extracting a model from experimental data related to dynamic behavior of the plant. The modeling problem consists in choosing an appropriate structure for the model, so that enough information about the dynamic behavior of the plant can be extracted efficiently from the experimental data. Once the structure was determined, there is the parameters estimation problem so that a quadratic cost function of the approximation error between the outputs of the plant and the model is minimized. This problem is known as *systems identification* and several techniques have been proposed for linear and nonlinear plant modeling. A limitation of this approach is that the structure and parameters of the obtained models usually do not have physical meaning and they are not associated to physical variables of the plant.

- `Gray box modeling`. In this case some information on the dynamic behavior of the plant is available, but the model structure and parameters must be determined from experimental data. This approach, also known as hybrid modeling, combines the features of the white box and black box approaches.

The area of mathematical modeling covers topics from linear regression up to sofisticated concepts related to qualitative information from expert, and great attention have been given to this issue in the academy and industry (Abonyi et al., 2000; Brown & Harris, 1994; Pedrycz & Gomide, 1998; Wang, 1996). A mathematical model can be used for:

- Analysis and better understanding of phenomena (models in engineering, economics, biology, sociology, physics and chemistry);

- Estimate quantities from indirect measurements, where no sensor is available;

- Hypothesis testing (fault diagnostics, medical diagnostics and quality control);

- Teaching through simulators for aircraft, plants in the area of nuclear energy and patients in critical conditions of health;

- Prediction of behavior (adaptive control of time-varying plants);

- Control and regulation around some operating point, optimal control and robust control;

- Signal processing (cancellation of noise, filtering and interpolation);

Modeling techniques are widely used in the control systems design, and successful applications have appeared over the past two decades. There are cases in which the identification procedure is implemented in real time as part of the controller design. This technique, known as adaptive control, is suitable for nonlinear and/or time varying plants. In adaptive control schemes, the plant model, valid in several operating conditions is identified on-line. The controller is designed in accordance to current identified model, in order to garantee the performance specifications. There is a vast literature on modeling and control design (Åström & Wittenmark, 1995; Keesman, 2011; Sastry & Bodson, 1989; Isermann & Münchhof, 2011; Zhu, 2011; Chalam, 1987; Ioannou, 1996; Lewis & Syrmos, 1995; Ljung, 1999; Söderström & Stoica, 1989; Van Overschee & De Moor, 1996; Walter & Pronzato, 1997). Most approaches have a focus on models and controllers described by linear differential or finite differences equations, based on transfer functions or state space representation. Moreover, motivated by the fact that all plant present some type of nonlinear behavior, there are several approaches to analysis, modeling and control of nonlinear plants (Tee et al., 2011; Isidori, 1995; Khalil, 2002; Sjöberg et al., 1995; Ogunfunmi, 2007; Vidyasagar, 2002), and one of the key elements for these applications are the fuzzy systems (Lee et al., 2011; Hellendoorn & Driankov, 1997; Grigorie, 2010; Vukadinovic, 2011; Michels, 2006; Serra & Ferreira, 2011; Nelles, 2011).

2. Fuzzy inference systems

The theory of fuzzy systems has been proposed by Lotfi A. Zadeh (Zadeh, 1965; 1973), as a way of processing vague, imprecise or linguistic information, and since 1970 presents wide industrial application. This theory provides the basis for knowledge representation and developing the essential mechanisms to infer decisions about appropriate actions to be taken on a real problem. Fuzzy inference systems are typical examples of techniques that make use of human knowledge and deductive process. Its structure allows the mathematical modeling of a large class of dynamical behavior, in many applications, and provides greater flexibility in designing high-performance control with a certain degree of transparency for interpretation and analysis, that is, they can be used to explain solutions or be built from expert knowledge in a particular field of interest. For example, although it does not know the exact mathematical model of an oven, one can describe their behavior as follows: " **IF** is applied more power on the heater **THEN** the temperature increases", where **more** and **increases** are linguistic terms that, while imprecise, they are important information about the behavior of the oven. In fact, for many control problems, an expert can determine a set of efficient control rules based on linguistic descriptions of the plant to be controlled. Mathematical models can not incorporate the traditional linguistic descriptions directly into their formulations. Fuzzy inference systems are powerful tools to achieve this goal, since the logical structure of its **IF** <antecedent proposition> **THEN** <consequent proposition> rules facilitates the understanding and analysis of the problem in question. According to consequent proposition, there are two types of fuzzy inference systems:

- *Mamdani Fuzzy Inference Systems*: In this type of fuzzy inference system, the antecedent and consequent propositions are linguistic informations.

- *Takagi-Sugeno Fuzzy Inference Systems*: In this type of fuzzy inference system, the antecedent proposition is a linguistic information and the consequent proposition is a functional expression of the linguistic variables defined in the antecedent proposition.

2.1 Mamdani fuzzy inference systems

The Mamdani fuzzy inference system was proposed by E. H. Mamdani (Mamdani, 1977) to capture the qualitative knowledge available in a given application. Without loss of generality, this inference system presents a set of rules of the form:

$$\Re^i : \textbf{IF } \tilde{x}_1 \text{ is } F_{j|\tilde{x}_1}^i \textbf{ AND } \ldots \textbf{ AND } \tilde{x}_n \text{ is } F_{j|\tilde{x}_n}^i \textbf{ THEN } \tilde{y} \text{ is } G_{j|\tilde{y}}^i \tag{1}$$

In each rule i $|^{[i=1,2,\ldots,l]}$, where l is the number of rules, $\tilde{x}_1, \tilde{x}_2, \ldots, \tilde{x}_n$ are the linguistic variables of the antecedent (input) and $\tilde{y}$ is the linguistic variable of the consequent (output), defined, respectively, in the own universe of discourse $\mathcal{U}_{\tilde{x}_1}, \ldots, \mathcal{U}_{\tilde{x}_n}$ e $\mathcal{Y}$. The fuzzy sets $F^i_{j|\tilde{x}_1}, F^i_{j|\tilde{x}_2}, \ldots, F^i_{j|\tilde{x}_n}$ e $G^i_{j|\tilde{y}}$, are the linguistic values (terms) used to partition the unierse of discourse of the linguistic variables of antecedent and consequent in the inference system, that is, $F^i_{j|\tilde{x}_t} \in \{F^i_{1|\tilde{x}_t}, F^i_{2|\tilde{x}_t}, \ldots, F^i_{p_{\tilde{x}_t}|\tilde{x}_t}\}^{t=1,2,\ldots,n}$ and $G^i_{j|\tilde{y}} \in \{G^i_{1|\tilde{y}}, G^i_{2|\tilde{y}}, \ldots, G^i_{p_{\tilde{y}}|\tilde{y}}\}$, where $p_{\tilde{x}_t}$ and $p_{\tilde{y}}$ are the partitions number of the universes of discourses associated to the linguistic variables $\tilde{x}_t$ and $\tilde{y}$, respectively. The variable $\tilde{x}_t$ belongs to the fuzzy set $F^i_{j|\tilde{x}_t}$ with a value $\mu^i_{F_{j|\tilde{x}_t}}$ defined by the membership function $\mu^i_{\tilde{x}_t} : R \to [0,1]$, where $\mu^i_{F_{j|\tilde{x}_t}} \in \{\mu^i_{F_{1|\tilde{x}_t}}, \mu^i_{F_{2|\tilde{x}_t}}, \ldots, \mu^i_{F_{p_{\tilde{x}_t}|\tilde{x}_t}}\}$. The variable $\tilde{y}$ belongs to the fuzzy set $G^i_{j|\tilde{y}}$ with a value $\mu^i_{G_{j|\tilde{y}}}$ defined by the membership function $\mu^i_{\tilde{y}} : R \to [0,1]$ where $\mu^i_{G_{j|\tilde{y}}} \in \{\mu^i_{G_{1|\tilde{y}}}, \mu^i_{G_{2|\tilde{y}}}, \ldots, \mu^i_{G_{p_{\tilde{y}}|\tilde{y}}}\}$. Each rule is interpreted by a fuzzy implication

$$\mathfrak{R}^i : \mu^i_{F_{j|\tilde{x}_1}} \star \mu^i_{F_{j|\tilde{x}_2}} \star \ldots \star \mu^i_{F_{j|\tilde{x}_n}} \to \mu^i_{G_{j|\tilde{y}}} \tag{2}$$

where $\star$ is a T-norm, $\mu^i_{F_{j|\tilde{x}_1}} \star \mu^i_{F_{j|\tilde{x}_2}} \star \ldots \star \mu^i_{F_{j|\tilde{x}_n}}$ is the fuzzy relation between the linguistic inputs, on the universes of discourses $\mathcal{U}_{\tilde{x}_1} \times \mathcal{U}_{\tilde{x}_2} \times \ldots \times \mathcal{U}_{\tilde{x}_n}$, and $\mu^i_{G_{j|\tilde{y}}}$ is the linguistic output defined on the universe of discourse $\mathcal{Y}$. The Mamdani inference systems can represent MISO (Multiple Input and Single Output) systems directly, and the set of implications correspond to a unique fuzzy relation in $\mathcal{U}_{\tilde{x}_1} \times \mathcal{U}_{\tilde{x}_2} \times \ldots \times \mathcal{U}_{\tilde{x}_n} \times \mathcal{Y}$ of the form

$$\mathfrak{R}_{MISO} : \bigvee_{i=1}^{l} [\mu^i_{F_{j|\tilde{x}_1}} \star \mu^i_{F_{j|\tilde{x}_2}} \star \ldots \star \mu^i_{F_{j|\tilde{x}_n}} \star \mu^i_{G_{j|\tilde{y}}}] \tag{3}$$

where $\bigvee$ is a S-norm.

The fuzzy output $m |^{[m=1,2,\ldots,r]}$ is given by

$$G(\tilde{y}_m) = \mathfrak{R}_{MISO} \circ (\mu^i_{F_{j|\tilde{x}_1^*}} \star \mu^i_{F_{j|\tilde{x}_2^*}} \star \ldots \star \mu^i_{F_{j|\tilde{x}_n^*}}) \tag{4}$$

where $\circ$ is a inference based composition operator, which can be of the type *max-min* or *max-product*, and $\tilde{x}_t^*$ is any point in $\mathcal{U}_{x_t}$. The Mamdani inference systems can represent MIMO (Multiple Input and Multple Output) systems of r outputs by a set of r MISO sub-rules coupled base $\mathfrak{R}^j_{MISO} |^{[j=1,2,\ldots,l]}$, that is,

$$\boldsymbol{G}(\tilde{\boldsymbol{y}}) = \mathfrak{R}_{MIMO} \circ (\mu^i_{F_{j|\tilde{x}_1^*}} \star \mu^i_{F_{j|\tilde{x}_2^*}} \star \ldots \star \mu^i_{F_{j|\tilde{x}_n^*}}) \tag{5}$$

with $\boldsymbol{G}(\tilde{\boldsymbol{y}}) = [G(\tilde{y}_1), \ldots, G(\tilde{y}_r)]^T$ and

$$\mathfrak{R}_{MIMO} : \bigcup_{m=1}^{r} \{\bigvee_{i=1}^{l} [\mu^i_{F_{j|\tilde{x}_1}} \star \mu^i_{F_{j|\tilde{x}_2}} \star \ldots \star \mu^i_{F_{j|\tilde{x}_n}} \star \mu^i_{G_{j|\tilde{y}_m}}]\} \tag{6}$$

where the operator $\bigcup$ represents the set of all fuzzy relations $\mathfrak{R}^j_{MISO}$ associated to each output $\tilde{y}_m$.

2.2 Takagi-Sugeno fuzzy inference systems

The Takagi-Sugeno fuzzy inference system uses in the consequent proposition, a functional expression of the linguistic variables defined in the antecedent proposition (Takagi & Sugeno, 1985). Without loss of generality, the i $|^{[i=1,2,\ldots,l]}$-th rule of this inference system, where l is the maximum number of rules, is given by:

$$R^i : \text{ IF } \tilde{x}_1 \text{ is } F^i_{j|\tilde{x}_1} \text{ AND } \ldots \text{ AND } \tilde{x}_n \text{ is } F^i_{j|\tilde{x}_n} \text{ THEN } \tilde{y}_i = f_i(\tilde{x}) \tag{7}$$

The vector $\tilde{x} \in \Re^n$ contains the linguistic variables of the antecedent proposition. Each linguistic variable has its own universe of discourse $\mathcal{U}_{\tilde{x}_1}, \ldots, \mathcal{U}_{\tilde{x}_n}$ partitioned by fuzzy sets which represent the linguistic terms. The variable $\tilde{x}_t$ $|^{t=1,2,\ldots,n}$ belongs to the fuzzy set $F^i_{j|\tilde{x}_t}$ with value $\mu^i_{F_{j|\tilde{x}_t}}$ defined by a membership function $\mu^i_{\tilde{x}_t} : R \rightarrow [0,1]$, with $\mu^i_{F_{j|\tilde{x}_t}} \in \{\mu^i_{F_{1|\tilde{x}_t}}, \mu^i_{F_{2|\tilde{x}_t}}, \ldots, \mu^i_{F_{p_{\tilde{x}_t}|\tilde{x}_t}}\}$, where $p_{\tilde{x}_t}$ is the partitions number of the universe of discourse associated to the linguistic variable $\tilde{x}_t$. The activation degree h_i of the rule i is given by:

$$h_i(\tilde{x}) = \mu^i_{F_{j|\tilde{x}^*_1}} \star \mu^i_{F_{j|\tilde{x}^*_2}} \star \ldots \star \mu^i_{F_{j|\tilde{x}^*_n}} \tag{8}$$

where $\tilde{x}^*_t$ is any point in $\mathcal{U}_{\tilde{x}_t}$. The normalized activation degree of the rule i is defined as:

$$\gamma_i(\tilde{x}) = \frac{h_i(\tilde{x})}{\sum_{r=1}^l h_r(\tilde{x})} \tag{9}$$

This normalization implies that

$$\sum_{i=1}^l \gamma_i(\tilde{x}) = 1 \tag{10}$$

The response of the Takagi-Sugeno fuzzy inference system is a weighted sum of the functional expressions defined on the consequent proposition of each rule, that is, a convex combination of local functions f_i:

$$y = \sum_{i=1}^l \gamma_i(\tilde{x}) f_i(\tilde{x}) \tag{11}$$

Such inference system can be seen as linear parameter varying system. In this sense, the Takagi-Sugeno fuzzy inference system can be considered as a mapping from antecedent space (input) to the convex region (polytope) defined on the local functional expressions in the consequent space. This property allows the analysis of the Takagi-Sugeno fuzzy inference system as a robust system which can be applied in modeling and controllers design for complex plants.

3. Fuzzy computational modeling based control

Many human skills are learned from examples. Therefore, it is natural establish this "didactic principle" in a computer program, so that it can learn how to provide the desired output as function of a given input. The Computational intelligence techniques, basically derived from the theory of Fuzzy Systems, associated to computer programs, are able to process numerical data and/or linguistic information, whose parameters can be adjusted from examples. The examples represent what these systems should respond when subjected to a particular input. These techniques use a numeric representation of knowledge, demonstrate adaptability and fault tolerance in contrast to the classical theory of artificial intelligence that uses symbolic representation of knowledge. The human knowledge, in turn, can be classified into two categories:

1. *Objective knowledge*: This kind of knowledge is used in the engineering problems formulation and is defined by mathematical equations (mathematical model of a submarine, aircraft or robot; statistics analysis of the communication channel behaviour; Newton's laws for motion analysis and Kirchhoff's Laws for circuit analysis).

2. *Subjective knowledge*: This kind of knowledge represents the linguistic informations defined through set of rules, knowledge from expert and design specifications, which are usually impossible to be described quantitatively.

Fuzzy systems are able to coordinate both types of knowledge to solve real problems. The necessity of expert and engineers to deal with increasingly complex control systems problems, has enabled via computational intelligence techniques, the identification and control of real plants with difficult mathematical modeling. The computational intelligence techniques, once related to classical and modern control techniques, allow the use of constraints in its formulation and satisfaction of robustness and stability requirements in an efficient and practical form. The implementation of intelligent systems, especially from 70's, has been characterized by the growing need to improve the efficiency of industrial control systems in the following aspects: increasing product quality, reduced losses, and other factors related to the improvement of the disabilities of the identification and control methods. The intelligent identification and control methodologies are based on techniques motivated by biological systems, human intelligence, and have been introduced exploring alternative representations schemes from the natural language, rules, semantic networks or qualitative models.

The research on fuzzy inference systems has been developed in two main directions. The first direction is the linguistic or qualitative information, in which the fuzzy inference system is developed from a collection of rules (propositions). The second direction is the quantitative information and is related to the theory of classical and modern systems. The combination of the qualitative and quantitative informations, which is the main motivation for the use of intelligent systems, has resulted in several contributions on stability and robustness of advanced control systems. In (Ding, 2011) is addressed the output feedback predictive control for a fuzzy system with bounded noise. The controller optimizes an infinite-horizon objective function respecting the input and state constraints. The control law is parameterized as a dynamic output feedback that is dependent on the membership functions, and the closed-loop stability is specified by the notion of quadratic boundedness. In (Wang et al., 2011) is considered the problem of fuzzy control design for a class of nonlinear distributed parameter systems that is described by first-order hyperbolic partial differential equations (PDEs), where the control actuators are continuously distributed in space. The goal of this methodology is to develop a fuzzy state-feedback control design methodology for these systems by employing a combination of PDE theory and concepts from Takagi-Sugeno fuzzy control. First, the Takagi-Sugeno fuzzy hyperbolic PDE model is proposed to accurately represent the nonlinear first-order hyperbolic PDE system. Subsequently, based on the Takagi-Sugeno fuzzy-PDE model, a Lyapunov technique is used to analyze the closed-loop exponential stability with a given decay rate. Then, a fuzzy state-feedback control design procedure is developed in terms of a set of spatial differential linear matrix inequalities (SDLMIs) from the resulting stability conditions. The developed design methodology is successfully applied to the control of a nonisothermal plug-flow reactor. In (Sadeghian & Fatehi, 2011) is used a nonlinear system identification method to predict and detect process fault of a cement rotary kiln from the White Saveh Cement Company. After selecting proper inputs and output, an inputŨoutput locally linear neuro-fuzzy (LLNF) model is identified for the plant in various operation points in the kiln. In (Li & Lee, 2011) an observer-based adaptive controller is developed from a hierarchical fuzzy-neural network (HFNN) is employed to solve the controller time-delay problem for a class of multi-input multi-output(MIMO) non-affine nonlinear systems under the constraint that only system outputs are available for measurement. By using the implicit function theorem and Taylor series expansion, the observer-based control law and the weight update law of the HFNN adaptive controller are derived. According to the design of the HFNN hierarchical fuzzy-neural network, the observer-based adaptive controller can alleviate the online computation burden and can guarantee that all signals involved are bounded and that the outputs of the closed-loop system track asymptotically the desired output trajectories.

Fuzzy inference systems are widely found in the following areas: Control Applications - aircraft (Rockwell Corp.), cement industry and motor/valve control (Asea Brown Boveri Ltd.), water treatment and robots control (Fuji Electric), subway system (Hitachi), board control (Nissan), washing machines (Matsushita, Hitachi), air conditioning system (Mitsubishi); Medical Technology - cancer diagnosis (Kawasaki medical School); Modeling and Optimization - prediction system for earthquakes recognition (Institute of Seismology Bureau of Metrology, Japan); Signal Processing For Adjustment and Interpretation - vibration compensation in video camera (Matsushita), video image stabilization (Matsushita / Panasonic), object and voice recognition (CSK, Hitachi Hosa Univ., Ricoh), adjustment of images on TV (Sony). Due to the development, the many practical possibilities and the commercial success of their applications, the theory of fuzzy systems have a wide acceptance in academic community as well as industrial applications for modeling and advanced control systems design.

4. Takagi-Sugeno fuzzy black box modeling

This section aims to illustrate the problem of black box modeling, well known as systems identification, addressing the use of Takagi-Sugeno fuzzy inference systems. The nonlinear input-output representation is often used for building TS fuzzy models from data, where the regression vector is represented by a finite number of past inputs and outputs of the system. In this work, the nonlinear autoregressive with exogenous input (NARX) structure model is used. This model is applied in most nonlinear identification methods such as neural networks, radial basis functions, cerebellar model articulation controller (CMAC), and also fuzzy logic. The NARX model establishes a relation between the collection of past scalar input-output data and the predicted output

$$y_{k+1} = F[y_k, \ldots, y_{k-n_y+1}, u_k, \ldots, \ldots, u_{k-n_u+1}] \tag{12}$$

where k denotes discrete time samples, n_y and n_u are integers related to the system's order. In terms of rules, the model is given by

$$R^i : \text{IF } y_k \text{ is } F_1^i \text{ AND } \cdots \text{ AND } y_{k-n_y+1} \text{ is } F_{n_y}^i \text{ AND } u_k \text{ is } G_1^i \text{ AND } \cdots \text{ AND } u_{k-n_u+1} \text{ is } G_{n_u}^i$$

$$\text{THEN } \hat{y}_{k+1}^i = \sum_{j=1}^{n_y} a_{i,j} y_{k-j+1} + \sum_{j=1}^{n_u} b_{i,j} u_{k-j+1} + c_i \tag{13}$$

where $a_{i,j}$, $b_{i,j}$ and c_i are the consequent parameters to be determined. The inference formula of the TS fuzzy model is a straightforward extension of (11) and is given by

$$y_{k+1} = \frac{\sum_{i=1}^{l} h_i(\boldsymbol{x}) \hat{y}_{k+1}^i}{\sum_{i=1}^{l} h_i(\boldsymbol{x})} \tag{14}$$

or

$$y_{k+1} = \sum_{i=1}^{l} \gamma_i(\boldsymbol{x}) \hat{y}_{k+1}^i \tag{15}$$

with

$$\boldsymbol{x} = [y_k, \ldots, y_{k-n_y+1}, u_k, \ldots, u_{k-n_u+1}] \tag{16}$$

and $h_i(\boldsymbol{x})$ is given as (8). This NARX model represents multiple input and single output (MISO) systems directly and multiple input and multiple output (MIMO) systems in a decomposed form as a set of coupled MISO models.

4.1 Antecedent parameters estimation problem

The experimental data based antecedent parameters estimation can be done by fuzzy clustring algorithms. A cluster is a group of similar objects. The term "similarity" should be understood as mathematical similarity measured in some well-define sense. In metric spaces, similarity is often defined by means of a distance norm. Distance can be measured from data vector to some cluster prototypical (center). Data can reveal clusters of different geometric shapes, sizes and densities. The clusters also can be characterized as linear and nonlinear subspaces of the data space.

The objective of clustering is partitioning the data set Z into c clusters. Assume that c is known, based on priori knowledge. The fuzzy partition of Z can be defined as a family of subsets $\{A_i | 1 \leq i \leq c\} \subset P(Z)$, with the following properties:

$$\bigcup_{i=1}^{c} A_i = Z \tag{17}$$

$$A_i \cap A_j = 0 \tag{18}$$

$$0 \subset A_i \subset Z_i \tag{19}$$

Equation (17) means that the subsets A_i collectively contain all the data in Z. The subsets must be disjoint, as stated by (18), and none off them is empty nor contains all the data in Z, as stated by (19). In terms of membership functions, μ_{A_i} is the membership function of A_i. To simplifly the notation, in this paper is used μ_{ik} instead of $\mu_i(z_k)$. The $c \times N$ matrix $U = [\mu_{ik}]$ represents a fuzzy partitioning space if and only if:

$$M_{fc} = \left\{ U \in \Re^{c \times N} | \mu_{ik} \in [0,1], \forall i, k; \sum_{i=1}^{c} \mu_{ik} = 1, \forall k; 0 < \sum_{k=1}^{N} \mu_{ik} < N, \forall i \right\} \tag{20}$$

The i-th row of the fuzzy partition matrix U contains values of the i-th membership function of the fuzzy subset A_i of Z. The clustering algorithm optimizes an initial set of centroids by minimizing a cost function J in an iterative process. This function is usually formulated as:

$$J(Z; U, V, A) = \sum_{i=1}^{c} \sum_{k=1}^{N} \mu_{ik}^{m} D_{ikA_i}^{2} \tag{21}$$

where, $Z = \{z_1, z_2, \cdots, z_N\}$ is a finite data set. $U = [\mu_{ik}] \in M_{fc}$ is a fuzzy partition of Z. $V = \{v_1, v_2, \cdots, v_c\}$, $v_i \in \Re^n$, is a vector of cluster prototypes (centers). A denote a c-tuple of the norm-induting matrices: $A = (A_1, A_2, \cdots, A_c)$. $D_{ikA_i}^{2}$ is a square inner-product distance norm. The $m \in [1, \infty)$ is a weighting exponent which determines the fuzziness of the clusters. The clustering algorithms differ in the choice of the norm distance. The norm metric influences the clustering criterion by changing the measure of dissimilarity. The Euclidean norm induces hiperspherical clusters. It's characterizes the FCM algorithm, where the norm-inducing matrix $A_{i_{FCM}}$ is equal to identity matrix ($A_{i_{FCM}} = I$), which strictly imposes a circular shape to all clusters. The Euclidean Norm is given by:

$$D_{ik_{FCM}}^{2} = (z_k - v_i)^T A_{i_{FCM}} (z_k - v_i) \tag{22}$$

An adaptative distance norm in order to detect clusters of different geometrical shapes in a data set characterizes the *GK algorithm*:

$$D_{ik_{GK}}^{2} = (z_k - v_i)^T A_{i_{GK}} (z_k - v_i) \tag{23}$$

In this algorithm, each cluster has its own norm-inducing matrix $A_{i_{GK}}$, where each cluster adapts the distance norm to the local topological structure of the data set. $A_{i_{GK}}$ is given by:

$$A_{i_{GK}} = [\rho_i det(F_i)]^{1/n} F_i^{-1}, \tag{24}$$

where ρ_i is cluster volume, usually fixed in 1. The n is data dimension. The F_i is fuzzy covariance matrix of the i-th cluster defined by:

$$F_i = \frac{\sum_{k=1}^{N} (\mu_{ik})^m (z_k - v_i)(z_k - v_i)^T}{\sum_{k=1}^{N} (\mu_{ik})^m} \tag{25}$$

The eigenstructure of the cluster covariance matrix provides information about the shape and orientation cluster. The ratio of the hyperellipsoid axes is given by the ratio of the square roots of the eigenvalues of F_i. The directions of the axes are given by the eigenvectores of F_i. The eigenvector corresponding to the smallest eigenvalue determines the normal to the hyperplane, and it can be used to compute optimal local linear models from the covariance matrix. The fuzzy maximum likelihood estimates (FLME) algorithm employs a distance norm based on maximum lekelihood estimates:

$$D_{ik_{FLME}} = \frac{\sqrt{G_{i_{FLME}}}}{P_i} \exp\left[\frac{1}{2}(z_k - v_i)^T F_{i_{FLME}}^{-1}(z_k - v_i)\right] \tag{26}$$

Note that, contrary to the GK algorithm, this distance norm involves an exponential term and decreases faster than the inner-product norm. The $F_{i_{FLME}}$ denotes the fuzzy covariance matrix of the i-th cluster, given by (25). When m is equal to 1, it has a strict algorithm FLME. If m is greater than 1, it has a extended algorithm FLME, or Gath-Geva (GG) algorithm. Gath and Geva reported that the FLME algorithm is able to detect clusters of varying shapes, sizes and densities. This is because the cluster covariance matrix is used in conjuncion with an "exponential" distance, and the clusters are not constrained in volume. P_i is the prior probability of selecting cluster i, given by:

$$P_i = \frac{1}{N} \sum_{k=1}^{N} (\mu_{ik})^m \tag{27}$$

4.2 Consequent parameters estimation problem

The inference formula of the TS fuzzy model in (15) can be expressed as

$$
\begin{aligned}
y_{k+1} = \ &\gamma_1(x_k)[a_{1,1}y_k + \ldots + a_{1,ny}y_{k-n_y+1} + b_{1,1}u_k + \ldots + b_{1,nu}u_{k-n_u+1} + c_1] + \\
&\gamma_2(x_k)[a_{2,1}y_k + \ldots + a_{2,ny}y_{k-n_y+1} + b_{2,1}u_k + \ldots + b_{2,nu}u_{k-n_u+1} + c_2] + \\
&\vdots \\
&\gamma_l(x_k)[a_{l,1}y_k + \ldots + a_{l,ny}y_{k-n_y+1} + b_{l,1}u_k + \ldots + b_{l,nu}u_{k-n_u+1} + c_l]
\end{aligned}
\tag{28}
$$

which is linear in the consequent parameters: a, b and c. For a set of N input-output data pairs $\{(x_k, y_k) | i = 1, 2, \ldots, N\}$ available, the following vetorial form is obtained

$$Y = [\psi_1 X, \psi_2 X, \ldots, \psi_l X]\theta + \Xi \tag{29}$$

where $\psi_i = diag(\gamma_i(x_k)) \in \Re^{N \times N}$, $X = [y_k, \ldots, y_{k-ny+1}, u_k, \ldots, u_{k-nu+1}, 1] \in \Re^{N \times (n_y + n_u + 1)}$, $Y \in \Re^{N \times 1}$, $\Xi \in \Re^{N \times 1}$ and $\theta \in \Re^{l(n_y + n_u + 1) \times 1}$ are the normalized membership degree matrix of (9), the data matrix, the output vector, the approximation error vector and the estimated parameters vector, respectively. If the unknown parameters associated variables are *exactly known* quantities, then the least squares method can be used efficiently. However, in practice, and in the present context, the elements of X are no exactly known quantities so that its value can be expressed as

$$y_k = \chi_k^T \theta + \eta_k \tag{30}$$

where, at the k-th sampling instant, $\chi_k^T = [\gamma_k^1(x_k + \xi_k), \ldots, \gamma_k^l(x_k + \xi_k)]$ is the vector of the data with error in variables, $x_k = [y_{k-1}, \ldots, y_{k-n_y}, u_{k-1}, \ldots, u_{k-n_u}, 1]^T$ is the vector of the data with exactly known quantities, e.g., free noise input-output data, ξ_k is a vector of noise associated with the observation of x_k, and η_k is a disturbance noise.

The normal equations are formulated as

$$[\sum_{j=1}^{k} \chi_j \chi_j^T]\hat{\theta}_k = \sum_{j=1}^{k} \chi_j y_j \tag{31}$$

and multiplying by $\frac{1}{k}$ gives

$$\{\frac{1}{k} \sum_{j=1}^{k} [\gamma_j^1(x_j + \xi_j), \ldots, \gamma_j^l(x_j + \xi_j)][\gamma_j^1(x_j + \xi_j), \ldots, \gamma_j^l(x_j + \xi_j)]^T\}\hat{\theta}_k =$$
$$\frac{1}{k} \sum_{j=1}^{k} [\gamma_j^1(x_j + \xi_j), \ldots, \gamma_j^l(x_j + \xi_j)]y_j \tag{32}$$

Noting that $y_j = \chi_j^T \theta + \eta_j$,

$$\{\frac{1}{k} \sum_{j=1}^{k} [\gamma_j^1(x_j + \xi_j), \ldots, \gamma_j^l(x_j + \xi_j)][\gamma_j^1(x_j + \xi_j), \ldots, \gamma_j^l(x_j + \xi_j)]^T\}\hat{\theta}_k =$$
$$\frac{1}{k} \sum_{j=1}^{k} [\gamma_j^1(x_j + \xi_j), \ldots, \gamma_j^l(x_j + \xi_j)][\gamma_j^1(x_j + \xi_j), \ldots, \gamma_j^l(x_j + \xi_j)]^T \theta + \frac{1}{k} \sum_{j=1}^{k} [\gamma_j^1(x_j +$$
$$\xi_j), \ldots, \gamma_j^l(x_j + \xi_j)]\eta_j \tag{33}$$

and

$$\tilde{\theta}_k = \{\frac{1}{k} \sum_{j=1}^{k} [\gamma_j^1(x_j + \xi_j), \ldots, \gamma_j^l(x_j + \xi_j)][\gamma_j^1(x_j + \xi_j), \ldots, \gamma_j^l(x_j + \xi_j)]^T\}^{-1} \frac{1}{k} \sum_{j=1}^{k} [\gamma_j^1(x_j +$$
$$\xi_j), \ldots, \gamma_j^l(x_j + \xi_j)]\eta_j \tag{34}$$

where $\tilde{\theta}_k = \hat{\theta}_k - \theta$ is the parameter error. Taking the probability in the limit as $k \to \infty$,

$$\text{p.lim } \tilde{\theta}_k = \text{p.lim } \{\frac{1}{k}C_k^{-1}\frac{1}{k}b_k\} \tag{35}$$

with

$$C_k = \sum_{j=1}^{k} [\gamma_j^1(x_j + \xi_j), \ldots, \gamma_j^l(x_j + \xi_j)][\gamma_j^1(x_j + \xi_j), \ldots, \gamma_j^l(x_j + \xi_j)]^T$$

$$b_k = \sum_{j=1}^{k} [\gamma_j^1(x_j + \xi_j), \ldots, \gamma_j^l(x_j + \xi_j)]\eta_j$$

Applying Slutsky's theorem and assuming that the elements of $\frac{1}{k}C_k$ and $\frac{1}{k}b_k$ converge in probability, we have

$$\text{p.lim } \tilde{\theta}_k = \text{p.lim } \frac{1}{k}C_k^{-1} \text{ p.lim } \frac{1}{k}b_k \tag{36}$$

Thus,

$$\text{p.lim } \frac{1}{k}C_k = \text{p.lim } \frac{1}{k}\sum_{j=1}^{k}[\gamma_j^1(x_j + \xi_j), \ldots, \gamma_j^l(x_j + \xi_j)][\gamma_j^1(x_j + \xi_j), \ldots, \gamma_j^l(x_j + \xi_j)]^T$$

$$\text{p.lim } \frac{1}{k}C_k = \text{p.lim } \frac{1}{k}\sum_{j=1}^{k}(\gamma_j^1)^2(x_j + \xi_j)(x_j + \xi_j)^T + \ldots + \text{p.lim } \frac{1}{k}\sum_{j=1}^{k}(\gamma_j^l)^2(x_j + \xi_j)(x_j + \xi_j)^T$$

Assuming x_j and ξ_j statistically independent,

$$\text{p.lim } \frac{1}{k}C_k = \text{p.lim } \frac{1}{k}\sum_{j=1}^{k}(\gamma_j^1)^2[x_jx_j^T + \xi_j\xi_j^T] + \ldots + \text{p.lim } \frac{1}{k}\sum_{j=1}^{k}(\gamma_j^l)^2[x_jx_j^T + \xi_j\xi_j^T]$$

$$\text{p.lim } \frac{1}{k}C_k = \text{p.lim } \frac{1}{k}\sum_{j=1}^{k}x_jx_j^T[(\gamma_j^1)^2 + \ldots + (\gamma_j^l)^2] + \text{p.lim } \frac{1}{k}\sum_{j=1}^{k}\xi_j\xi_j^T[(\gamma_j^1)^2 + \ldots + (\gamma_j^l)^2] \tag{37}$$

with $\sum_{i=1}^{l}\gamma_j^i = 1$. Hence, the asymptotic analysis of the TS fuzzy model consequent parameters estimation is based in a weighted sum of the fuzzy covariance matrices of x and ξ. Similarly,

$$\text{p.lim } \frac{1}{k}b_k = \text{p.lim } \frac{1}{k}\sum_{j=1}^{k}[\gamma_j^1(x_j + \xi_j), \ldots, \gamma_j^l(x_j + \xi_j)]\eta_j$$

$$\text{p.lim } \frac{1}{k}b_k = \text{p.lim } \frac{1}{k}\sum_{j=1}^{k}[\gamma_j^1\xi_j\eta_j, \ldots, \gamma_j^l\xi_j\eta_j] \tag{38}$$

Substituting from (37) and (38) in (36), results

$$\text{p.lim } \tilde{\theta}_k = \{\text{p.lim } \frac{1}{k}\sum_{j=1}^{k}x_jx_j^T[(\gamma_j^1)^2 + \ldots + (\gamma_j^l)^2] + \text{p.lim } \frac{1}{k}\sum_{j=1}^{k}\xi_j\xi_j^T[(\gamma_j^1)^2 + \ldots$$

$$+ (\gamma_j^l)^2]\}^{-1}\text{p.lim } \frac{1}{k}\sum_{j=1}^{k}[\gamma_j^1\xi_j\eta_j, \ldots, \gamma_j^l\xi_j\eta_j] \tag{39}$$

with $\sum_{i=1}^{l}\gamma_j^i = 1$. For the case of only one rule ($l = 1$), the analysis is simplified to the linear one, with $\gamma_j^i \mid_{j=1,\ldots,k}^{i=1} = 1$. Thus, this analysis, which is a contribution of this article, is an extension of the standard linear one, from which can result several studies for fuzzy filtering and modeling in a noisy environment, fuzzy signal enhancement in communication channel, and so forth.

Provided that the input u_k continues to excite the process and, at the same time, the coefficients in the submodels from the consequent are not all zero, then the output y_k will exist for all k observation intervals. As a result, the fuzzy covariance matrix $\sum_{j=1}^{k} x_j x_j^T [(\gamma_j^1)^2 + \ldots + (\gamma_j^l)^2]$ will also be non-singular and its inverse will exist. Thus, the only way in which the asymptotic error can be zero is for $\xi_j \eta_j$ identically zero. But, in general, ξ_j and η_j are correlated, the asymptotic error will not be zero and the least squares estimates will be asymptotically biased to an extent determined by the relative ratio of noise to signal variances. In other words, least squares method is not appropriate to estimate the TS fuzzy model consequent parameters in a noisy environment because the estimates will be inconsistent and the bias error will remain no matter how much data can be used in the estimation.

As a consequence of this analysis, the definition of the vector $[\beta_j^1 z_j, \ldots, \beta_j^l z_j]$ as *fuzzy instrumental variable vector* or simply the *fuzzy instrumental variable* (FIV) is proposed. Clearly, with the use of the FIV vector in the form suggested, becomes possible to eliminate the asymptotic bias while preserving the existence of a solution. However, the statistical efficiency of the solution is dependent on the degree of correlation between $[\beta_j^1 z_j, \ldots, \beta_j^l z_j]$ and $[\gamma_j^1 x_j, \ldots, \gamma_j^l x_j]$. In particular, the lowest variance estimates obtained from this approach occur only when $z_j = x_j$ and $\beta_j^i |_{j=1,\ldots,k}^{i=1,\ldots,l} = \gamma_j^i |_{j=1,\ldots,k}^{i=1,\ldots,l}$, i.e., when the z_j are equal to the dynamic system "free noise" variables, which are unavailable in practice. According to situation, several fuzzy instrumental variables can be chosen. An effective choice of FIV would be the one based on the delayed input sequence

$$z_j = [u_{k-\tau}, \ldots, u_{k-\tau-n}, u_k, \ldots, u_{k-n}]^T$$

where τ is chosen so that the elements of the fuzzy covariance matrix C_{zx} are maximized. In this case, the input signal is considered persistently exciting, e.g., it continuously perturbs or excites the system. Another FIV would be the one based on the delayed input-output sequence

$$z_j = [y_{k-1-dl}, \cdots, y_{k-n_y-dl}, u_{k-1-dl}, \cdots, u_{k-n_u-dl}]^T$$

where dl is the applied delay. Other FIV could be the one based in the input-output from a "fuzzy auxiliar model" with the same structure of the one used to identify the nonlinear dynamic system. Thus,

$$z_j = [\hat{y}_{k-1}, \cdots, \hat{y}_{k-n_y}, u_{k-1}, \cdots, u_{k-n_u}]^T$$

where $\hat{y}_k$ is the output of the fuzzy auxiliar model, and u_k is the input of the dynamic system. The inference formula of this fuzzy auxiliar model is given by

$$
\begin{aligned}
\hat{y}_{k+1} = &\ \beta_1(z_k)[\alpha_{1,1}\hat{y}_k + \ldots + \alpha_{1,ny}\hat{y}_{k-n_y+1} + \rho_{1,1}u_k + \ldots + \rho_{1,nu}u_{k-n_u+1} + \delta_1] + \\
&\ \beta_2(z_k)[\alpha_{2,1}\hat{y}_k + \ldots + \alpha_{2,ny}\hat{y}_{k-n_y+1} + \rho_{2,1}u_k + \ldots + \rho_{2,nu}u_{k-n_u+1} + \delta_2] + \\
&\ \vdots \\
&\ \beta_l(z_k)[\alpha_{l,1}\hat{y}_k + \ldots + \alpha_{l,ny}\hat{y}_{k-n_y+1} + \rho_{l,1}u_k + \ldots + \rho_{l,nu}u_{k-n_u+1} + \delta_l]
\end{aligned}
\tag{40}
$$

which is also linear in the consequent parameters: α, ρ and δ. The closer these parameters are to the actual, but unknown, system parameters (a, b, c), more correlated z_k and x_k will be, and the obtained FIV estimates closer to the optimum.

4.2.1 Batch processing scheme

The normal equations are formulated as

$$\sum_{j=1}^{k}[\beta_j^1 z_j,\ldots,\beta_j^l z_j][\gamma_j^1(x_j+\xi_j),\ldots,\gamma_j^l(x_j+\xi_j)]^T\hat{\theta}_k - \sum_{j=1}^{k}[\beta_j^1 z_j,\ldots,\beta_j^l z_j]y_j = 0 \quad (41)$$

or, with $\zeta_j = [\beta_j^1 z_j,\ldots,\beta_j^l z_j]$,

$$[\sum_{j=1}^{k}\zeta_j\chi_j^T]\hat{\theta}_k - \sum_{j=1}^{k}\zeta_j y_j = 0 \quad (42)$$

so that the FIV estimate is obtained as

$$\hat{\theta}_k = \{\sum_{j=1}^{k}[\beta_j^1 z_j,\ldots,\beta_j^l z_j][\gamma_j^1(x_j+\xi_j),\ldots,\gamma_j^l(x_j+\xi_j)]^T\}^{-1}\sum_{j=1}^{k}[\beta_j^1 z_j,\ldots,\beta_j^l z_j]y_j \quad (43)$$

and, in vectorial form, the interest problem may be placed as

$$\hat{\theta} = (\Gamma^T\Sigma)^{-1}\Gamma^T Y \quad (44)$$

where $\Gamma^T \in \Re^{l(n_y+n_u+1)\times N}$ is the fuzzy extended instrumental variable matrix with rows given by ζ_j, $\Sigma \in \Re^{N\times l(n_y+n_u+1)}$ is the fuzzy extended data matrix with rows given by χ_j and $Y \in \Re^{N\times 1}$ is the output vector and $\hat{\theta} \in \Re^{l(n_y+n_u+1)\times 1}$ is the parameters vector. The models can be obtained by the following two approaches:

- *Global approach* : In this approach all linear consequent parameters are estimated simultaneously, minimizing the criterion:

$$\hat{\theta} = \arg\min \| \Gamma^T\Sigma\theta - \Gamma^T Y \|_2^2 \quad (45)$$

- *Local approach* : In this approach the consequent parameters are estimated for each rule i, and hence independently of each other, minimizing a set of weighted local criteria ($i = 1,2,\ldots,l$):

$$\hat{\theta}_i = \arg\min \| Z^T\Psi_i X\theta_i - Z^T\Psi_i Y \|_2^2 \quad (46)$$

where Z^T has rows given by z_j and Ψ_i is the normalized membership degree diagonal matrix according to z_j.

Example 1. So that the readers can understand the definitions of global and local fuzzy modeling estimations, consider the following second-order polynomial given by

$$y = 2u_k^2 - 4u_k + 3 \quad (47)$$

where u_k is the input and y_k is the output, respectively. The TS fuzzy model used to approximate this polynomial has the following structure with 2 rules:

$$R^i : \text{IF } u_k \text{ is } F_i \text{ THEN } \hat{y}_k = a_0 + a_1 u_k$$

where $i = 1, 2$. It was choosen the points $u_k = -0.5$ and $u_k = 0.5$ to analysis the consequent models obtained by global and local estimation, and it was defined triangular membership functions for $-0.5 \leq u_k \leq 0.5$ in the antecedent. The following rules were obtained:

Local estimation:

$$R^1 : \text{IF } u_k \text{ is } -0.5 \text{ THEN } \hat{y} = 3.1000 - 4.4012 u_k$$
$$R^2 : \text{IF } u_k \text{ is } +0.5 \text{ THEN } \hat{y} = 3.1000 - 3.5988 u_k$$

Global estimation:

$$R^1 : \text{IF } u_k \text{ is } -0.5 \text{ THEN } \hat{y} = 4.6051 - 1.7503 u_k$$
$$R^2 : \text{IF } u_k \text{ is } +0.5 \text{ THEN } \hat{y} = 1.3464 + 0.3807 u_k$$

The application of local and global estimation to the TS fuzzy model results in the consequent models given in Fig. 1. The consequent models obtained by local estimation describe properly the local behavior of the function and the fuzzy model can easily be interpreted in terms of the local behavior (the rule consequents). The consequent models obtained by global estimation are not relevant for the local behavior of the nonlinear function. The fit of the function is

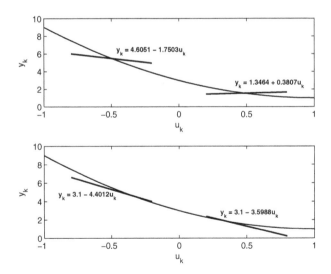

Fig. 1. The nonlinear function and the result of global (top) and local (bottom) estimation of the consequent parameters of the TS fuzzy models.

shown in Fig. 2. The global estimation gives a good fit and a minimal prediction error, but it bias the estimates of the consequent as parameters of local models. In the local estimation a larger prediction error is obtained than with global estimation, but it gives locally relevant parameters of the consequent. This is the tradeoff between local and global estimation. All the results of the Example 1 can be extended for any nonlinear estimation problem and they would be considered for computational and experimental results analysis in this paper.

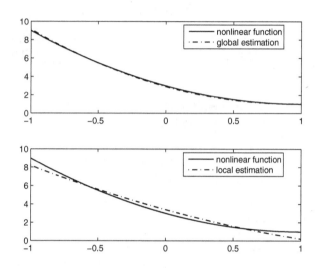

Fig. 2. The nonlinear function approximation result by global (top) and local (bottom) estimation of the consequent parameters of the TS fuzzy models.

4.2.2 Recursive processing scheme

An on line FIV scheme can be obtained by utilizing the recursive solution to the FIV equations and then updating the fuzzy auxiliar model continuously on the basis of these recursive consequent parameters estimates. The FIV estimate in (43) can take the form

$$\hat{\theta}_k = P_k b_k \tag{48}$$

where

$$P_k = \sum_{j=1}^{k} [\beta_j^1 z_j, \dots, \beta_j^l z_j][\gamma_j^1(x_j + \xi_j), \dots, \gamma_j^l(x_j + \xi_j)]^T\}^{-1}$$

and

$$b_k = \sum_{j=1}^{k} [\beta_j^1 z_j, \dots, \beta_j^l z_j] y_j$$

which can be expressed as

$$P_k^{-1} = P_{k-1}^{-1} + [\beta_k^1 z_k, \dots, \beta_j^l z_k][\gamma_j^1(x_k + \xi_k), \dots, \gamma_k^l(x_k + \xi_k)]^T \tag{49}$$

and

$$b_k = b_{k-1} + [\beta_k^1 z_k, \dots, \beta_k^l z_k] y_k \tag{50}$$

respectively. Pre-multiplying (49) by P_k and post-multiplying by P_{k-1} gives

$$P_{k-1} = P_k + P_k[\beta_k^1 z_k, \ldots, \beta_j^l z_k][\gamma_j^1(x_k + \xi_k), \ldots, \gamma_k^l(x_k + \xi_k)]^T P_{k-1} \tag{51}$$

then post-multiplying (51) by the FIV vector $[\beta_j^1 z_j, \ldots, \beta_j^l z_j]$, results

$$P_{k-1}[\beta_k^1 z_k, \ldots, \beta_j^l z_k] = P_k[\beta_k^1 z_k, \ldots, \beta_j^l z_k] + P_k[\beta_k^1 z_k, \ldots, \beta_j^l z_k][\gamma_j^1(x_k + \xi_k), \ldots, \gamma_k^l(x_k+$$
$$\xi_k)]^T P_{k-1}[\beta_k^1 z_k, \ldots, \beta_j^l z_k] \tag{52}$$

$$P_{k-1}[\beta_k^1 z_k, \ldots, \beta_j^l z_k] = P_k[\beta_k^1 z_k, \ldots, \beta_j^l z_k]\{1 + [\gamma_j^1(x_k + \xi_k), \ldots, \gamma_k^l(x_k + \xi_k)]^T P_{k-1}$$
$$[\beta_k^1 z_k, \ldots, \beta_j^l z_k]\} \tag{53}$$

Then, post-multiplying by

$$\{1 + [\gamma_j^1(x_k + \xi_k), \ldots, \gamma_k^l(x_k + \xi_k)]^T P_{k-1}[\beta_k^1 z_k, \ldots, \beta_j^l z_k]\}^{-1}[\gamma_j^1(x_k + \xi_k), \ldots,$$
$$\gamma_k^l(x_k + \xi_k)]^T P_{k-1} \tag{54}$$

we obtain

$$P_{k-1}[\beta_k^1 z_k, \ldots, \beta_j^l z_k]\{1 + [\gamma_j^1(x_k + \xi_k), \ldots, \gamma_k^l(x_k + \xi_k)]^T P_{k-1}[\beta_k^1 z_k, \ldots, \beta_j^l z_k]\}^{-1}$$
$$[\gamma_j^1(x_k + \xi_k), \ldots, \gamma_k^l(x_k + \xi_k)]^T P_{k-1} =$$
$$P_k[\beta_k^1 z_k, \ldots, \beta_j^l z_k][\gamma_j^1(x_k + \xi_k), \ldots, \gamma_k^l(x_k + \xi_k)]^T P_{k-1} \tag{55}$$

Substituting (51) in (55), we have

$$P_k = P_{k-1} - P_{k-1}[\beta_k^1 z_k, \ldots, \beta_j^l z_k]\{1 + [\gamma_j^1(x_k + \xi_k), \ldots, \gamma_k^l(x_k + \xi_k)]^T P_{k-1}$$
$$[\beta_k^1 z_k, \ldots, \beta_j^l z_k]\}^{-1}[\gamma_j^1(x_k + \xi_k), \ldots, \gamma_k^l(x_k + \xi_k)]^T P_{k-1} \tag{56}$$

Substituting (56) and (50) in (48), the recursive consequent parameters estimates will be:

$$\hat{\theta}_k = \{P_{k-1} - P_{k-1}[\beta_k^1 z_k, \ldots, \beta_j^l z_k]\{1 + [\gamma_j^1(x_k + \xi_k), \ldots, \gamma_k^l(x_k + \xi_k)]^T P_{k-1}$$
$$[\beta_k^1 z_k, \ldots, \beta_j^l z_k]\}^{-1}[\gamma_j^1(x_k + \xi_k), \ldots, \gamma_k^l(x_k + \xi_k)]^T P_{k-1}\}\{b_{k-1} + [\beta_k^1 z_k, \ldots, \beta_k^l z_k]y_k\} \tag{57}$$

so that finally,

$$\hat{\theta}_k = \hat{\theta}_{k-1} - K_k\{[\gamma_j^1(x_k + \xi_k), \ldots, \gamma_k^l(x_k + \xi_k)]^T \hat{\theta}_{k-1} - y_k\} \tag{58}$$

where

$$K_k = P_{k-1}[\beta_k^1 z_k, \ldots, \beta_k^l z_k]\{1 + [\gamma_j^1(x_k + \xi_k), \ldots, \gamma_k^l(x_k + \xi_k)]^T P_{k-1}[\beta_k^1 z_k, \ldots, \beta_j^l z_k]\}^{-1} \tag{59}$$

Equations (56)-(59) compose the recursive algorithm to be implemented so the consequent parameters of a Takagi-Sugeno fuzzy model can be estimated from experimental data.

5. Results

In the sequel, some results will be presented to demonstrate the effectiveness of black box fuzzy modeling for advanced control systems design.

5.1 Computational results

5.1.1 Stochastic nonlinear SISO system identification

The plant to be identified consists on a second order highly nonlinear discrete-time system

$$y_{k+1} = \frac{y_k y_{k-1}(y_k + 2.5)}{1 + y_k^2 + y_{k-1}^2} + u_k + e_k \tag{60}$$

which is a benchmark problem in neural and fuzzy modeling, where y_k is the output and $u_k = \sin(\frac{2\pi k}{25})$ is the applied input. In this case e_k is a white noise with zero mean and variance σ^2. The TS model has two inputs y_k and y_{k-1} and a single output y_{k+1}, and the antecedent part of the fuzzy model (the fuzzy sets) is designed based on the evolving clustering method (ECM). The model is composed of rules of the form:

$$R^i : \text{ IF } y_k \text{ is } F_1^i \text{ AND } y_{k-1} \text{ is } F_2^i \text{ THEN}$$

$$\hat{y}_{k+1}^i = a_{i,1} y_k + a_{i,2} y_{k-1} + b_{i,1} u_k + c_i \tag{61}$$

where $F_{1,2}^i$ are gaussian fuzzy sets.

Experimental data sets of N points each are created from (60), with $\sigma^2 \in [0, 0.20]$. This means that the noise applied take values between 0 and $\pm 30\%$ of the output nominal value, which is an acceptable practical percentage of noise. These data sets are presented to the proposed algorithm, for obtaining an IV fuzzy model, and to the LS based algorithm, for obtaining a LS fuzzy model. The models are obtained by the global and local approaches as in (45) and (46), repetively. The noise influence is analized according to the difference between the outputs of the fuzzy models, obtained from the noisy experimental data, and the output of the plant without noise. The antecedent parameters and the structure of the fuzzy models are the same in the experiments, while the consequent parameters are obtained by the proposed method and by the LS method. Thus, the obtained results are due to these algorithms and accuracy conclusions will be derived about the proposed algorithm performance in the presence of noise. Two criteria, widely used in experimental data analysis, are applied to avaliate the obtained fuzzy models fit: Variance Accounted For (VAF)

$$\text{VAF}(\%) = 100 \times \left[1 - \frac{var(\mathbf{Y} - \hat{\mathbf{Y}})}{var(\mathbf{Y})} \right] \tag{62}$$

where $\mathbf{Y}$ is the nominal plant output, $\hat{\mathbf{Y}}$ is the fuzzy model output and var means signal variance, and Mean Square Error (MSE)

$$\text{MSE} = \frac{1}{N} \sum_{k=1}^{N} (y_k - \hat{y}_k)^2 \tag{63}$$

where y_k is the nominal plant output, $\hat{y}_k$ is the fuzzy model output and N is the number of points. Once obtained these values, a comparative analysis will be established between the proposed algorithm, based on IV, and the algorithm based on LS according to the approaches presented above. In the performance of the TS models obtained off-line according to (45) and (46), the number of points is 500, the proposed algorithm used λ equal to 0.99; the number of rules is 4, the structure is the presented in (61) and the antecedent parameters are obtained by the ECM method for both algorithms. The proposed algorithm performs better than the LS algorithm for the two approaches as it is more robust to noise. This is due to the chosen instrumental variable matrix, with $dl = 1$, to satisfy the convergence conditions as well as possible. In the global approach, for low noise variance, both algorithms presented similar performance with VAF and MSE of 99.50% and 0.0071 for the proposed algorithm and of 99.56% and 0.0027 for the LS based algorithm, respectively. However, when the noise variance increases, the chosen instrumental variable matrix satisfies the convergence conditions, and as a consequence the proposed algorithm becomes more robust to the noise with VAF and MSE of 98.81% and 0.0375. On the other hand the LS based algorithm presented VAF and MSE of 82.61% and 0.4847, respectively, that represents a poor performance. Similar analysis can be applied to the local approach: increasing the noise variance, both algorithms present good performances where the VAF and MSE values increase too. This is due to the polytope property, where the obtained models can represent local approximations giving more flexibility curves fitting. The proposed algorithm presented VAF and MSE values of 93.70% and 0.1701 for the worst case and of 96.3% and 0.0962 for the better case. The LS based algorithm presented VAF and MSE values of 92.4% and 0.2042 for the worst case and of 95.5% and 0.1157 for the better case. The worst case of noisy data set was still used by the algorithm proposed in (Wang & Langari, 1995), where the VAF and MSE values were of 92.6452% and 0.1913, and by the algorithm proposed in (Pedrycz, 2006) where the VAF and MSE values were of 92.5216% and 0.1910, respectively. These results, considering the local approach, show that they have an intermediate performance between the proposed method in this paper and the LS based algorithm. For the global approach, the VAF and MSE values are 96.5% and 0.09 for the proposed method and of 81.4% and 0.52 for the LS based algorithm, respectively. For the local approach, the VAF and MSE values are 96.0% and 0.109 for the proposed method and of 95.5% and 0.1187 for the LS based algorithm, respectively. In sense to be clear to the reader, the results of local and global estimation to the TS fuzzy model from the stochastic SISO nonlinear system identification, it has the following conclusions: When interpreting TS fuzzy models obtained from data, one has to be aware of the tradeoffs between local and global estimation. The TS fuzzy models estimated by local approach describe properly the local behavior of the nonlinear system, but not give a good fit; for the global approach, the opposite holds - a perfect fit is obtained, but the TS fuzzy models are not relevant for the local behavior of the nonlinear system. This is the tradeoffs between local and global estimation. To illustrate the robustness of the FIV algorithm, it was performed a numerical experiment based on 300 different realizations of noise. The numerical experiment followed a particular computational pattern:

• Define a domain with 300 different sequences of noise;

• Generate a realization of noise randomly from the domain, and perform the identification procedure for the IV and LS based algorithms;

• Aggregate the results of IV and LS algorithms according to VAF and MSE criteria into the final result from histograms, indicating the number of its occurences (frequency) during the numerical experiment.

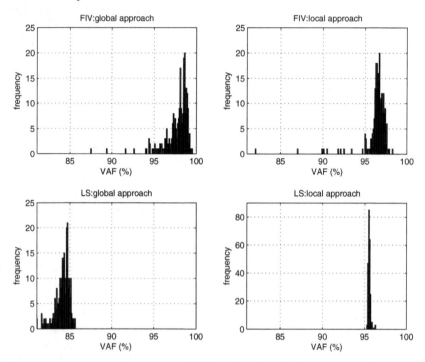

Fig. 3. Robustness analysis: Histogram of VAF for the IV and LS based algorithms.

The IV and LS based algorithms were submitted to these different conditions of noise at same time and the efficiency was observed through VAF and MSE criteria according to the histograms shown on Fig. 3 and Fig. 4, respectively. Clearly, the proposed method presented the best results compared with LS based algorithm. For the global approach, the results of VAF and MSE values are of $98.60 \pm 1.25\%$ and 0.037 ± 0.02 for the proposed method and of $84.70 \pm 0.65\%$ and 0.38 ± 0.15 for the LS based algorithm, respectively. For the local approach, the results of VAF and MSE values are of $96.70 \pm 0.55\%$ and 0.07 ± 0.015 for the proposed method and of $95.30 \pm 0.15\%$ and 0.1150 ± 0.005 for the LS based algorithm, respectively. In general, from the results shown in Tab. 1, it can conclude that the proposed method has favorable results compared with existing techniques and good robustness properties for identification of stochastic nonlinear systems.

5.2 Experimental results

In this section, the experimental results on adaptive model based control of a multivariable (two inputs and one output) nonlinear pH process, commonly found in industrial environment, are presented.

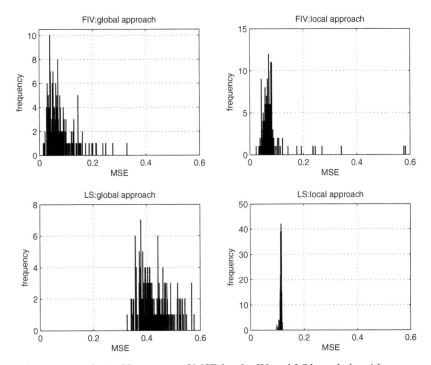

Fig. 4. Robustness analysis: Histogram of MSE for the IV and LS based algorithms.

5.2.1 Fuzzy adaptive black box fuzzy model based control of pH neutralization process

The input-output experimental data set of the nonlinear plant were obtained from DAISY[1] (Data Acquisition For Identification of Systems) plataform.

This plant presents the following input-output variables:

- $u_1(t)$: acid flow (l);
- $u_2(t)$: base flow (l);
- $y(t)$: level of pH in the tank.

Figure 5 shows the open loop temporal response of the plant, considering a sampling time of 10 seconds. These data will be used for modeling of the process. The obtained fuzzy model will be used for indirect multivariable adaptive fuzzy control design. The TS fuzzy inference system uses a functional expression of the pH level in the tank. The i $|^{i=1,2,\dots,l}$-th rule of the multivariable TS fuzzy model, where l is the number of rules is given by:

$$R^i : \textbf{IF } \tilde{Y}(z)z^{-1} \text{ is } F^i_{j|\tilde{Y}(z)z^{-1}} \textbf{ THEN}$$

$$Y^i(z) = \frac{b_1^i}{1 - a_1^i z^{-1} - a_2^i z^{-2}} U_1(z) + \frac{b_2^i}{1 - a_1^i z^{-1} - a_2^i z^{-2}} U_2(z) \tag{64}$$

---

[1] accessed in http://homes.esat.kuleuven.be/ smc/daisy.

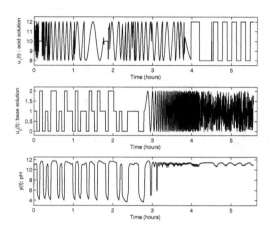

Fig. 5. Open loop temporal response of the nonlinear pH process

The C-means fuzzy clustering algorithm was used to estimate the antecedent parameters of the TS fuzzy model. The fuzzy recursive instrumental variable algorithm based on QR factorization, was used to estimate the consequent submodels parameters of the TS fuzzy model. For initial estimation was used 100 points, the number of rules was $l = 2$, and the fuzzy frequency response validation method was used for fuzzy controller design based on the inverse model (Serra & Ferreira, 2011).

The parameters of the submodels in the consequent proposition of the multivariable TS fuzzy model are shown in Figure 6. It is observed that in addition to nonlinearity, the pH neutralization process presents uncertainty behavior in order to commit any application of fix control design.

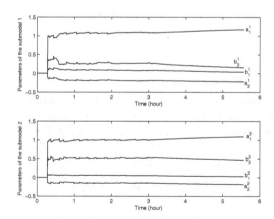

Fig. 6. TS fuzzy model parameters estimated by fuzzy instrumental variable algortihm based on QR factoration

The TS multivariable fuzzy model, at last sample, is given by:

$$R^1 : \textbf{IF } \tilde{y}(k-1) \text{ is } F^1 \textbf{THEN}$$
$$y^1(k) = 1.1707y(k-1) - 0.2187y(k-2) + 0.0372u_1(k) + 0.1562u_2(k)$$
$$R^2 : \textbf{IF } \tilde{y}(k-1) \text{ is } F^2 \textbf{THEN}$$
$$y^2(k) = 1.0919y(k-1) - 0.1861y(k-2) + 0.0304u_1(k) + 0.4663u_2(k) \tag{65}$$

The validation of the TS fuzzy model, according to equation (65) via fuzzy frequency response

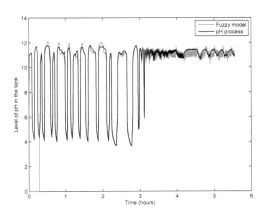

Fig. 7. Recursive estimation processing for submodels parameters in the TS multivariable fuzzy model consequent proposition.

is shown in Figure 8. It can be observed the efficiency of the proposed identification algorithm to track the output variable of pH neutralization process. This result has fundamental importance for multivariable adaptive fuzzy controller design step. The region of uncertainty defined by fuzzy frequency response for the identified model contains the frequency response of the pH process. It means that the fuzzy model represents the dynamic behavior perfectly, considering the uncertainties and nonlinearities of the pH neutralization process. Consequently, the model based control design presents robust stability characteristic. The adaptive control design methodology adopted in this paper consists of a control action based on the inverse model. Once the plant model becomes known precisely by the rules of multivariable TS fuzzy model, considering the fact that the submodels are stable, one can develop a strategy to control the flow of acid and base, in order to maintain the pH level of 7. Thus, the multivariable fuzzy controller is designed so that the control system closed-loop presents unity gain and the output is equal to the reference. So, it yields:

$$G_{MF}(z) = \frac{R(z)}{Y(z)} = \frac{G^i_{c_1} G^i_{p_1} + G^i_{c_2} G^i_{p_2}}{1 + G^i_{c_1} G^i_{p_1} + G^i_{c_2} G^i_{p_2}} \tag{66}$$

where $G^i_{c_1}$ e $G^i_{c_2}$ are the transfer functions of the controllers in the i-th rule, as $G^i_{p_1}$ and $G^i_{p_2}$ are submodels in the consequent proposition from the output $Y(z)$ to inputs $U_1(z)$ and $U_2(z)$,

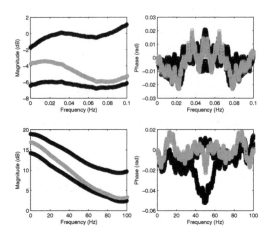

Fig. 8. Validation step of the multivariable TS fuzzy model: (a) - (b) Fuzzy frequency response of the TS fuzzy model (black curve) representing the dynamic behavior of the pH level and the flow of acid solution (red curve), (c) - (d) Fuzzy frequency response of the TS fuzzy model (black curve) representing the dynamic behavior of the pH level and flow of the base (red curve).

respectively. Considering

$$G_{c_1}^i = \frac{1}{G_{p_1}^i}$$

and

$$G_{c_2}^i = \frac{1}{G_{p_2}^i}$$

results:

$$G_{MF}(z) = \frac{R(z)}{Y(z)} = \frac{2}{3} \tag{67}$$

this is,

$$Y(z) = \frac{2}{3}R(z) \tag{68}$$

For compensation this closed loop gain of the control system, it is necessary generate a reference signal so that $Y(z) = R(z)$. Therefore, adopting the new reference signal $R'(z) = \frac{3}{2}R(z)$, it yields:

$$Y(z) = \frac{2}{3}R'(z) \tag{69}$$

$$Y(z) = \frac{2}{3}\frac{3}{2}R(z) \tag{70}$$

$$Y(z) = R(z) \tag{71}$$

For the inverse model based indirect multivariable fuzzy control design, one adopte a new reference signal given by $R'(z) = \frac{3}{2}R(z)$. The TS fuzzy multivariable controller presents the folowing structure:

$$R^i : \textbf{IF } \tilde{Y}(z)z^{-1} \text{ is } F^i_{j|\tilde{Y}(z)z^{-1}} \textbf{ THEN}$$

$$G^i_{c_1} = \frac{1 - \hat{a}^i_1 z^{-1} - \hat{a}^i_2 z^{-2}}{\hat{b}^i_1} E(z)$$

$$G^i_{c_2} = \frac{1 - \hat{a}^i_1 z^{-1} - \hat{a}^i_2 z^{-2}}{\hat{b}^i_2} E(z) \tag{72}$$

The temporal response of the TS fuzzy multivariable adaptive control is shown in Fig. 9. It can be observed the control system track the reference signal, $pH = 7$, because the controller can tune itself based on the identified TS fuzzy multivariable model.

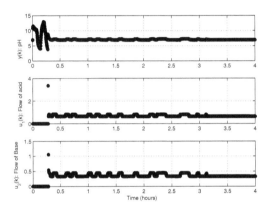

Fig. 9. Performance of the TS fuzzy multivariable adaptive control system.

6. References

Abonyi, J., Babuska, R., Ayala Botto, M., Szeifert, F. and Nagy, L. Identification and Control of Nonlinear Systems Using Fuzzy Hammerstein Models. *Industrial Engineering and Chemistry Research*, Vol. 39, No 11, November 2000, 4302-4314, ISSN 0888-5885

Tee, K. P., Ren, B., Ge, S. S. Control of nonlinear systems with time-varying output constraints. *Automatica*, Vol. 47, No. 11, November 2011, 2331-2552, ISSN 0005-1098

Åström, K.J., Albertos, P., Blamke, M., Isidori, A., Schaufelberger, W. and Sanz, R. (2001). *Control of Complex Systems*. Springer-Verlag, 1st *Edition*, ISBN 1-85233-324-3, Berlin-Heidelberg

Åström, K. J. and Wittenmark, B. (1995). *Adaptive Control*. Addison-Wesley, 2nd Edition, ISBN 0-20-155866-1, United States of America

Brown, M. and Harris, C. (1994). *Neurofuzzy Adaptive Modelling and Control*. Prentice-Hall International Series in Systems and Control Engineering, 1st Edition, ISBN 0-13-134453-6, Upper Saddle River, New Jersey

Keesman, K. J. (2011). *System Identification: An Introduction*. Springer-Verlag, 1st Edition, ISBN 0-85-729521-7, Berlin-Heidelberg

Sastry, S. and Bodson, M. (1989). *Adaptive Control: Stability, Convergence, and Robustness.* Prentice Hall Advanced Reference Series, ISBN 0-13-004326-5, Englewood Cliffs, New Jersey

Isermann, R. and Münchhof, M. (2011). *Identification of Dynamic Systems: An Introduction with Applications.* Springer-Verlag, 1st Edition, ISBN 978-3-540-78878-2, Berlin-Heidelberg

Li, I. and Lee, L.W. Hierarchical Structure of Observer-Based Adaptive Fuzzy-Neural Controller for MIMO systems. *Fuzzy Sets and Systems,* Vol. 185, No. 1, December 2011, 52-82, ISSN 0165-0114

Zhu, Y. (2011). *Multivariable System Identification For Process Control.* Elsevier, 1st Edition, ISBN 978-0-08-043985-3

Lee, D. H., Park, J. B. and Joo, Y. H. Approaches to extended non-quadratic stability and stabilization conditions for discrete-time Takagi-Sugeno fuzzy systems. *Automatica,* Vol. 47, No. 3, March 2011, 534-538, ISSN 0005-1098

Castillo-Toledo, B. and Meda-Campaña, A. The Fuzzy Discrete-Time Robust Regulation Problem: An LMI Approach. *IEEE Transactions on Fuzzy Systems,* Vol.12, No.3, June 2004, 360–367, ISSN 1063-6706

Chalam, V.V. (1987). *Adaptive Control Systems: Techniques and Applications.* Marcel Dekker, ISBN 0-82-477650-X, New York, United States of America

Ding, B. Dynamic Output Feedback Predictive Control for Nonlinear Systems Represented by a Takagi-Sugeno Model. *IEEE Transactions on Fuzzy Systems,* Vol. 19, No. 5, October 2011, 831-843, ISSN 1063-6706

Hellendoorn, H. and Driankov, D. *Fuzzy Model Identification: Selected Approaches.* Springer Verlag, ISBN 978-3540627210, Berlin-Heidelberg

Ioannou, P.A. and Sun, J. (1996). *Robust Adaptive Control.* Prentice Hall , ISBN 978-0134391007, Upper Saddle River, New Jersey

Isidori, A. (1995). *Nonlinear Control Systems.* Springer Verlag, 3rd Edition, ISBN 978-3540199168 , Berlin-Heidelberg

Wang, J.W., Wu, H.N. and Li, H.X. Distributed Fuzzy Control Design of Nonlinear Hyperbolic PDE Systems With Application to Nonisothermal Plug-Flow Reactor. *IEEE Transactions on Fuzzy Systems,* Vol. 19, No. 3, June 2011, 514-526, ISSN 1063-6706

Kadmiry, B. and Driankov, D. A Fuzzy Gain-Scheduler for the Attitude Control of an Unmanned Helicopter. *IEEE Transactions on Fuzzy Systems,* Vol.12, No.3, August 2004, 502-515, ISSN 1063-6706

Khalil, H. *Nonlinear Systems.* Prentice Hall, 3rd Edition, ISBN 0-13-067389-7, Upper Saddle River, New Jersey

Sadeghian, M. and Fatehi, A. Identification, prediction and detection of the process fault in a cement rotary kiln by locally linear neuro-fuzzy technique. *Journal of Process Control,* Vol. 21, No. 2, February 2011, 302-308, ISSN 0959-1524

Grigorie, L. (2010). *Fuzzy Controllers, Theory and Applications.* Intech, ISBN 978-953-307-543-3

Vukadinovic, D. (2011). *Fuzzy Control Systems.* Nova Science Publishers, ISSN 978-1-61324-488-3

Lewis, F.L. and Syrmos, V.L. (1995). *Optimal Control.* John Wiley & Sons - IEEE, $2^{n}d$ Edition, ISBN 0-471-03378-2, United States of America

Ljung, L. (1999). *System Identification: Theory for the User.* Prentice Hall, 2nd Edition, ISBN 0-13-656695-2, Upper Saddle River, New Jersey

Mamdani, E.H. Application of Fuzzy Logic to Approximate Reasoning Using Linguistic Systems. *IEEE Transactions on Computers*, Vol. 26, No. 12, December 1977, 1182-1191, ISSN 0018-9340

Michels, K., Klawonn, F., Kruse, R., Nürnberger, A. (2006). *Fuzzy Control: Fundamentals, Stability and Design of Fuzzy Controllers*. Springer Verlag, Studies in Fuzziness and Soft Computing Series, Vol. 200, ISBN 978-3-540-31765-4, Berlin-Heidelberg

Pedrycz, W. and Gomide, F.C. (1998). *An Introduction to Fuzzy Sets - Analysis and Design*. A Bradford Book, 1st Edition, ISBN 0-262-16171-0, Massachusetts Institute of Technology

Pedrycz, W. OR/AND neurons and the development of interpretable logic models. *IEEE Transactions on Neural Networks*, Vol. 17, No. 3, May 2006, 636–658, ISSN 1045-9227

Ren, T. and Chen, T. A Robust Model Reference Fuzzy Control for Nonlinear Systems. *Proceedings of IEEE International Conference on Control Applications*, pp. 165-170, ISBN 0-7803-8633-7, Taiwan, September 2004, Taipei

Serra, G.L.O. and Ferreira, C.C.T. Fuzzy Frequency Response: Proposal and Application for Uncertain Dynamic Systems. *Engineering Applications of Artificial Intelligence*, Vol. 24, No. 7, October 2011, 1186-1194, ISSN 0952-1976

Sjöberg, J., Zhang, Q., Ljung, L., Benveniste, A., Delyon, B., Glorennec, P., Hjalmarsson, H. and Juditsky, A. Nonlinear Black-box Modeling in System Identification: A Unified Overview. *Automatica*, Vol. 31, No. 12, December 1995, 1691-1724, ISSN 0005-1098

Söderström, T. and Stoica, P. (1989). *System Identification*. Prentice Hall, ISBN 0-13-881236-5, Upper Saddle River, New Jersey

Takagi, T. and Sugeno, M. Fuzzy Identification of Systems and its Application to Modeling and Control. *IEEE Transactions on Systems, Man and Cibernetics*, Vol. 15, No. 1, January/February 1985, 116-132, ISSN 0018-9472

Tong, S. and Li, H. Observer-based Robust Fuzzy Control of Nonlinear Systems with Parametric Uncertainties. *Fuzzy Sets and Systems*, Vol. 131, No. 2, October 2002, 165-184, ISSN 0165-0114

Van Overschee, P. and De Moor, B. (1996). *Subspace Identification for Linear Systems, Theory, Implementation, Applications*. Kluwer Academic Publishers, ISBN 0-7923-9717-7, The Netherlands

Ogunfunmi, T. (2007). *Adaptive Nonlinear System Identification: The Volterra and Wiener Model Approaches*. Springer Verlag, 1st Edition, ISBN 978-0387263281, Berlin-Heidelberg

Vidyasagar, M. (2002). *Nonlinear Systems Analysis*. SIAM: Society for Industrial and Applied Mathematics, 2nd Edition, ISBN 0-89871-526-1

Walter, E. and Pronzato, L. (1997). *Identification of Parametric Models: From Experimental Data*. Springer-Verlag, 1st Edition, ISBN 3-540-76119-5, Berlin-Heidelberg

Nelles, O. (2011). *Nonlinear System Identification: From Classical Approaches to Neural Networks and Fuzzy Models*. Springer Verlag, 3rd Edition, ISBN 978-3642086748, Berlin-Heidelberg

Wang, L. (1996). *A Course in Fuzzy Systems and Control*. Prentice Hall, 1st Edition, ISBN 0-13-540882-2, Englewood Cliffs, New Jersey

Wang, W.J. and Luoh, L. Stability and Stabilization of Fuzzy Large-Scale Systems. *IEEE Transactions on Fuzzy Systems*, Vol.12, No.3, June 2004, 309–315, ISSN 1063-6706

Wang, L. and Langari, R. Building Sugeno Type Models Using Fuzzy Discretization and Orthogonal Parameter Estimation Techniques. *IEEE Transactions on Fuzzy Systems*, Vol. 3, No. 4, November 1995, 454–458, ISSN 1063-6706

Yoneyama, J. H_∞ Control for Fuzzy Time-Delay Systems via Descriptor System. *Proceedings of IEEE International Symposium on Intelligent Control*, pp. 407-412, ISBN 0-7803-8635-3, Taiwan, September 2004, Taipei

Zadeh, L.A. Fuzzy Sets. *Information and Control*, Vol. 8, No. 3, June 1965, 338-353, ISSN 0019-9958

Zadeh, L.A. Outline of a New Approach to the Analysis of Complex Systems and Decision Processes. *IEEE Transactions on Systems, Man and Cybernetics*, Vol. 3, No. 1, January 1973, 28-44, ISSN 0018-9472

2

New Techniques for Optimizing the Norm of Robust Controllers of Polytopic Uncertain Linear Systems

L. F. S. Buzachero, E. Assunção, M. C. M. Teixeira and E. R. P. da Silva
FEIS - School of Electrical Engineering,
UNESP, 15385-000, Ilha Solteira, São Paulo
Brazil

1. Introduction

The history of linear matrix inequalities (LMIs) in the analysis of dynamical systems dates from over 100 years. The story begins around 1890 when Lyapunov published his work introducing what is now called the Lyapunov's theory (Boyd et al., 1994). The researches and publications involving the Lyapunov's theory have grown up a lot in recent decades (Chen, 1999), opening a very wide range for various approaches such as robust stability analysis of linear systems (Montagner et al., 2009), LMI optimization approach (Wang et al., 2008), $\mathcal{H}_2$ (Apkarian et al., 2001; Assunção et al., 2007a; Ma & Chen, 2006) or $\mathcal{H}_\infty$ (Assunção et al., 2007b; Chilali & Gahinet, 1996; Lee et al., 2004) robust control, design of controllers for systems with state feedback (Montagner et al., 2005), and design of controllers for systems with state-derivative feedback (Cardim et al., 2009). The design of robust controllers can also be applied to nonlinear systems.

In addition to the various current controllers design techniques, the design of robust controllers (or controller design by quadratic stability) using LMI stands out for solving problems that previously had no known solution. These designs use specialized computer packages (Gahinet et al., 1995), which made the LMIs important tools in control theory.

Recent publications have found a certain conservatism inserted in the analysis of quadratic stability, which led to a search for solutions to eliminate this conservatism (de Oliveira et al., 1999). Finsler's lemma (Skelton et al., 1997) has been widely used in control theory for the stability analysis by LMIs (Montagner et al., 2009; Peaucelle et al., 2000), with better results than the quadratic stability of LMIs, but with extra matrices, which allows a certain relaxation in the stability analysis (here called extended stability), by obtaining a larger feasibility region. The advantage found in its application to design of state feedback is the fact that the synthesis of gain K becomes decoupled from Lyapunov's matrix P (Oliveira et al., 1999), leaving Lyapunov's matrix free as it is necessarily symmetric and positive defined to meet the initial restrictions.

The reciprocal projection lemma used in robust control literature $\mathcal{H}_2$ (Apkarian et al., 2001), can also be used for the synthesis of robust controllers, eliminating in a way the existing conservatism, as it makes feasible dealing with multiple Lyapunov's matrices, as in the case of extended stability point, allowing extra matrices through a relaxation in the case of extended stability, making feasible a relaxation in the stability analysis (here called projective stability) through extra matrices. The synthesis of the controller K is depending now on an auxiliary matrix V, not necessarily symmetrical, and in this situation it becomes completely decoupled from Lyapunov's matrix P, leaving it free.

Two critical points in the design of robust controllers are explored here. One of them is the magnitude of the designed controllers that are often high, affect their practical implementation and therefore require a minimization of the gains of the controller to facilitate its implementation (optimization of the norm of K).The other one is the fact that the system settling time can be larger than the required specifications of the project, thus demanding restrictions on LMIs to limit the decay rate, formulated with the inclusion of the parameter γ in LMIs.

The main focus of this work is to propose new methods for optimizing the controller's norm, through a different approach from that found in (Chilali & Gahinet, 1996), and compare it with the optimization method presented in (Assunção et al., 2007c) considering the different criteria of stability, aiming at the advantages and disadvantages of each method, as well as the inclusion of a decay rate (Boyd et al., 1994) in LMIs formulation.

In (Šiljak & Stipanovic, 2000) an optimization of the controllers's norm was proposed for decentralized control, but without the decay rate, so no comparisons were made with this work due to the necessity to insert this parameter to improve the performance of the system response.

The LMIs of optimization that will be used for new design techniques, had to be reformulated because the matrix controller synthesis does not depend more on a symmetric matrix, a necessary condition for the formulation of the existing LMI optimization. Comparisons will be made through a practical implementation in the Quanser's 3-DOF helicopter (Quanser, 2002) and a general analysis involving 1000 randomly generated polytopic uncertain systems.

2. Quadratic stability of continuous time linear systems

Consider (1) an autonomous linear dynamic system without state feedback. Lyapunov proved that the system

$$\dot{x}(t) = Ax(t) \tag{1}$$

with $x(t) \in \mathbb{R}^n$ e $A \in \mathbb{R}^{n \times n}$ a known matrix, is asymptotically stable (i.e., all trajectories converge to zero) if and only if there exists a matrix $P = P' \in \mathbb{R}^{nxn}$ such that the LMIs (2) and (3) are met (Boyd et al., 1994).

$$A'P + PA < 0 \tag{2}$$

$$P > 0 \tag{3}$$

Consider in equation (2) that A is not precisely known, but belongs to a politopic bounded uncertainty domain $\mathcal{A}$. In this case, the matrix A within the domain of uncertainty can be written as convex combination of vertexes A_j, $j = 1, ..., N$, of the convex bounded uncertainty domain (Boyd et al., 1994), i.e. $A(\alpha) \in \mathcal{A}$ and $\mathcal{A}$ shown in (4).

$$\mathcal{A} = \{A(\alpha) \in \mathbb{R}^{n \times n} \ : \ A(\alpha) = \sum_{j=1}^{N} \alpha_j A_j \ , \ \sum_{j=1}^{N} \alpha_j = 1 \ , \ \alpha_j \geq 0 \ , \ j = 1...N\} \tag{4}$$

A sufficient condition for stability of the convex bounded uncertainty domain $\mathcal{A}$ (now on called polytope) is given by the existence of a Lyapunov's matrix $P = P' \in \mathbb{R}^{n \times n}$ such that the LMIs (5) and (6)

$$A(\alpha)'P + PA(\alpha) < 0 \tag{5}$$
$$P > 0 \tag{6}$$

are checked for every $A(\alpha) \in \mathcal{A}$ (Boyd et al., 1994). This stability condition is known as quadratic stability and can be easily verified in practice thanks to the convexity of Lyapunov's inequality that turns the conditions (5) and (6) equivalent to checking the existence of $P = P' \in \mathfrak{R}^{n \times n}$ such that conditions (7) and (8) are met with $j = 1, ..., N$.

$$A'_j P + PA_j < 0 \tag{7}$$
$$P > 0 \tag{8}$$

It can be observed that (5) can be obtained multiplying by $\alpha_j \geq 0$ and adding in j of $j = 1$ to $j = N$.

Due to being a sufficient condition for stability of the polytope $\mathcal{A}$, conservative results are generated, nevertheless this quadratic stability has been widely used for robust controllers's synthesis.

3. Decay rate restriction for closed-loop systems

Consider a linear time invariant controllable system described in (9)

$$\dot{x}(t) = Ax(t) + Bu(t), \quad x(0) = x_0 \tag{9}$$

with $A \in \mathbb{R}^{n \times n}$, $B \in \mathbb{R}^{n \times m}$ the matrix of system input, $x(t) \in \mathbb{R}^n$ the state vector and $u(t) \in \mathbb{R}^m$ the input vector. Assuming that all state are available for feedback, the control law for the same feedback is given by (10)

$$u(t) = -Kx(t) \tag{10}$$

being $K \in \mathbb{R}^{m \times n}$ a constant elements matrix. Often the norm of the controller K can be high, leading to saturation of amplifiers and making the implementation in analogic systems difficult. Thus it is necessary to reduce the norm of the controllers elements to facilitate its implementation.

Considering the controlled system (9) - (10), the decay rate (or largest Lyapunov's exponent) is defined as the largest positive constant γ, such that (11)

$$\lim_{t \to \infty} e^{\gamma t} ||x(t)|| = 0 \tag{11}$$

remains for all trajectories $x(t)$, $t > 0$.

From the quadratic Lyapunov's function (12),

$$V(x(t)) = x(t)'Px(t) \tag{12}$$

to establish a lower limit on the decay rate of (9), with (13)

$$\dot{V}(x(t)) \leq -2\gamma V(x(t)) \tag{13}$$

for all trajectories (Boyd et al., 1994).

From (12) and (9), (14) can be found.

$$\dot{V}(x(t)) = \dot{x}(t)'Px(t) + x(t)'P\dot{x}(t)$$
$$= x(t)'(A - BK)'Px(t) + x(t)'P(A - BK)x(t) \tag{14}$$

Adding the restriction on the decay rate (13) in the equation (14) and making the appropriate simplifications, (15) and (16) are met.

$$(A - BK)'P + P(A - BK) < -2\gamma P \tag{15}$$
$$P > 0 \tag{16}$$

As the inequality (15) became a bilinear matrix inequality (BMI) it is necessary to perform manipulations to fit them back into the condition of LMIs. Multiplying the inequalities (17) and (18) on the left and on the right by P^{-1}, making $X = P^{-1}$ and $G = KX$ results:

$$AX - BG + XA' - G'B' + 2\gamma X < 0 \tag{17}$$
$$X > 0 \tag{18}$$

If the LMIs (17) and (18) are feasible, a controller that stabilizes the closed-loop system can be given by $K = GX^{-1}$.

Consider the linear uncertain time-invariant system (19).

$$\dot{x}(t) = A(\alpha)x(t) + B(\alpha)u(t) \tag{19}$$

This system can be described as convex combination of the polytope's vertexes shown in (20).

$$\dot{x}(t) = \sum_{j=1}^{r} \alpha_j A_j x(t) + \sum_{j=1}^{r} \alpha_j B_j u(t) \tag{20}$$

with A and B belonging to the uncertainty polytope (21)

$$(\mathcal{A}, \mathcal{B}) = \{(A, B)(\alpha) \in \mathbb{R}^{n \times n} : (A, B)(\alpha) = \sum_{j=1}^{N} \alpha_j (A, B)_j, \sum_{j=1}^{N} \alpha_j = 1, \alpha_j \geq 0, j = 1...N\} \tag{21}$$

being r the number of vertexes (Boyd et al., 1994).

Knowing the existing theory for uncertain systems, Theorem 3.1 theorem can be enunciated (Boyd et al., 1994):

Theorem 3.1. *A sufficient condition which guarantees the stability of the uncertain system (20) subject to decay rate γ is the existence of matrices $X = X' \in \mathbb{R}^{n \times n}$ and $G \in \mathbb{R}^{m \times n}$, such that (22) and (23) are met.*

$$A_j X - B_j G + XA_j' - G'B_j' + 2\gamma X < 0 \tag{22}$$
$$X > 0 \tag{23}$$

with $j = 1, ..., r$.

When the LMIs (22) and (23) are feasible, a state feedback matrix which stabilizes the system can be given by (24).

$$K = GX^{-1} \tag{24}$$

Proof. The proof can be found at (Boyd et al., 1994). □

Thus, it can be feedback into the uncertain system shown in (19) being (22) and (23) sufficient conditions for the polytope asymptotic stability, now for a closed-loop system subject to decay rate.

4. Optimization of the K matrix norm of the closed-loop system

In many situations the norm of the state feedback matrix is high, precluding its practical implementation. Thus Theorem 4.1 was proposed in order to limit the norm of K (Assunção et al., 2007c; Faria et al., 2010).

Theorem 4.1. *Given an fixed constant $\mu_0 > 0$, that enables to find feasible results, it can be obtained a constraint for the $K \in \mathbb{R}^{m \times n}$ matrix norm from the state feedback, with $K = GX^{-1}$, $X = X' > 0 \in \mathbb{R}^{n \times n}$ and $G \in \mathbb{R}^{m \times n}$ finding the minimum value β, $\beta > 0$ such that $KK' < \frac{\beta}{\mu_0^2} I_m$. The optimum value for β can be found solving the optimization problem with the LMIs (25), (26) and (27).*

$$\min \beta$$
$$s.t. \begin{bmatrix} \beta I_m & G \\ G' & I_n \end{bmatrix} > 0 \tag{25}$$

$$X > \mu_0 I_n \tag{26}$$

$$A_j X - B_j G + X A_j' - G' B_j' + 2\gamma X < 0 \tag{27}$$

where I_m and I_n are the identity matrices of m and n order respectively.

Proof. The proof can be found at (Assunção et al., 2007c). □

5. New optimization of the K matrix norm of the closed-loop system

It can be verified that the LMIs given in Theorem 4.1 can produce conservative results, so in order to find better results, new methodologies are proposed.

Using the theory presented in (Assunção et al., 2007c) for the optimization of the norm of robust controllers subject to failures, it is proposed an alternative approach for the same problem grounded in Lemma (5.1).

The approach of the optimum norm used was modified to fit to the new structures of LMIs that will be given in sequence. At first, this new approach has produced better results comparing to the existing ones for the optimization stated in Theorem 4.1 using the set of LMIs (22) and (23).

Lemma 5.1. *Consider $L \in \mathbb{R}^{n \times m}$ a a given matrix and $\beta \in \mathbb{R}$, $\beta > 0$. The conditions*

1. $L'L \leq \beta I_m$

2. $LL' \leq \beta I_n$

are equivalent.

Proof. Note that if $L = 0$ the lemma conditions are verified. Then consider the case where $L \neq 0$.

Note that in the first statement of the lemma, (28) is met

$$L'L \leq \beta I_m \Leftrightarrow x'(L'L)x \leq \beta x'x \tag{28}$$

for all $x \in \mathbb{R}^m$.

Knowing that (29) is true

$$x'(L'L)x \leq \lambda_{max}(L'L)x'x \tag{29}$$

and $\lambda_{max}(L'L)$ the maximum eigenvalue of $L'L$, which is real (every symmetric matrix has only real eigenvalues). Besides, when x is equal to the eigenvector of $L'L$ associated to the eigenvalue $\lambda_{max}(L'L)$, and $x'(L'L)x = \lambda_{max}(L'L)x'x$. Thus, from (28) and (29), $\beta \geq \lambda_{max}(L'L)$.

Similarly, for every $z \in \mathbb{R}^n$, the second assertion of the lemma results in (30).

$$LL' \leq \beta I_n \Leftrightarrow z'(LL')z \leq \lambda_{max}(LL')z'z \leq \beta z'z \tag{30}$$

and then, $\beta \geq \lambda_{max}(LL')$.

Now, note that the condition (31) is true (Chen, 1999).

$$\lambda^m det(\lambda I_n - L'L) = \lambda^n det(\lambda I_m - LL') \tag{31}$$

Consequently, every non-zero eigenvalue of $L'L$ is also an eigenvalue of LL'. Therefore, $\lambda_{max}(L'L) = \lambda_{max}(LL')$, and from (29) and (30) the lemma is proved . $\qquad \square$

Knowing that $P = X^{-1}$ is the matrix used to define Lyapunov's quadratic function, Theorem 5.1 is proposed.

Theorem 5.1. *Given a constant $\mu_0 > 0$, a constraint for the state feedback $K \in \mathbb{R}^{m \times n}$ matrix norm is obtained, with $K = GX^{-1}$, $X = X' > 0$, $X \in \mathbb{R}^{n \times n}$ and $G \in \mathbb{R}^{m \times n}$ by finding the minimum of β, $\beta > 0$ such that $K'K < \frac{\beta}{\mu_0} I_n$. You can get the minimum β solving the optimization problem with the LMIs (32), (33) and (34).*

$$\min \beta$$

$$s.t. \begin{bmatrix} X & G' \\ G & \beta I_m \end{bmatrix} > 0 \tag{32}$$

$$X > \mu_0 I_n \tag{33}$$

$$A_j X - B_j G + XA_j' - G'B_j' + 2\gamma X < 0 \tag{34}$$

where I_m and I_n are the identity matrices of m and n order respectively.

Proof. Applying the Schur complement for the first inequality of (32) results in (35).

$$\beta I_m > 0 \quad e \quad X - G'(\beta I_m)^{-1}G > 0 \tag{35}$$

Thus, from (35), (36) is found.

$$X > \frac{1}{\beta}G'G \Rightarrow G'G < \beta X \tag{36}$$

Replacing $G = KX$ in (36) results in (37)

$$XK'KX < \beta X \Rightarrow K'K < \beta X^{-1} \tag{37}$$

So from (33), (37) and (33), (38) is met.

$$K'K < \frac{\beta}{\mu_0}I_n \tag{38}$$

on which K is the optimal controller associated with (22).   □

It follows that minimizing the norm of a matrix is equivalent to the minimization of a $\beta > 0$ variable such that $K'K < \frac{\beta}{\mu_0}I_n$, with $\mu_0 > 0$. Note that the position of the transposed matrix was replaced in this condition, comparing to that used in Theorem 4.1.

A comparison will be shown between the optimization methods, using the robust LMIs with decay rate (22) and (23) in the results section. Since the new method may suit the relaxed LMIs listed below, it was used in the comparative analysis for the control design for extended stability and projective stability.

Finsler's lemma shown in Lemma (5.2) can be used to express stability conditions referring to matrix inequalities, with advantages over existing Lyapunov's theory (Boyd et al., 1994), because it introduces new variables and generate new degrees of freedom in the analysis of uncertain systems with the possibility of nonlinearities elimination.

Lemma 5.2 (Finsler). *Consider $w \in \mathbb{R}^{n_x}$, $\mathcal{L} \in \mathbb{R}^{n_x \times n_x}$ and $\mathcal{B} \in \mathbb{R}^{m_x \times n_x}$ with $rank(\mathcal{B}) < n_x$ e $\mathcal{B}^{\perp}$ a basis for the null space of $\mathcal{B}$ (i.e., $\mathcal{B}\mathcal{B}^{\perp} = 0$). Then the following conditions are equivalent:*

1. $w'\mathcal{L}w < 0, \ \forall \ w \neq 0 \ : \ \mathcal{B}w = 0$
2. $\mathcal{B}^{\perp '}\mathcal{L}\mathcal{B}^{\perp} < 0$
3. $\exists \mu \in \mathbb{R} \ : \ \mathcal{L} - \mu \mathcal{B}'\mathcal{B} < 0$
4. $\exists \mathcal{X} \in \mathbb{R}^{n_x \times m_x} \ : \ \mathcal{L} + \mathcal{X}\mathcal{B} + \mathcal{B}'\mathcal{X}' < 0$

Proof. Finsler's lemma proof can be found at (Oliveira & Skelton, 2001; Skelton et al., 1997).   □

5.1 Stability of systems using Finsler's lemma restricted by the decay rate

Consider the closed-loop system (9). Defining $w = \begin{bmatrix} x \\ \dot{x} \end{bmatrix}$, $\mathcal{B} = \begin{bmatrix} (A - BK) & -I \end{bmatrix}$, $\mathcal{B}^{\perp} = \begin{bmatrix} I \\ (A - BK) \end{bmatrix}$ and $\mathcal{L} = \begin{bmatrix} 2\gamma P & P \\ P & 0 \end{bmatrix}$. Note that $\mathcal{B}w = 0$ corresponds to (9) and $w'\mathcal{L}w < 0$ corresponds to stability constraint with decay rate given by (12) and (13). In this case the dimensions of the lemma's variables (5.2) are: $n_x = 2n$ and $m_x = n$. Considering that P is the matrix used to define the quadratic Lyapunov's function (12), the properties 1 and 2 of Finsler's lemma can be written as:

1. $\exists P = P' > 0$ such that
$$\begin{bmatrix} x \\ \dot{x} \end{bmatrix}' \begin{bmatrix} 2\gamma P & P \\ P & 0 \end{bmatrix} \begin{bmatrix} x \\ \dot{x} \end{bmatrix} < 0 \; \forall x, \dot{x} \neq 0 \; : \; \begin{bmatrix} (A - BK) & -I \end{bmatrix} \begin{bmatrix} x \\ \dot{x} \end{bmatrix} = 0$$

2. $\exists P = P' > 0$ such that
$$\begin{bmatrix} I \\ (A - BK) \end{bmatrix}' \begin{bmatrix} 2\gamma P & P \\ P & 0 \end{bmatrix} \begin{bmatrix} I \\ (A - BK) \end{bmatrix} < 0$$

which results in the equations of stability, according to Lyapunov, including decay rate:

1. $x(t)'P\dot{x}(t) + \dot{x}(t)'Px(t) + 2\gamma x(t)'Px(t) < 0 \; \forall x, \dot{x} \neq 0 \; : \; \dot{x}(t) = (A - BK)x(t)$
2. $P(A - BK) + (A - BK)'P + 2\gamma P < 0$

Thus, it is possible to characterize stability through Lyapunov's quadratic function $(V(x(t)) = x(t)'Px(t))$, generating new degrees of freedom for the synthesis of controllers.

From Finsler's lemma proof follows that if the properties 1 and 2 are true, then properties 3 and 4 will also be true. Thus, the fourth propriety can be written as (39).

4. $\exists \mathcal{X} \in \mathbb{R}^{2n \times n}, P = P' > 0$ such that

$$\begin{bmatrix} 2\gamma P & P \\ P & 0 \end{bmatrix} + \mathcal{X} \begin{bmatrix} (A - BK) & -I \end{bmatrix} + \begin{bmatrix} (A - BK)' \\ -I \end{bmatrix} \mathcal{X}' < 0. \tag{39}$$

Choosing conveniently the matrix of variables $\mathcal{X} = \begin{bmatrix} Z \\ aZ \end{bmatrix}$, with $Z \in \mathbb{R}^{n \times n}$ invertible and not necessarily symmetric and $a > 0$ a fixed relaxation constant of the LMI (Pipeleers et al., 2009). Developing the equation (39) and applying the congruence transformation $\begin{bmatrix} Z^{-1} & 0 \\ 0 & Z^{-1} \end{bmatrix}$ on the left and $\begin{bmatrix} Z^{-1} & 0 \\ 0 & Z^{-1} \end{bmatrix}'$ on the right, is found (40).

$$\begin{bmatrix} AZ'^{-1} + Z^{-1}A' - BKZ'^{-1} - Z^{-1}K'B' + 2\gamma Z^{-1}PZ'^{-1} & Z^{-1}PZ'^{-1} + aZ^{-1}A' - aZ^{-1}K'B' - Z'^{-1} \\ Z^{-1}PZ'^{-1} + aAZ'^{-1} - aBKZ'^{-1} - Z^{-1} & -aZ'^{-1} - aZ^{-1} \end{bmatrix} < 0$$

Making $Y = Z'^{-1}$; $G = KY$ and $Q = Y'PY$, there were found LMIs (40) and (41) subject to decay rate γ.

$$\begin{bmatrix} AY + Y'A' - BG - G'B' + 2\gamma Q & Q + aY'A' - aG'B' - Y \\ Q + aAY - aBG - Y' & -aY - aY' \end{bmatrix} < 0, \tag{40}$$

$$Q > 0 \tag{41}$$

with $Y \in \mathbb{R}^{n \times n}, Y \neq Y', G \in \mathbb{R}^{m \times n}$ and $Q \in \mathbb{R}^{n \times n}, Q = Q' > 0$, for some $a > 0$.

These LMIs meet the restrictions for the asymptotic stability (Feron et al., 1996) of the system described in (9) with state feedback given by (10). It can be checked that the first principal minor of the LMI (40) has the structure of the result found in the theorem of quadratic stability with decay rate (Faria et al., 2009). Nevertheless, there is also, as stated in the Finsler's lemma, a greater degree of freedom because the matrix of variables Y, responsible for the synthesis of the controller, doesn't need to be symmetric and the Lyapunov's matrix now turned into Q, which remains restricted to positive definite, is partially detached from the controller synthesis, since that $Q = Y'PY$.

The stability of the LMIs derived from Finsler's lemma stability is commonly called extended stability and it will be designated this way now on.

5.2 Robust stability of systems using Finsler's lemma restricted by the decay rate

As discussed for the condition of quadratic stability, the stability analysis can be performed for a robust stability condition considering the continuous time linear system as a convex combination of r vertexes of the polytope described in (20). The advantage of using the Finsler's lemma for robust stability analysis is the freedom of Lyapunov's function, now defined as $Q(\alpha) = \sum_{j=1}^{r} \alpha_j Q_j$, $\sum_{j=1}^{r} \alpha_j = 1$, $\alpha_j \geq 0$ e $j = 1...r$, i.e., it can be defined a Lyapunov's function Q_j for each vertex j. As $Q(\alpha)$ depends on α, the Lyapunov matrix use fits to time-invariant polytopic uncertainties, being permitted rate of variation sufficiently small. To verify this, Theorem 5.2 is proposed.

Theorem 5.2. *A sufficient condition which guarantees the stability of the uncertain system (20) is the existence of matrices* $Y \in \mathbb{R}^{n \times n}$, $Q \in \mathbb{R}^{n \times n}$, $Q_j = Q_j' > 0$ e $G \in \mathbb{R}^{m \times n}$, *decay rate greater than* γ *and a fixed constant* $a > 0$ *such that the LMIs (42) and (43) are met.*

$$\begin{bmatrix} A_j Y + Y' A_j' - B_j G - G' B_j' + 2\gamma Q_j & Q_j + aY' A_j' - aG' B_j' - Y \\ Q_j + aA_j Y - aB_j G - Y' & -aY - aY' \end{bmatrix} < 0 \qquad (42)$$

$$Q_j > 0 \qquad (43)$$

with $j = 1, ..., r$. *When the LMIs (42) and (43) are feasible, a state feedback matrix which stabilizes the system can be given by (44).*

$$K = GY^{-1} \qquad (44)$$

Proof. Multiplying (42) and (43) by $\alpha_j \geq 0$, and adding in j, for $j = 1$ to $j = N$, LMIs (45) and (46) are found.

$$\begin{bmatrix} (\sum_{j=1}^{r} \alpha_j A_j)Y + Y'(\sum_{j=1}^{r} \alpha_j A_j)' - (\sum_{j=1}^{r} \alpha_j B_j)G - G'(\sum_{j=1}^{r} \alpha_j B_j)' + 2\gamma(\sum_{j=1}^{r} \alpha_j Q_j) \\ (\sum_{j=1}^{r} \alpha_j Q_j) + a(\sum_{j=1}^{r} \alpha_j A_j)Y - a(\sum_{j=1}^{r} \alpha_j B_j)G - Y' \end{bmatrix}$$

$$\begin{bmatrix} (\sum_{j=1}^{r} \alpha_j Q_j) + aY'(\sum_{j=1}^{r} \alpha_j A_j)' - aG'(\sum_{j=1}^{r} \alpha_j B_j)' - Y \\ -aY - aY' \end{bmatrix} < 0$$

$$\sum_{j=1}^{r} \alpha_j Q_j > 0$$

$$\begin{bmatrix} A(\alpha)Y + Y'A(\alpha)' - B(\alpha)G - G'B(\alpha)' + 2\gamma Q(\alpha) & Q(\alpha) + aY'A(\alpha)' - aG'B(\alpha)' - Y \\ Q(\alpha) + aA(\alpha)Y - aB(\alpha)G - Y' & -aY - aY' \end{bmatrix} < 0 \qquad (45)$$

$$Q(\alpha) > 0 \qquad (46)$$

with $Q(\alpha) = \sum\limits_{j=1}^{r} \alpha_j Q_j$, $\sum\limits_{j=1}^{r} \alpha_j = 1$, $\alpha_j \geq 0$ and $j = 1...r$. □

Thus, the uncertain system shown can be fed back in (19) with (45) and (46) sufficient conditions for asymptotic stability of the polytope.

Observation 1. *In the LMIs (42) and (43), the constant "a" has to be fixed for all vertexes and to satisfy the LMIs and it can be found through a one-dimensional search.*

5.3 Optimization of the K matrix norm using Finsler's lemma

The motivation for the study of an alternative optimization of the K matrix norm of state feedback control was due to less conservative results obtained with Finsler's lemma. This way expecting to find, for some situations, controllers with lower gains, thus being easier to implement than those designed using the existing quadratic stability theory (Faria et al., 2010), avoiding the signal control saturations.

Some difficulty in applying the existing theorem (Faria et al., 2010) was found to the new structure of LMIs, as the controller synthesis matrix Y is not symmetric, a condition that was necessary for the development of Theorem (4.1) when the controller synthesis matrix was $X = P^{-1}$. Thus, Theorem 5.3 is proposed.

Theorem 5.3. *A constraint for the $K \in \mathbb{R}^{m \times n}$ matrix norm of state feedback can be obtained, with $K = GY^{-1}$ and $Q_j = Y'P_jY$, being $Y \in \mathbb{R}^{n \times n}$, $G \in \mathbb{R}^{m \times n}$ and $P \in \mathbb{R}^{n \times n}$, $P_j = P'_j > 0$ finding the minimum β, $\beta > 0$, such that $K'K < \beta P_j$, $j = 1...N$. You can get the optimal value of β solving the optimization problem with the LMIs (47) and (48).*

$$\min \beta$$
$$s.t. \begin{bmatrix} Q_j & G' \\ G & \beta I_m \end{bmatrix} > 0 \tag{47}$$

$$\begin{bmatrix} A_jY + Y'A_j' - B_jG - G'B_j' + 2\gamma Q_j & Q_j + aY'A_j' - aG'B_j' - Y \\ Q_j + aA_jY - aB_jG - Y' & -aY - aY' \end{bmatrix} < 0 \tag{48}$$

where I_m denotes the identity matrix of m order.

Proof. Applying the Schur complement for (47) results in (49).

$$\beta I_m > 0 \ and \ Q_j - G'(\beta I_m)^{-1}G > 0 \tag{49}$$

Thus, from (49), (50) is found.

$$Q_j > \frac{1}{\beta}G'G \Rightarrow G'G < \beta Q_j \tag{50}$$

Replacing $G = KY$ and $Q_j = Y'P_jY$ in (50), (51) is met.

$$Y'K'KY < \beta Y'P_jY \Rightarrow K'K < \beta P_j \tag{51}$$

on which K is the optimal controller associated with (42) and (43). □

Thus it was possible the adequacy of the proposed optimization method with the minimization of a scalar β, using the inequality of minimization $K'K < \beta P_j$ with P_j the Lyapunov's matrix, to the new relaxed parameters.

5.4 Stability of systems using reciprocal projection lemma restricted by the decay rate

Another tool that can be used for stability analysis using LMIs is the reciprocal projection lemma (Apkarian et al., 2001) set out in Lemma 5.3.

Lemma 5.3 (reciprocal projection lemma). *Consider* $Y = Y' > 0$ *a given matrix. The following statements are equivalent*

1. $\psi + S + S' < 0$
2. *The following LMI is feasible for W*

$$\begin{bmatrix} \psi + Y - (W + W') & S' + W' \\ S + W & -Y \end{bmatrix} < 0$$

Proof. Reciprocal projection lemma proof can be found at (Apkarian et al., 2001).   □

Consider the Lyapunov's inequality subject to a decay rate given by (15) and (16), which can be rewritten as (52) and (53).

$$(A - BK)X + X(A - BK)' + 2\gamma X < 0 \tag{52}$$

$$X > 0 \tag{53}$$

where $X \triangleq P^{-1}$ e P is the Lyapunov's matrix. The original Lyapunov's inequality (15) can be recovered by multiplying the inequality (52) on the left and on the right by P.

Assuming $\psi \triangleq 0$ e $S' = (A - BK)X + \gamma X$, it will be verified that the first claim of the reciprocal projection lemma will be exactly Lyapunov's inequality subject to the decay rate described in (52):

$$\psi + S + S' = (A - BK)X + X(A - BK)' + 2\gamma X < 0$$

From the reciprocal projection lemma, if the first statement is true, then the second one will also be true as (54) shows.

$$\begin{bmatrix} Y - (W + W') & (A - BK)X + \gamma X + W' \\ X(A - BK)' + \gamma X + W & -Y \end{bmatrix} < 0 \tag{54}$$

Multiplying (54) on the left and on the right by $\begin{bmatrix} I & 0 \\ 0 & X^{-1} \end{bmatrix}$ with $P \triangleq X^{-1}$ results in (55).

$$\begin{bmatrix} Y - (W + W') & (A - BK) + \gamma I + W'P \\ (A - BK)' + \gamma I + PW & -PYP \end{bmatrix} < 0 \tag{55}$$

Multiplying (55) on the left and on the right by $\begin{bmatrix} W'^{-1} & 0 \\ 0 & I \end{bmatrix}$ and $\begin{bmatrix} W^{-1} & 0 \\ 0 & I \end{bmatrix}$ respectively with $V \triangleq W^{-1}$, (56) is found.

$$\begin{bmatrix} V'YV - (V + V') & V'(A - BK) + \gamma V' + P \\ (A - BK)'V + \gamma V + P & -PYP \end{bmatrix} < 0 \tag{56}$$

Applying the Schur complement in $V'YV$, (57) is found.

$$\begin{bmatrix} -(V+V') & V'(A-BK)+\gamma V'+P & V' \\ (A-BK)'V+\gamma V+P & -PYP & 0 \\ V & 0 & -Y^{-1} \end{bmatrix} < 0 \qquad (57)$$

performing the linearizing variable change $Y \triangleq P^{-1}$ results in (58).

$$\begin{bmatrix} -(V+V') & V'(A-BK)+\gamma V'+P & V' \\ (A-BK)'V+\gamma V+P & -P & 0 \\ V & 0 & -P \end{bmatrix} < 0 \qquad (58)$$

In literature it can be found a formulation close to the insertion of the decay rate but with different positioning of the parameter of decay rate (Shen et al., 2006). It is easy to verify that some conservatism was introduced with the choice of $Y \triangleq P^{-1}$, but the state feedback matrix is unrelated to the Lyapunov's matrix P, which results in relaxation of Lyapunov's LMI. Using the dual form $(A-BK) \rightarrow (A-BK)'$ (Apkarian et al., 2001) results in inequality (59).

$$\begin{bmatrix} -(V+V') & V'(A-BK)'+\gamma V'+P & V' \\ (A-BK)V+\gamma V+P & -P & 0 \\ V & 0 & -P \end{bmatrix} < 0 \qquad (59)$$

Performing the change of variable $Z \triangleq KV$ and inserting the constraint $P > 0$, the LMIs (60) and (60) that guarantee system stability can be found.

$$\begin{bmatrix} -(V+V') & V'A' - Z'B' + \gamma V' + P & V' \\ AV - BZ + \gamma V + P & -P & 0 \\ V & 0 & -P \end{bmatrix} < 0 \qquad (60)$$

$$P > 0 \qquad (61)$$

The inequalities (60) and (61) are LMIs, and being feasible, it is deduced a state feedback matrix that can stabilize the system (9) - (10) given by (62).

$$K = ZV^{-1} \qquad (62)$$

The result of relaxation of LMIs is interesting in the design of robust controllers, proposed below.

5.5 Robust stability of systems using reciprocal projection lemma restricted by the decay rate

A stability analysis for a robust stability condition can be performed considering the continuous time linear system an convex combination of r vertexes of the polytope described in (20). As in the extended stability case, the advantage of using the reciprocal projection lemma for robust stability analysis is the Lyapunov's function degree of freedom, now defined as $P(\alpha) = \sum_{j=1}^{r} \alpha_j P_j$, $\sum_{j=1}^{r} \alpha_j = 1$, $\alpha_j \geq 0$ e $j = 1...r$, i.e., it is defined a Lyapunov's function P_j for each vertex j. As described before Theorem 5.2, the use of $P(\alpha)$ fits to time-invariant polytopic uncertainties, being permitted rate of variation sufficiently small. To verify this, Theorem 5.4 is proposed.

Theorem 5.4. *A sufficient condition which guarantees the stability of the uncertain system (20) is the existence of matrices $V \in \mathbb{R}^{n \times n}$, $P_j = P_j' \in \mathbb{R}^{n \times n}$ and $Z \in \mathbb{R}^{m \times n}$, such that LMIs (63) and (64) are met.*

$$
\begin{bmatrix}
-(V+V') & V'A_j' - Z'B_j' + \gamma V' + P_j & V' \\
A_j V - B_j Z + \gamma V + P_j & -P_j & 0 \\
V & 0 & -P_j
\end{bmatrix} < 0 \tag{63}
$$

$$
P_j > 0 \tag{64}
$$

with $j = 1, ..., r$.

When the LMIs (63) and (64) are feasible, a state feedback matrix which stabilizes the system can be given by (65).

$$
K = ZV^{-1} \tag{65}
$$

Proof. Multiplying (63) and (64) by $\alpha_j \geq 0$, and adding in j, for $j = 1$ to $j = N$, (65) and (66) are found.

$$
\begin{bmatrix}
-(V+V') & V'(\sum_{j=1}^{r} \alpha_j A'_j) - Z'(\sum_{j=1}^{r} \alpha_j B'_j) + \gamma V' + (\sum_{j=1}^{r} \alpha_j P_j) & V' \\
(\sum_{j=1}^{r} \alpha_j A_j)V - (\sum_{j=1}^{r} \alpha_j B_j)Z + \gamma V + (\sum_{j=1}^{r} \alpha_j P_j) & -(\sum_{j=1}^{r} \alpha_j P_j) & 0 \\
V & 0 & -(\sum_{j=1}^{r} \alpha_j P_j)
\end{bmatrix} < 0
$$

$$
(\sum_{j=1}^{r} \alpha_j P_j) > 0
$$

$$
\begin{bmatrix}
-(V+V') & V'A'(\alpha) - Z'B'(\alpha) + \gamma V' + P(\alpha) & V' \\
A(\alpha)V - B(\alpha)Z + \gamma V + P(\alpha) & -P(\alpha) & 0 \\
V & 0 & -P(\alpha)
\end{bmatrix} < 0 \tag{65}
$$

$$
P(\alpha) > 0 \tag{66}
$$

with $P(\alpha) = \sum_{j=1}^{r} \alpha_j P_j$, $\sum_{j=1}^{r} \alpha_j = 1$, $\alpha_j \geq 0$ and $j = 1...r$. $\qquad \square$

It appears that K is unique and there are r Lyapunov's matrices P_j, generating a relaxation in the LMIs. The same trend was observed in the formulation via Finsler's lemma in which variables were the Lyapunov's matrices Q_j, but in (65) and (66) there is a greater degree of freedom with the inclusion of V in the design of the control matrix K, V being totally disconnected from P_j, $j = 1, ..., n$.

5.6 Optimization of the K matrix norm using reciprocal projection lemma

A study was carried out to fit the LMIs to the new relaxed parameters once the state feedback matrix K is completely detached from the Lyapunov's matrix $P(\alpha)$. Therefore, relevant changes took place in the optimization proposed in this study to suit the reciprocal projection lemma. This optimization has provided interesting results in practice.

Due to the lack of relations to assemble LMI able to optimize the module of K it was proposed a minimization procedure similar to the optimization procedure for redesign presented in (Chang et al., 2002) inserting an extra restriction to the LMIs (63) and (64).

Thus Theorem 5.5 was proposed.

Theorem 5.5. *A constraint for the $K \in \mathbb{R}^{m \times n}$ matrix norm of state feedback is obtained, with $K = ZV^{-1}$, $V \in \mathbb{R}^{n \times n}$ and $Z \in \mathbb{R}^{m \times n}$ finding the minimum β, $\beta > 0$, such that $K'K < \beta M$, being $M = V'^{-1}V^{-1}$ and therefore $M = M' > 0$. You can get the optimal value of β solving the optimization problem with the LMIs (67) and (68).*

$$\min \beta$$

$$s.t. \begin{bmatrix} I_n & Z' \\ Z & \beta I_m \end{bmatrix} > 0 \tag{67}$$

$$(Set\ of\ LMIs\ (63)\ and\ (64)) \tag{68}$$

which I_m and I_n denote the identity matrices of m and n order respectively.

Proof. Applying the Schur complement in (67) results in (69).

$$\beta I_m > 0 \ e \ I_n - Z'(\beta I_m)^{-1}Z > 0 \tag{69}$$

Thus, from (69), (70) is found.

$$I_n > \frac{1}{\beta}Z'Z \Rightarrow Z'Z < \beta I_n \tag{70}$$

Replacing $Z = KV$ in (70) results in (71).

$$V'K'KV < \beta I_n \tag{71}$$

Multiplying on the left and on the right (71) for V'^{-1} e V^{-1} respectively and naming $V'^{-1}V^{-1} = M$ (72) is met.

$$V'K'KV < \beta I_n \Rightarrow K'K < \beta M \tag{72}$$

where K is the optimal controller associated with (63) and (64). $\square$

Due to M being defined as $M = V'^{-1}V^{-1}$ and so $M = M' > 0$, it is possible to find a relationship that optimizes the matrix K minimizing a scalar β, with the relation of minimizing $K'K < \beta M$.

6. Practical application in the 3-DOF helicopter

Consider the schematic model in Figure (2) of the 3-DOF helicopter (Quanser, 2002) shown in Figure (1). Two DC motors are mounted at the two ends of a rectangular frame and drive two propellers. The motors axis are parallel and the thrust vector is normal to the frame. The helicopter frame is suspended from the instrumented joint mounted at the end of a long arm and is free to pitch about its center (Quanser, 2002).

The arm is gimbaled on a 2-DOF instrumented joint and is free to pitch and yaw. The other end of the arm carries a counterweight such that the effective mass of the helicopter is light enough for it to be lifted using the thrust from the motors. A positive voltage applied to the front motor causes a positive pitch while a positive voltage applied to the back motor causes a negative pitch (angle *pitch* (ρ)). A positive voltage to either motor also causes an elevation of the body (i.e., pitch of the arm). If the body pitches, the thrust vectors result in a travel of the body (i.e., yaw (ε) of the arm) as well. If the body pitches, the impulsion vector results in the displacement of the system (i.e., travel (λ) of the system).

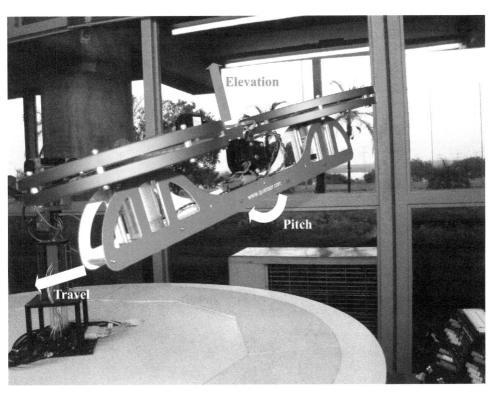

Fig. 1. Quanser's 3-DOF helicopter of UNESP - Campus Ilha Solteira.

The objective of this experiment is to design a control system to track and regulate the elevation and travel of the 3-DOF Helicopter.

The 3-DOF Helicopter can also be fitted with an active mass disturbance system that will not be used in this work.

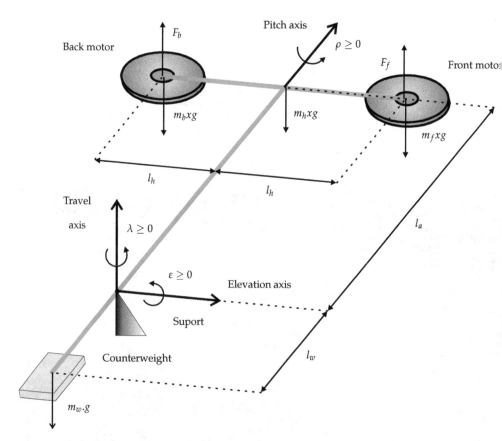

Fig. 2. Schematic drawing of 3-DOF Helicopter

The state space model that describes the helicopter is (Quanser, 2002) shown in (73).

$$\begin{bmatrix} \dot{\varepsilon} \\ \dot{\rho} \\ \dot{\lambda} \\ \ddot{\varepsilon} \\ \ddot{\rho} \\ \ddot{\lambda} \\ \dot{\varsigma} \\ \dot{\gamma} \end{bmatrix} = A \begin{bmatrix} \varepsilon \\ \rho \\ \lambda \\ \dot{\varepsilon} \\ \dot{\rho} \\ \dot{\lambda} \\ \varsigma \\ \gamma \end{bmatrix} + B \begin{bmatrix} V_f \\ V_b \end{bmatrix} \qquad (73)$$

The variables ζ and γ represent the integrals of the angles ε of yaw and λ of travel, respectively. The matrices A and B are presented in sequence.

$$
A = \begin{bmatrix}
0 & 0 & 0\ 1\ 0\ 0\ 0\ 0 \\
0 & 0 & 0\ 0\ 1\ 0\ 0\ 0 \\
0 & 0 & 0\ 0\ 0\ 1\ 0\ 0 \\
0 & 0 & 0\ 0\ 0\ 0\ 0\ 0 \\
0 & 0 & 0\ 0\ 0\ 0\ 0\ 0 \\
0 & \dfrac{2m_f l_a - m_w l_w g}{2m_f l_a{}^2 + 2m_f l_h{}^2 + m_w l_w{}^2} & 0\ 0\ 0\ 0\ 0\ 0 \\
1 & 0 & 0\ 0\ 0\ 0\ 0\ 0 \\
0 & 0 & 1\ 0\ 0\ 0\ 0\ 0
\end{bmatrix}
\quad \text{and} \quad
B = \begin{bmatrix}
0 & 0 \\
0 & 0 \\
0 & 0 \\
\dfrac{l_a k_f}{m_w l_w^2 + 2m_f l_a^2} & \dfrac{l_a k_f}{m_w l_w^2 + 2m_f l_a^2} \\
\dfrac{1}{2}\dfrac{k_f}{m_f l_h} & -\dfrac{1}{2}\dfrac{k_f}{m_f l_h} \\
0 & 0 \\
0 & 0 \\
0 & 0
\end{bmatrix}
$$

The values used in the project were those that appear in the MATLAB programs for implementing the original design manufacturer, to maintain fidelity to the parameters. The constants used are described in Table (1).

Power constant of the propeller (found experimentally)	k_f	0.1188
Mass of the helicopter body (kg)	m_h	1.15
Mass of counterweight (kg)	m_w	1.87
Mass of the whole front of the propeller (kg)	m_f	$m_h/2$
Mass of the whole back of the propeller (kg)	m_b	$m_h/2$
Distance between each axis of pitch and motor (m)	l_h	0.1778
Distance between the lift axis and the body of the helicopter (m)	l_a	0.6604
Distance between the axis of elevation and the counterweight (m)	l_w	0.4699
Gravitational constant (m/s^2)	g	9.81

Table 1. Helicopter parameters

Practical implementations of the controllers were carried out in order to view the controller acting in real physical systems subject to failures.

The trajectory of the helicopter was divided into three stages. The first stage is to elevate the helicopter 27.5° reaching the yaw angle $\varepsilon = 0°$. In the second stage the helicopter travels 120°, keeping the same elevation i.e., the helicopter reaches $\lambda = 120°$ with reference to the launch point. In the third stage the helicopter performs the landing recovering the initial angle $\varepsilon = -27.5°$.

During the landing stage, more precisely in the instant 22 s, the helicopter loses 30% of the power back motor. The robust controller should maintain the stability of the helicopter and have small oscilation in the occurrence of this failure.

To add robustness to the system without any physical change, a 30% drop in power of the back motor is forced by inserting a timer switch connected to an amplifier with a gain of 0.7 in tension acting directly on engine, and thus being constituted a polytope of two vertexes with an uncertainty in the input matrix of the system acting on the helicopter voltage between $0, 7V_b$ and V_b. The polytope described as follows.

Vertex 1 (100% of V_b):

$$A_1 = \begin{bmatrix} 0 & 0 & 0\,1\,0\,0\,0\,0 \\ 0 & 0 & 0\,0\,1\,0\,0\,0 \\ 0 & 0 & 0\,0\,0\,1\,0\,0 \\ 0 & 0 & 0\,0\,0\,0\,0\,0 \\ 0 & 0 & 0\,0\,0\,0\,0\,0 \\ 0 & -1.2304 & 0\,0\,0\,0\,0\,0 \\ 1 & 0 & 0\,0\,0\,0\,0\,0 \\ 0 & 0 & 1\,0\,0\,0\,0\,0 \end{bmatrix} \quad \text{and} \quad B_1 = \begin{bmatrix} 0 & 0 \\ 0 & 0 \\ 0 & 0 \\ 0.0858 & 0.0858 \\ 0.5810 & -0.5810 \\ 0 & 0 \\ 0 & 0 \\ 0 & 0 \end{bmatrix}$$

Vertex 2 (70% of V_b):

$$A_2 = \begin{bmatrix} 0 & 0 & 0\,1\,0\,0\,0\,0 \\ 0 & 0 & 0\,0\,1\,0\,0\,0 \\ 0 & 0 & 0\,0\,0\,1\,0\,0 \\ 0 & 0 & 0\,0\,0\,0\,0\,0 \\ 0 & 0 & 0\,0\,0\,0\,0\,0 \\ 0 & -1.2304 & 0\,0\,0\,0\,0\,0 \\ 1 & 0 & 0\,0\,0\,0\,0\,0 \\ 0 & 0 & 1\,0\,0\,0\,0\,0 \end{bmatrix} \quad \text{and} \quad B_2 = \begin{bmatrix} 0 & 0 \\ 0 & 0 \\ 0 & 0 \\ 0.0858 & 0.0601 \\ 0.5810 & -0.4067 \\ 0 & 0 \\ 0 & 0 \\ 0 & 0 \end{bmatrix}$$

Fixing the decay rate equal to 0.8, there were designed: a controller with quadratic stability using the existing optimization (Assunção et al., 2007c), a controller with quadratic stability with the proposed optimization and controllers with extended stability and projective stability also with the proposed optimization to perform the practical implementation.

The controller designed by quadratic stability with existing optimization (Theorem 4.1) is shown in (74) (Assunção et al., 2007c).

$$K = \begin{bmatrix} -46.4092 & -15.6262 & 21.3173 & -24.7541 & -3.9269 & 23.5800 & -27.4973 & 7.4713 \\ -70.3091 & 13.3795 & -10.1982 & -37.5960 & 4.3357 & -15.1521 & -41.5328 & -2.7935 \end{bmatrix} \tag{74}$$

where $||K|| = 107.83$.

This controller was implemented in helicopter and the results are shown in Figure 3.

In (75) follows the quadratic stability controller design with the proposed optimization follows (Theorem 5.1).

$$K = \begin{bmatrix} -18.8245 & -12.2370 & 10.9243 & -13.9612 & -4.4480 & 14.6213 & -9.1334 & 3.2483 \\ -27.9219 & 10.6586 & -7.6096 & -20.1096 & 4.5602 & -11.0774 & -13.7202 & -2.2629 \end{bmatrix} \tag{75}$$

where $||K|| = 44.88$.

This controller was implemented in the helicopter and the results are shown in Figure 4.

In (76) follows the extended stability controller design with the proposed optimization follows (Theorem 5.3). For this LMIs an $a = 10^{-6}$ solves the problem. Though the Theorem 5.3

and Theorem 5.5 hypothesis establishes a sufficiently low time variation of α. However, for comparison purposes of Theorem 5.3 and Theorem 5.5 with Theorem 5.1, the same abrupt loss of power test was done with controllers (76) and (77).

$$K = \begin{bmatrix} -23.7152 & -12.9483 & 9.8587 & -18.7322 & -4.9737 & 14.3283 & -10.7730 & 2.6780 \\ -33.8862 & 15.2923 & -11.6132 & -25.4922 & 6.0776 & -16.5503 & -15.8350 & -3.4475 \end{bmatrix} \qquad (76)$$

where $||K|| = 56.47$.

This controller was implemented in the helicopter and the results are shown in Figure (5).

In (77) follows the projective stability controller design with the proposed optimization follows (Theorem 5.5).

$$K = \begin{bmatrix} -50.7121 & -28.7596 & 35.1829 & -29.8247 & -7.9563 & 41.0906 & -28.8974 & 11.7405 \\ -66.5405 & 31.9853 & -34.7642 & -38.3173 & 9.9376 & -42.0298 & -38.3418 & -11.8207 \end{bmatrix} \qquad (77)$$

where $||K|| = 110.46$.

This controller was implemented in the helicopter and the results are shown in Figure 6.

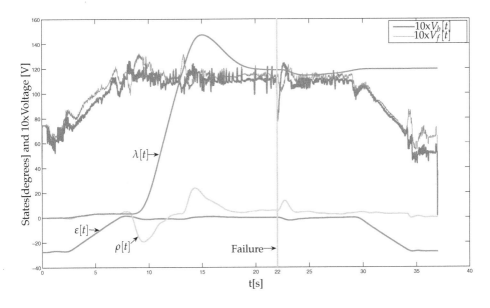

Fig. 3. Practical implementation of the designed K by quadratic stability with the optimization method presented in (Assunção et al., 2007c).

The graphics of Figures 3, 4, 5 and 6, refer to the actual data of the angles and voltages on the front motor (V_f) and back motor (V_b) measured with the designed controllers acting on the plant during the trajectory described as a failure in the instant 22 s. Tensions (V_f) and (V_b) on the motors were multiplied by 10 to match the scales of the two graphics.

Note that the variations of the amplitudes of (V_f) and (V_b) using optimized controllers proposed (75) and (76) in Figures 4 and 5 are smaller than those obtained with the existing

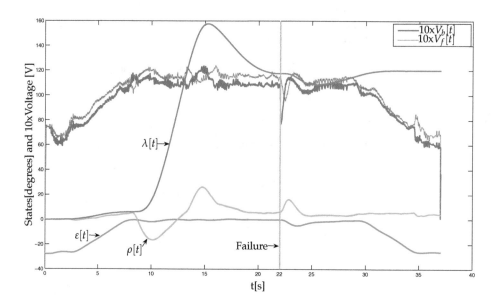

Fig. 4. Practical implementation of the K designed by quadratic stability with the proposed optimization method.

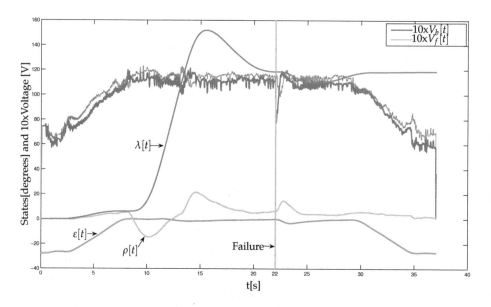

Fig. 5. Practical implementation of the K designed by extended stability with the proposed optimization method.

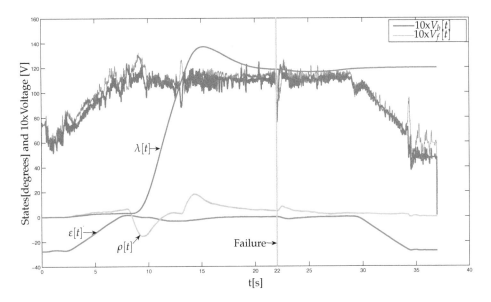

Fig. 6. Practical implementation of the K designed by projective stability with the proposed optimization method.

controller in the literature (74) shown in Figure 3. This is due to the fact that our proposed controllers (75) and (76) have lower gains then (74). For this implementation the projective stability designed controllers with proposed optimization (77) obtained the worst results as Figure 6.

It was checked that the γ used in the implementation of robust controllers, if higher, forces the system to have a quick and efficient recovery, with small fluctuations.

7. General comparison of the two optimization methods

In order to obtain more satisfactory results on which would be the best way to optimize the norm of K, a more general comparison has been made between the two methods as Theorems 4.1 and 5.1.

There were randomly generated 1000 uncertain polytopes of second order systems, with only one uncertain parameter (two vertexes) and after that, 1000 uncertain polytopes of fourth order uncertain systems, with two uncertain parameter (four vertexes). The 1000 uncertain polytopes were generated feasible in at least one case of optimization for $\gamma = 0.5$, and the consequences of γ increase were analyzed and plotted in a bar charts showing the number of controllers with lower norm due to the increase of γ, shown in Figure 7 for second-order systems and in Figure 8 to fourth-order systems.

The controllers designed with elevated values of γ do not have much practical application due to the fact that the increase of γ affect the increasing of the norm and make higher peaks of the transient oscillation, used here only for the purpose of analyzing feasibility and better

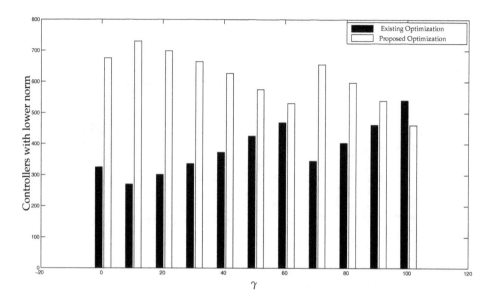

Fig. 7. Number of controllers with lower norm for 1000 uncertain politopic systems of second-order randomly generated.

results for the norm of K, so comparisons were closed in $\gamma = 100.5$, because this γ is already considered high.

In Figure 8 can be seen that the proposed optimization method produces better results for all cases analyzed. Due to the complexity of the polytopes used in this case (fourth-order uncertain systems with two uncertainties (four vertexes)), is natural a loss of feasibility with the increase of γ, and yet the proposed method shows very good results.

8. General comparison of the new design and optimization methods

A generic comparison between the three methods of design and optimization of K was also carried out: design by quadratic stability with proposed optimization shown in Theorem 5.1, design and proposed optimization with extended stability shown in Theorem 5.3 (using the parameter $a = 10^{-6}$ in the LMIs) and projective stability design with proposed optimization shown in Theorem 5.5.

Initially 1000 polytopes of second order uncertain systems were randomly generated, with only one uncertain parameter (two vertexes) and after that, fourth order uncertain systems, with two uncertain parameter (four vertexes). The 1000 polytopes were generated feasible in at least one case of optimization for $\gamma = 0.5$ and the consequences of γ increase were analyzed. In fourth-order uncertain systems, the 1000 polytopes were generated feasible in at least one case of optimization for $\gamma = 0.2$ and then, the consequences of γ of 0.2 in 0.2 increase were analyzed. This comparison was carried out with the intention of examining feasibility and better results for the norm of K. So, a bar graphics showing the number of controllers with lower norm with the increase of γ was plotted, and is shown in Figures 9 and 10.

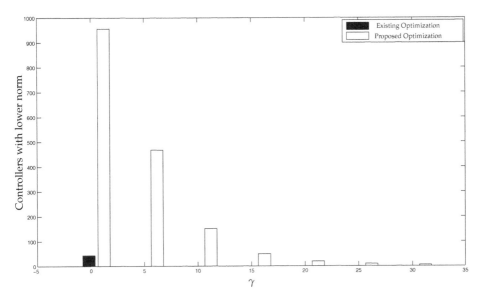

Fig. 8. Number of controllers with lower norm for 1000 uncertain politopic systems of fourth-order randomly generated.

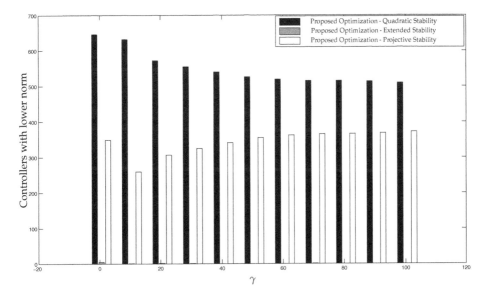

Fig. 9. Number of controllers with lower norm for 1000 uncertain politopic systems of second-order randomly generated. All these methods are proposed in this work.

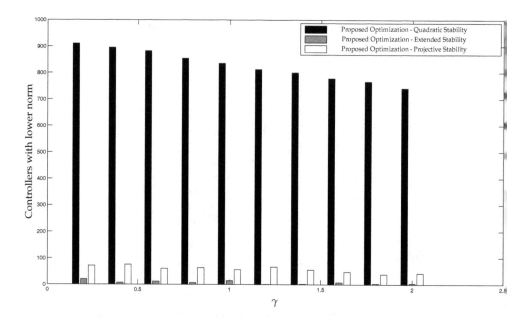

Fig. 10. Number of controllers with lower norm for 1000 uncertain polytopic systems of fourth-order randomly generated. All these methods are proposed in this work.

Both figures 9 and 10 show that the proposed optimization method using quadratic stability showed better results for the controller norm with the increase of γ, due to optimization this method no longer depend on the matrices that guarantee system stability as it can be seen in equation (22). In contrast, using the proposed optimizations with extended stability and projective stability, they still depend on the matrices that guarantee system stability as seen in equations (51) and (72) and this is the obstacle to finding better results for these methods.

9. Conclusions

At the 3-DOF helicopter practical application, the controllers designed with the proposed optimization showed lower values of the controller's norm designed by the existing optimization with quadratic stability, except the design for projective stability which had the worst value of the norm for this case, thus showing the advantage of the proposed method regarding implementation cost and required effort on the motors. These characteristics of optimality and robustness make our design methodology attractive from the standpoint of practical applications for systems subject to structural failure, guaranteeing robust stability and small oscillations in the occurrence of faults.

It is clear that the design of K via the optimization proposed here achieved better results than the existing optimizing K (Assunção et al., 2007c), using the LMI quadratic stability for second order polytopes with one uncertainty. The proposed optimization project continued to show better results even when the existing optimization has become totally infeasible for fourth order polytopes with two uncertainties.

By comparing the three optimal design methods proposed here (quadratic stability, extended stability, and projective stability) it can be concluded that the design using quadratic stability had a better performance for both analysis: 1000 second order polytopes with one uncertainty and for the 1000 fourth order polytopes with two uncertainties, showing so that the proposed optimization ensures best results when used with the quadratic stability.

10. References

Apkarian, P., Tuan, H. D. & Bernussou, J. (2001). Continuous-time analysis, eigenstructure assignment, and $\mathcal{H}_2$ synthesis with enhanced linear matrix inequalities (LMI) characterizations, *IEEE Transactions on Automatic Control* 46(12): 1941–1946.

Assunção, E., Andrea, C. Q. & Teixeira, M. C. M. (2007a). $\mathcal{H}_2$ and $\mathcal{H}_\infty$-optimal control for the tracking problem with zero variation, *IET Control Theory Applications* 1(3): 682–688.

Assunção, E., Marchesi, H. F., Teixeira, M. C. M. & Peres, P. L. D. (2007b). Global optimization for the $\mathcal{H}_\infty$-norm model reduction problem, *International Journal of Systems Science* 38(2): 125–138.

Assunção, E., Teixeira, M. C. M., Faria, F. A., Silva, N. A. P. D. & Cardim, R. (2007c). Robust state-derivative feedback LMI-based designs for multivariable linear systems., *International Journal of Control* 80(8): 1260–1270.

Boyd, S., Ghaoui, L. E., Feron, E. & Balakrishnan, V. (1994). *Linear Matrix Inequalities in Systems and Control Theory*, Studies in Applied Mathematics, 15, 2nd edn, SIAM Studies in Applied Mathematics.

Cardim, R., Teixeira, M., Faria, F. & Assuncao, E. (2009). Lmi-based digital redesign of linear time-invariant systems with state-derivative feedback, *Control Applications, (CCA) Intelligent Control, (ISIC), 2009 IEEE*, pp. 745 –749.

Chang, W., H.J., J. P., Lee & Joo, Y. (2002). LMI approach to digital redesign of linear time-invariant systems, *Control Theory and Applications, IEEE Proceedings* 149(4): 297–302.

Chen, C. (1999). *Linear system theory and design*, Oxford Series in Electrical and Computer Engineering, 3rd edn, Oxford, New York.

Chilali, M. & Gahinet, P. (1996). $\mathcal{H}_\infty$ design with pole placement constraints: an LMI approach, *IEEE Transactions on Automatic Control* 41(3): 358–367.

de Oliveira, M. C., Geromel, J. C. & Hsu, L. (1999). Lmi characterization of structural and robust stability: the discrete-time case, *Linear Algebra and its Applications* 296(1-3): 27 – 38.

Faria, F. A., Assunção, E., Teixeira, M. C. M. & Cardim, R. (2010). Robust state-derivative feedback LMI-based designs for linear descriptor systems, *Mathematical Problems in Engineering* 2010: 1–15.

Faria, F. A., Assunção, E., Teixeira, M. C. M., Cardim, R. & da Silva, N. A. P. (2009). Robust state-derivative pole placement LMI-based designs for linear systems, *International Journal of Control* 82(1): 1–12.

Feron, E., Apkarian, P. & Gahinet, P. (1996). Analysis and synthesis of robust control systems via parameter-dependent lyapunov functions, *IEEE Transactions on Automatic Control* 41(7): 1041–1046.

Gahinet, P., Nemirovski, A., Laub, A. J. & Chilali, M. (1995). *LMI control toolbox - for use with MATLAB*, The Math Works Inc., Natick.

Lee, K. H., Lee, J. H. & Kwon, W. H. (2004). Sufficient lmi conditions for the h infin; output feedback stabilization of linear discrete-time systems, *Decision and Control, 2004. CDC. 43rd IEEE Conference on*, Vol. 2, pp. 1742 – 1747 Vol.2.

Ma, M. & Chen, H. (2006). *Constrained H2 control of active suspensions using LMI optimization*, IEEE.

Montagner, V. F., Oliveira, R. C. L. F., Peres, P. L. D. & Bliman, P.-A. (2009). Stability analysis and gain-scheduled state feedback control for continuous-time systems with bounded parameter variations, *International Journal of Control* 82(6): 1045–1059.

Montagner, V., Oliveira, R., Leite, V. & Peres, P. (2005). Lmi approach for h infin; linear parameter-varying state feedback control, *Control Theory and Applications, IEE Proceedings -* 152(2): 195 – 201.

Oliveira, M. C., Bernussou, J. & Geromel, J. C. (1999). A new discrete-time robust stability condition, *Systems Control Letters* 37(4): 261–265.

Oliveira, M. C. & Skelton, R. E. (2001). *Perspectives in robust control*, 1st edn, Springer Berlin / Heidelberg, Berlin, chapter Stability tests for constrained linear systems, pp. 241–257.

Peaucelle, D., Arzelier, D., Bachelier, O. & Bernussou, J. (2000). A new robust d-stability condition for real convex polytopic uncertainty, *Systems and Control Letters* 40(1): 21–30.

Pipeleers, G., Demeulenaere, B., Swevers, J. & Vandenberghe, L. (2009). Extended LMI characterizations for stability and performance of linear systems, *Systems & Control Letters* 58(7): 510 – 518.

Quanser (2002). *3-DOF helicopter reference manual*, Quanser Inc., Markham.

Shen, Y., Shen, W. & Gu, J. (2006). A new extended lmis approach for continuous-time multiobjective controllers synthesis, *Mechatronics and Automation, Proceedings of the 2006 IEEE International Conference on*, pp. 1054–1059.

Šiljak, D. D. & Stipanovic, D. M. (2000). Robust stabilization of nonlinear systems: The LMI approach, *Mathematical Problems in Engineering* 6(5): 461–493.

Skelton, R. E., Iwasaki, T. E. & Grigoriadis, K. (1997). *A unified algebraic approach to control design*, Taylor & Francis, Bristol.

Wang, J., Li, X., Ge, Y. & Jia, G. (2008). An LMI optimization approach to lyapunov stability analysis for linear time-invariant systems, *Control and Decision Conference, 2008. CCDC 2008. Chinese*, pp. 3044 –3048.

3

On Control Design of Switched Affine Systems with Application to DC-DC Converters

E. I. Mainardi Júnior[1], M. C. M. Teixeira[1], R. Cardim[1],
M. R. Moreira[1], E. Assunção[1] and Victor L. Yoshimura[2]
[1]UNESP - Univ Estadual Paulista
[2]IFMT - Federal Institute of Education
Brazil

1. Introduction

In last years has been a growing interest of researchers on theory and applications of switched control systems, widely used in the area of power electronics (Cardim et al., 2009), (Deaecto et al., 2010), (Yoshimura et al., 2011), (Batlle et al., 1996), (Mazumder et al., 2002), (He et al., 2010) and (Cardim et al., 2011). The switched systems are characterized by having a switching rule which selects, at each instant of time, a dynamic subsystem among a determined number of available subsystems (Liberzon, 2003). In general, the main goal is to design a switching strategy of control for the asymptotic stability of a known equilibrium point, with adequate assurance of performance (Decarlo et al., 2000), (Sun & Ge, 2005) and (Liberzon & Morse, 1999). The techniques commonly used to study this class of systems consist of choosing an appropriate Lyapunov function, for instance, the quadratic (Feron, 1996), (Ji et al., 2005) and (Skafidas et al., 1999). However, in switched affine systems, it is possible that the modes do not share a common point of equilibrium. Therefore, sometimes the concept of stability should be extended using the ideas contained in (Bolzern & Spinelli, 2004) and (Xu et al., 2008). Problems involving stability analysis can many times be reduced to problems described by Linear Matrix Inequalities, also known as LMIs (Boyd et al., 1994) that, when feasible, are easily solved by some tools available in the literature of convex programming (Gahinet et al., 1995) and (Peaucelle et al., 2002). The LMIs have been increasingly used to solve various types of control problems (Faria et al., 2009), (Teixeira et al., 2003) and (Teixeira et al., 2006). This paper is structured as follows: first, a review of previous results in the literature for stability of switched affine systems with applications in power electronics is described (Deaecto et al., 2010). Next, the main goal of this paper is presented: a new theorem, which conditions hold when the conditions of the two theorems proposed in (Deaecto et al., 2010) hold. Later, in order to obtain a design procedure more general than those available in the literature (Deaecto et al., 2010), it was considered a new performance indice for this control system: bounds on output peak in the project based on LMIs. The theory developed in this paper is applied to DC-DC converters: Buck, Boost, Buck-Boost and Sepic. It is also the first time that this class of controller is used for controlling a Sepic DC-DC converter. The notation used is described below. For real matrices or vectors $(')$ indicates transpose. The set composed by the first N positive integers, $1, ..., N$ is denoted by $\mathbb{K}$. The set of all vectors $\lambda = (\lambda_1, ..., \lambda_N)'$ such that $\lambda_i \geq 0$, $i = 1, 2, ..., N$ and $\lambda_1 + \lambda_2 + ... + \lambda_N = 1$ is denoted by Λ. The convex combination of a set of matrices $(A_1, \ldots, A_N)$ is denoted by $A_\lambda = \sum_{i=1}^{N} \lambda_i A_i$, where $\lambda \in \Lambda$. The trace of a matrix P is denoted by $Tr(P)$.

2. Switched affine systems

Consider the switched affine system defined by the following state space realization:

$$\dot{x} = A_{\sigma(t)}x + B_{\sigma(t)}w, \quad x(0) = x_0 \tag{1}$$
$$y = C_{\sigma(t)}x, \tag{2}$$

as presented in (Deaecto et al., 2010), were $x(t) \in \mathbb{R}^n$ is the state vector, $y(t) \in \mathbb{R}^p$ is the controlled output, $w \in \mathbb{R}^m$ is the input supposed to be constant for all $t \geq 0$ and $\sigma(t)$: $t \geq 0 \rightarrow \mathbb{K}$ is the switching rule. For a known set of matrices $A_i \in \mathbb{R}^{n \times n}$, $B_i \in \mathbb{R}^{n \times m}$ and $C_i \in \mathbb{R}^p$, $i = 1, \ldots, N$, such that:

$$A_{\sigma(t)} \in \{A_1, A_2, \ldots, A_N\}, \tag{3}$$
$$B_{\sigma(t)} \in \{B_1, B_2, \ldots, B_N\}, \tag{4}$$
$$C_{\sigma(t)} \in \{C_1, C_2, \ldots, C_N\}, \tag{5}$$

the switching rule $\sigma(t)$ selects at each instant of time $t \geq 0$, a known subsystem among the N subsystems available. The control design problem is to determine a function $\sigma(x(t))$, for all $t \geq 0$, such that the switching rule $\sigma(t)$, makes a known equilibrium point $x = x_r$ of (1), (2) globally asymptotically stable and the controlled system satisfies a performance index, for instance, a guaranteed cost. The paper (Deaecto et al., 2010) proposed two solutions for these problems, considering a quadratic Lyapunov function and the guaranteed cost:

$$\min_{\sigma \in \mathbb{K}} \int_0^\infty (y - C_\sigma x_r)'(y - C_\sigma x_r)dt = \min_{\sigma \in \mathbb{K}} \int_0^\infty (x - x_r)'Q_\sigma(x - x_r)dt, \tag{6}$$

where $Q_\sigma = C_\sigma'C_\sigma \geq 0$ for all $\sigma \in \mathbb{K}$.

2.1 Previous results

Theorem 1. *(Deaecto et al., 2010) Consider the switched affine system (1), (2) with constant input* $w(t) = w$ *for all* $t \geq 0$ *and let the equilibrium point* $x_r \in \mathbb{R}^n$ *be given. If there exist* $\lambda \in \Lambda$ *and a symmetric positive definite matrix* $P \in \mathbb{R}^{n \times n}$ *such that*

$$A_\lambda'P + PA_\lambda + Q_\lambda < 0, \tag{7}$$
$$A_\lambda x_r + B_\lambda w = 0, \tag{8}$$

then the switching strategy

$$\sigma(x) = \arg\min_{i \in \mathbb{K}} \xi'(Q_i\xi + 2P(A_ix + B_iw)), \tag{9}$$

where $Q_i = C_i'C_i$ *and* $\xi = x - x_r$, *makes the equilibrium point* $x_r \in \mathbb{R}^n$ *globally asymptotically stable and from (6) the guaranteed cost*

$$J = \int_0^\infty (y - C_\sigma x_r)'(y - C_\sigma x_r)dt < (x_0 - x_r)'P(x_0 - x_r), \tag{10}$$

holds.

Proof. See (Deaecto et al., 2010). □

Remembering that similar matrices have the same trace, it follows the minimization problem (Deaecto et al., 2010):

$$\inf_{P>0} \left\{ Tr(P) : A'_\lambda P + PA_\lambda + Q_\lambda < 0, \lambda \in \Lambda \right\}. \tag{11}$$

The next theorem provides another strategy of switching, more conservative, but easier and simpler to implement.

Theorem 2. *(Deaecto et al., 2010) Consider the switched affine system (1), (2) with constant input $w(t) = w$ for all $t \geq 0$ and let the equilibrium point $x_r \in \mathbb{R}^n$ be given. If there exist $\lambda \in \Lambda$, and a symmetric positive definite matrix $P \in \mathbb{R}^{n \times n}$ such that*

$$A'_i P + PA_i + Q_i < 0, \tag{12}$$
$$A_\lambda x_r + B_\lambda w = 0, \tag{13}$$

for all $i \in \mathbb{K}$, then the switching strategy

$$\sigma(x) = arg \min_{i \in \mathbb{K}} \xi' P(A_i x_r + B_i w), \tag{14}$$

where $\xi = x - x_r$, makes the equilibrium point $x_r \in \mathbb{R}^n$ globally asymptotically stable and the guaranteed cost (10) holds.

Proof. See (Deaecto et al., 2010). □

Theorem 2 gives us the following minimization problem (Deaecto et al., 2010):

$$\inf_{P>0} \left\{ Tr(P) : A'_i P + PA_i + Q_i < 0, i \in \mathbb{K} \right\}. \tag{15}$$

Note that (12) is more restrictive than (7), because it must be satisfied for all $i \in \mathbb{K}$. However, the switching strategy (14) proposed in Theorem 2 is simpler to implement than the strategy (9) proposed in Theorem 1, because it uses only the product of ξ by constant vectors.

2.2 Main results

The new theorem, proposed in this paper, is presented below.

Theorem 3. *Consider the switched affine system (1), (2) with constant input $w(t) = w$ for all $t \geq 0$ and let $x_r \in \mathbb{R}^n$ be given. If there exist $\lambda \in \Lambda$, symmetric matrices $N_i, i \in \mathbb{K}$ and a symmetric positive definite matrix $P \in \mathbb{R}^{n \times n}$ such that*

$$A'_i P + PA_i + Q_i - N_i < 0, \tag{16}$$
$$A_\lambda x_r + B_\lambda w = 0, \tag{17}$$
$$N_\lambda = 0, \tag{18}$$

for all $i \in \mathbb{K}$, where $Q_i = Q'_i$, then the switching strategy

$$\sigma(x) = arg \min_{i \in \mathbb{K}} \xi' \left(N_i \xi + 2P(A_i x_r + B_i w) \right), \tag{19}$$

where $\xi = x - x_r$, makes the equilibrium point $x_r \in \mathbb{R}^n$ globally asymptotically stable and from (10), the guaranteed cost $J < (x_0 - x_r)' P(x_0 - x_r)$ holds.

Proof. Adopting the quadratic Lyapunov candidate function $V(\xi) = \xi'P\xi$ and from (1), (16), (17) and (18) note that for $\xi \neq 0$:

$$\dot{V}(\xi) = \dot{x}'P\xi + \xi'P\dot{x} = 2\xi'P(A_\sigma x + B_\sigma w) = \xi'(A_\sigma'P + PA_\sigma)\xi + 2\xi'P(A_\sigma x_r + B_\sigma w)$$

$$< \xi'(-Q_\sigma + N_\sigma)\xi + 2\xi'P(A_\sigma x_r + B_\sigma w) = \xi'(N_\sigma\xi + 2P(A_\sigma x_r + B_\sigma w)) - \xi'Q_\sigma\xi$$

$$= \min_{i \in \mathbb{K}} \left\{ \xi'(N_i\xi + 2P(A_i x_r + B_i w)) \right\} - \xi'Q_\sigma\xi$$

$$= \min_{\lambda \in \Lambda} \left\{ \xi'(N_\lambda\xi + 2P(A_\lambda x_r + B_\lambda w)) \right\} - \xi'Q_\sigma\xi$$

$$\leq -\xi'Q_\sigma\xi \leq 0. \tag{20}$$

Since $\dot{V}(\xi) < 0$ for all $\xi \neq 0 \in \mathbb{R}^n$, and $\dot{V}(0) = 0$, then $x_r \in \mathbb{R}^n$ is an equilibrium point globally asymptotically stable. Now, integrating (20) from zero to infinity and taking into account that $\dot{V}(\xi(\infty)) = 0$, we obtain (10). The proof is concluded. $\qquad\square$

Theorem 3 gives us the following minimization problem:

$$\inf_{P>0} \left\{ Tr(P) : A_i'P + PA_i + Q_i - N_i < 0, \quad N_\lambda = 0, \quad i \in \mathbb{K} \right\}. \tag{21}$$

The next theorem compares the conditions of Theorems 1, 2 and 3.

Theorem 4. *The following statements hold:*

(i) if the conditions of Theorem 1 are feasible, then the conditions of Theorem 3 are also feasible;

(ii) if the conditions of Theorem 2 are feasible, then the conditions of Theorem 3 are also feasible.

Proof. (*i*) Consider the symmetric matrices N_i, $i \in \mathbb{K}$, as described below:

$$N_i = (A_i'P + PA_i + Q_i) - (A_\lambda'P + PA_\lambda + Q_\lambda). \tag{22}$$

Then, multiplying (22) by λ_i and taking the sum from 1 to N it follows that

$$N_\lambda = \sum_{i=1}^{N} \lambda_i N_i = \sum_{i=1}^{N} \lambda_i(A_i'P + PA_i + Q_i) - \sum_{i=1}^{N} \lambda_i(A_\lambda'P + PA_\lambda + Q_\lambda)$$

$$= (A_\lambda'P + PA_\lambda + Q_\lambda) - (A_\lambda'P + PA_\lambda + Q_\lambda) = 0. \tag{23}$$

Now, from (16), (18) and (22) observe that

$$A_i'P + PA_i + Q_i - N_i = A_i'P + PA_i + Q_i - \left((A_i'P + PA_i + Q_i) - (A_\lambda'P + PA_\lambda + Q_\lambda)\right)$$

$$= A_\lambda'P + PA_\lambda + Q_\lambda < 0, \quad \forall i \in \mathbb{K}. \tag{24}$$

(*ii*) It follows considering that $N_i = 0$ in (16):

$$A_i'P + PA_i + Q_i - N_i = A_i'P + PA_i + Q_i < 0, \quad \forall i \in \mathbb{K}. \tag{25}$$

Thus, the proof of Theorem 4 is completed. $\qquad\square$

2.3 Bounds on output peak

Considering the limitations imposed by practical applications of control systems, often must be considered constraints in the design. Consider the signal:

$$s = H\xi, \tag{26}$$

where $H \in \mathbb{R}^{q \times n}$ is a known constant matrix, and the following constraint:

$$\max_{t \geq 0} \|s(t)\| \leq \psi_o, \tag{27}$$

where $\|s(t)\| = \sqrt{s(t)'s(t)}$ and ψ_o is a known positive constant, for a given initial condition $\xi(0)$. In (Boyd et al., 1994), for an arbitrary control law were presented two LMIs for the specification of these restrictions, supposing that there exists a quadractic Lyapunov function $V(\xi) = \xi'P\xi$, with negative derivative defined for all $\xi \neq 0$. For the particular case, where $s(t) = y(t)$, with $y(t) \in \mathbb{R}^p$ defined in (2), is proposed the following lemma:

Lemma 1. *For a given constant $\psi_o > 0$, if there exist $\lambda \in \Lambda$, and a symmetric positive definite matrix $P \in \mathbb{R}^{n \times n}$, solution of the following optimization problem, for all $i \in \mathbb{K}$:*

$$\begin{bmatrix} P & C_i' \\ C_i & \psi_o^2 I_n \end{bmatrix} > 0, \tag{28}$$

$$\begin{bmatrix} I_n & \xi(0)'P \\ P\xi(0) & P \end{bmatrix} > 0, \tag{29}$$

$$(Set \ of \ LMIs), \tag{30}$$

where (Set of LMIs) can be equal to (7)-(8), (12)-(13) or (16)-(18) then the equilibrium point $\xi = x - x_r = 0$ is globally asymptotically stable, the guaranteed cost (10) and the constraint (27) hold.

Proof. It follows from Theorems 1, 2 and the condition for bounds on output peak given in (Boyd et al., 1994). □

The next section presents applications of Theorem 3 in the control design of three DC-DC converters: Buck, Boost and Buck-Boost.

3. DC-DC converters

Consider that $i_L(t)$ denotes the inductor current and $V_c(t)$ the capacitor voltage, that were adopted as state variables of the system:

$$x(t) = [x_1(t) \ x_2(t)]' = [i_L(t) \ V_c(t)]'. \tag{31}$$

Define the following operating point $x_r = [x_{1r} \ x_{2r}]' = [i_{Lr} \ V_{cr}]'$. Consider the DC-DC power converters: Buck, Boost and Buck-Boost, illustrated in Figures 1, 3 and 5, respectively. The DC-DC converters operate in continuous conduction mode. For theoretical analysis of DC-DC converters, no limit is imposed on the switching frequency because the trajectory of the system evolves on a sliding surface with infinite frequency. Simulation results are presented below.

The used solver was the LMILab from the software MATLAB interfaced by YALMIP (Lofberg, 2004) (Yet Another LMI Parser). Consider the following design parameters (Deaecto et al., 2010): $V_g = 100[V]$, $R = 50[\Omega]$, $r_L = 2[\Omega]$, $L = 500[\mu H]$, $C = 470[\mu F]$ and

$$Q_i = Q = \begin{bmatrix} \rho_1 r_L & 0 \\ 0 & \rho_2/R \end{bmatrix},$$

is the performance index matrix associated with the guaranteed cost:

$$\int_0^{\infty} (\rho_2 R^{-1}(V_c - V_{cr})^2 + \rho_1 r_L (i_L - i_{Lr})^2 dt,$$

where ρ_1 and $\rho_2 \in \mathbb{R}_+$ are design parameters. Note that $\rho_i \in \mathbb{R}_+$ plays an important role with regard to the value of peak current and duration of the transient voltage. Adopt $\rho_1 = 0$ and $\rho_2 = 1$.

3.1 Buck converter

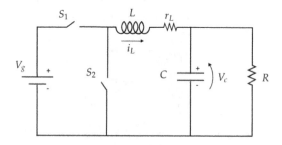

Fig. 1. Buck DC-DC converter.

Figure 1 shows the structure of the Buck converter, which allows only output voltage of magnitude smaller than the input voltage. The converter is modeled with a parasitic resistor in series with the inductor. The switched system state-space (1) is defined by the following matrices (Deaecto et al., 2010):

$$A_1 = \begin{bmatrix} -r_L/L & -1/L \\ 1/C & -1/RC \end{bmatrix}, \quad A_2 = \begin{bmatrix} -r_L/L & -1/L \\ 1/C & -1/RC \end{bmatrix},$$

$$B_1 = \begin{bmatrix} 1/L \\ 0 \end{bmatrix}, \quad B_2 = \begin{bmatrix} 0 \\ 0 \end{bmatrix}. \tag{32}$$

In this example, adopt $\lambda_1 = 0.52$ and $\lambda_2 = 0.48$. Using the minimization problems (11) and (15), corresponding to Theorems 1 and 2, respectively, we obtain the following matrix quadratic Lyapunov function

$$P = 1 \times 10^{-4} \begin{bmatrix} 0.0253 & 0.0476 \\ 0.0476 & 0.1142 \end{bmatrix},$$

needed for the implementation of the switching strategies (9) and (14). Maintaining the same parameters, from minimization problem of Theorem 3, we found the matrices below as a solution, and from (10) the guaranteed cost $J < (x_0 - x_r)'P(x_0 - x_r) = 0.029$:

$$P = 1 \times 10^{-4} \begin{bmatrix} 0.0253 & 0.0476 \\ 0.0476 & 0.1142 \end{bmatrix},$$

$$N_1 = -1 \times 10^{-6} \begin{bmatrix} 0.2134 & 0.0693 \\ 0.0693 & 0.0685 \end{bmatrix}, \quad N_2 = 1 \times 10^{-6} \begin{bmatrix} 0.2312 & 0.0751 \\ 0.0751 & 0.0742 \end{bmatrix}.$$

The results are illustrated in Figure 2. The initial condition was the origin $x = [i_L \ V_c]' = [0 \ 0]'$ and the equilibrium point is equal to $x_r = [1 \ 50]'$.

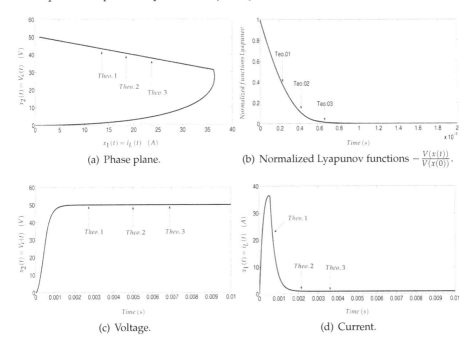

(a) Phase plane.

(b) Normalized Lyapunov functions $-\dfrac{V(x(t))}{V(x(0))}$.

(c) Voltage.

(d) Current.

Fig. 2. Buck dynamic.

Observe that Theorem 3 presented the same convergence rate and cost by applying Theorems 1 and 2. This effect is due to the fact that for this particular converter, the gradient of the switching surface does not depend on the equilibrium point (Deaecto et al., 2010). Table 1 presents the obtained results.

3.2 Boost converter

In order to compare the results from the previous theorems, designs and simulations will be also done for a DC-DC converter, Boost. The converter is modeled with a parasitic resistor

	Overshoot [A]	Time [ms]	Cost (6)
Theo. 1	36.5	2	0.029
Theo. 2	36.5	2	0.029
Theo. 3	36.5	2	0.029

Table 1. Buck results.

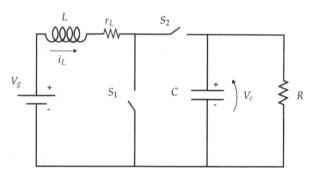

Fig. 3. Boost DC-DC converter.

in series with the inductor. The switched system state-space (1) is defined by the following matrices (Deaecto et al., 2010):

$$A_1 = \begin{bmatrix} -r_L/L & 0 \\ 0 & -1/RC \end{bmatrix}, \quad A_2 = \begin{bmatrix} -r_L/L & -1/L \\ 1/C & -1/RC \end{bmatrix},$$

$$B_1 = \begin{bmatrix} 1/L \\ 0 \end{bmatrix}, \quad B_2 = \begin{bmatrix} 1/L \\ 0 \end{bmatrix}. \tag{33}$$

In this example, $\lambda_1 = 0.4$ and $\lambda_2 = 0.6$. Using the minimization problems (11) of Theorem 1 and (15) of Theorem 2, the matrices of the quadratic Lyapunov functions are

$$P = 1 \times 10^{-4} \begin{bmatrix} 0.0237 & 0.0742 \\ 0.0742 & 0.2573 \end{bmatrix}, \quad P = 1 \times 10^{-3} \begin{bmatrix} 0.1450 & 0.0088 \\ 0.0088 & 0.2478 \end{bmatrix},$$

respectively. Now, from minimization problem of Theorem 3, we found the matrices below as a solution, and from (10) the guaranteed cost $J < (x_0 - x_r)'P(x_0 - x_r) = 0.59$:

$$P = 1 \times 10^{-4} \begin{bmatrix} 0.0237 & 0.0742 \\ 0.0742 & 0.2573 \end{bmatrix},$$

$$N_1 = \begin{bmatrix} -0.018 & -0.030 \\ -0.030 & 0.0178 \end{bmatrix}, \quad N_2 = \begin{bmatrix} 0.012 & 0.020 \\ 0.020 & -0.012 \end{bmatrix}.$$

The initial condition is the origin and the equilibrium point is $x_r = [5 \ 150]'$. The results are illustrated in Figure 4 and Table 2 presents the obtained results.

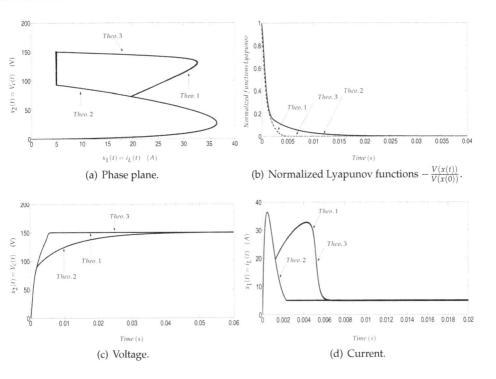

(a) Phase plane.

(b) Normalized Lyapunov functions $-\dfrac{V(x(t))}{V(x(0))}$.

(c) Voltage.

(d) Current.

Fig. 4. Boost dynamic.

	Overshoot $[A]$	Time $[ms]$	Cost (6)
Theo. 1	36.5	7	0.59
Theo. 2	36.5	40	5.59
Theo. 3	36.5	7	0.59

Table 2. Boost results.

3.3 Buck-Boost converter

Figure 5 shows the structure of the Buck-Boost converter. The switched system state-space (1) is defined by the following matrices (Deaecto et al., 2010):

$$A_1 = \begin{bmatrix} -r_L/L & 0 \\ 0 & -1/RC \end{bmatrix}, \quad A_2 = \begin{bmatrix} -r_L/L & -1/L \\ 1/C & -1/RC \end{bmatrix},$$

$$B_1 = \begin{bmatrix} 1/L \\ 0 \end{bmatrix}, \quad B_2 = \begin{bmatrix} 0 \\ 0 \end{bmatrix}. \tag{34}$$

The initial condition was the origin $x = [i_L \quad V_c]' = [0 \quad 0]'$, $\lambda_1 = 0.6$, $\lambda_2 = 0.4$ and the equilibrium point is equal to $x_r = [6 \quad 120]'$. Moreover, the optimal solutions of minimization

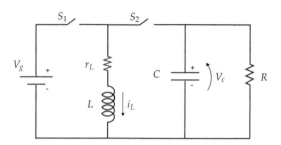

Fig. 5. Buck-Boost DC-DC converter.

problems (11) of Theorem 1 and (15) of Theorem 2, are

$$P = 1 \times 10^{-4} \begin{bmatrix} 0.0211 & 0.0989 \\ 0.0989 & 0.4898 \end{bmatrix}, \quad P = 1 \times 10^{-3} \begin{bmatrix} 0.1450 & 0.0088 \\ 0.0088 & 0.2478 \end{bmatrix},$$

respectively. Maintaining the same parameters, the optimal solution of minimization problem (21) are the matrices below and from (10) the guaranteed cost $J < (x_0 - x_r)'P(x_0 - x_r) = 0.72$:

$$P = 1 \times 10^{-4} \begin{bmatrix} 0.0211 & 0.0990 \\ 0.0990 & 0.4898 \end{bmatrix},$$

$$N_1 = \begin{bmatrix} -0.0168 & -0.0400 \\ -0.0400 & 0.0158 \end{bmatrix}, \quad N_2 = \begin{bmatrix} 0.0253 & 0.0600 \\ 0.0600 & -0.0237 \end{bmatrix}.$$

The results are illustrated in Figure 6. Table 3 presents the obtained results. The next section

	Overshoot [A]	Time [ms]	Cost (6)
Theo. 1	37.5	10	0.72
Theo. 2	7.5	70	3.59
Theo. 3	37.5	10	0.72

Table 3. Buck-Boost results.

is devoted to extend the theoretical results obtained in Theorems 1 (Deaecto et al., 2010) and 2 (Deaecto et al., 2010) for the model Sepic DC-DC converter.

4. Sepic DC-DC converter

A Sepic converter (Single-Ended Primary Inductor Converter) is characterized by being able to operate as a step-up or step-down, without suffering from the problem of polarity reversal. The Sepic converter consists of an active power switch, a diode, two inductors and two capacitors and thus it is a nonlinear fourth order. The converter is modeled with parasitic resistances in series with the inductors. The switched system (1) is described by the following matrices:

$$A_1 = \begin{bmatrix} -r_{L1}/L_1 & 0 & 0 & 0 \\ 0 & -r_{L2}/L_2 & -1/L_2 & 0 \\ 0 & 1/C_1 & 0 & 0 \\ 0 & 0 & 0 & -1/(RC_2) \end{bmatrix}, \quad B_1 = \begin{bmatrix} 1/L_1 \\ 0 \\ 0 \\ 0 \end{bmatrix},$$

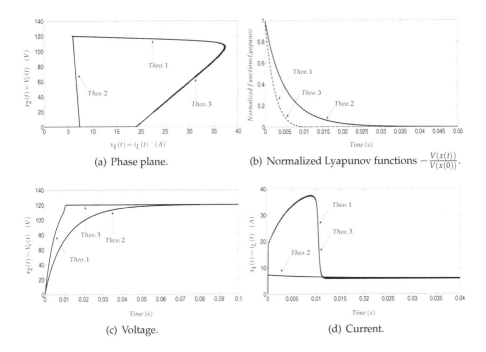

(a) Phase plane.

(b) Normalized Lyapunov functions $-\frac{V(x(t))}{V(x(0))}$.

(c) Voltage.

(d) Current.

Fig. 6. Buck-Boost dynamic.

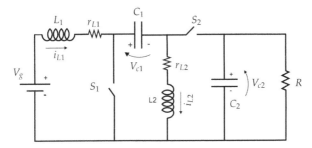

Fig. 7. Sepic DC-DC converter.

$$A_2 = \begin{bmatrix} -r_{L1}/L_1 & 0 & -1/L_1 & -1/L_1 \\ 0 & -r_{L2}/L_2 & 0 & 1/L_2 \\ 1/C_1 & 0 & 0 & 0 \\ 1/C_2 & -1/C_2 & 0 & -1/(RC_2) \end{bmatrix}, \quad B_2 = \begin{bmatrix} 1/L_1 \\ 0 \\ 0 \\ 0 \end{bmatrix}. \tag{35}$$

For this converter, consider that $i_{L1}(t)$, $i_{L2}(t)$ denote the inductors currents and $V_{c1}(t)$, $V_{c2}(t)$ the capacitors voltages, that again were adopted as state variables of the system:

$$x(t) = [x_1(t)\ x_2(t)\ x_3(t)\ x_4(t)]' = [i_{L1}(t)\ i_{L2}(t)\ V_{c1}(t)\ V_{c2}(t)]'. \tag{36}$$

Adopt the following operating point,

$$x_r = \left[x_{1r}(t)\ x_{2r}(t)\ x_{3r}(t)\ x_{4r}(t) \right]' = \left[i_{L1r}(t)\ i_{L2r}(t)\ V_{c1r}(t)\ V_{c2r}(t) \right]'. \tag{37}$$

The DC-DC converter operates in continuous conduction mode. The used solver was the LMILab from the software MATLAB interfaced by YALMIP (Lofberg, 2004) . The parameters are the following: $V_g = 100[V]$, $R = 50[\Omega]$, $r_{L1} = 2[\Omega]$, $r_{L2} = 3[\Omega]$, $L_1 = 500[\mu H]$, $L_2 = 600[\mu H]$, $C_1 = 800[\mu F]$, $C_2 = 470[\mu F]$ and

$$Q_i = Q = \begin{bmatrix} \rho_1 r_{L1} & 0 & 0 & 0 \\ 0 & \rho_2 r_{L2} & 0 & 0 \\ 0 & 0 & 0 & 0 \\ 0 & 0 & 0 & \rho_3/R \end{bmatrix}, \tag{38}$$

is the performance index matrix associated with the guaranteed cost

$$\int_0^\infty \left(\rho_1 r_{L1}(i_{L1} - i_{Lr1})^2 + \rho_2 r_{L2}(i_{L2} - i_{Lr2})^2 + \rho_3 R^{-1}(V_{c2} - V_{c2r})^2 \right) dt, \tag{39}$$

where $\rho_i \in \mathbb{R}_+$ are design parameters. Before of all, the set of all attainable equilibrium point is calculated considering that

$$x_r = \{ [i_{L1r}\ i_{L2r}\ V_{c1r}\ V_{c2r}]' : V_{c1r} = V_g,\ 0 \le V_{c2r} \le Ri_{L2r} \}. \tag{40}$$

The initial condition was the origin $x = [i_{L1}\ i_{L2}\ V_{c1}\ V_{c2}]' = [0\ 0\ 0\ 0]'$. Figure 8 shows the phase plane of the Sepic converter corresponding to the following values of load voltage $V_{c2r} = \{50, 60, \ldots, 150\}$.

In this case, Theorem 1 presented a voltage setting time smaller than $30[ms]$ and the maximum current peak $i_{L1} = 34[A]$ and $i_{L2} = 9[A]$. However, Theorem 2 showed a voltage setting time smaller than $80[ms]$, with currents peaks $i_{L1} = 34[A]$ and $i_{L2} = 13.5[A]$. Now, in order to compare the results from the proposed Theorem 3, adopt origin as initial condition, $\lambda_1 = 0.636$, $\lambda_2 = 0.364$ and the equilibrium point equal to $x_r = [5.24\ -3\ 100\ 150]'$. From the optimal solutions of minimization problems (11) and (15), we obtain respectively

$$P = 1 \times 10^{-4} \begin{bmatrix} 0.0141 & -0.0105 & 0.0037 & 0.0707 \\ -0.0105 & 0.0078 & -0.0026 & -0.0533 \\ 0.0037 & -0.0026 & 0.0016 & 0.0172 \\ 0.0707 & -0.0533 & 0.0172 & 0.3805 \end{bmatrix},$$

$$P = 1 \times 10^{-3} \begin{bmatrix} 0.0960 & -0.0882 & 0.0016 & 0.0062 \\ -0.0882 & 0.0887 & 0.0184 & -0.0034 \\ 0.0016 & -0.0184 & 0.0940 & 0.0067 \\ 0.0062 & -0.0034 & 0.0067 & 0.2449 \end{bmatrix}.$$

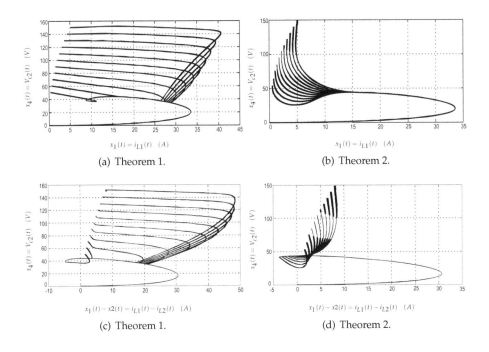

(a) Theorem 1.

(b) Theorem 2.

(c) Theorem 1.

(d) Theorem 2.

Fig. 8. Sepic DC-DC converter phase plane.

Maintaining the same parameters, the optimal solution of minimization problem (21) are the matrices below and from (10) the guaranteed cost $J < (x_0 - x_r)'P(x_0 - x_r) = 0.93$:

$$P = 1 \times 10^{-4} \begin{bmatrix} 0.0141 & -0.0105 & 0.0037 & 0.0707 \\ -0.0105 & 0.0078 & -0.0026 & -0.0533 \\ 0.0037 & -0.0026 & 0.0016 & 0.0172 \\ 0.0707 & -0.0533 & 0.0172 & 0.3805 \end{bmatrix},$$

$$N_1 = \begin{bmatrix} -0.0113 & 0.0099 & 0.0003 & -0.0286 \\ 0.0099 & -0.0085 & 0.0002 & 0.0290 \\ 0.0003 & 0.0002 & 0.0009 & 0.0088 \\ -0.0286 & 0.0290 & 0.0088 & 0.0168 \end{bmatrix},$$

$$N_2 = \begin{bmatrix} 0.0197 & -0.0173 & -0.0005 & 0.0500 \\ -0.0173 & 0.0148 & -0.0003 & -0.0507 \\ -0.0005 & -0.0003 & -0.0015 & -0.0154 \\ 0.0500 & -0.0507 & -0.0154 & -0.0293 \end{bmatrix}.$$

The results are illustrated in Figure 9 and Table 4 presents the obtained results from the simulations.

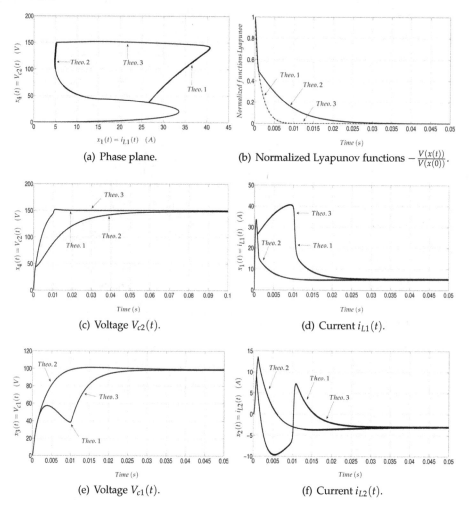

(a) Phase plane.

(b) Normalized Lyapunov functions $-\frac{V(x(t))}{V(x(0))}$.

(c) Voltage $V_{c2}(t)$.

(d) Current $i_{L1}(t)$.

(e) Voltage $V_{c1}(t)$.

(f) Current $i_{L2}(t)$.

Fig. 9. Sepic dynamic.

	Overshoot [A]	Time [ms]	Cost (6)
Theo. 1	34	30	0.93
Theo. 2	34	80	6.66
Theo. 3	34	30	0.93

Table 4. Sepic results.

Remark 1. *From the simulations results, note that the proposed Theorem 3 presented the same results obtained by applying Theorem 1. Theorem 3 is an interesting theoretical result, as described in Theorem 1, and the authors think that it can be useful in the design of more general switched controllers.*

5. Conclusions

This paper presented a study about the stability and control design for switched affine systems. Theorems proposed in (Deaecto et al., 2010) and later modified to include bounds on output peak on the control project were presented. A new theorem for designing switching affine control systems, with a flexibility that generalises Theorems 1 and 2 from (Deaecto et al., 2010) was proposed. Finally, simulations involving four types of converters namely Buck, Boost, Buck-Boost and Sepic illustrate the simplicity, quality and usefulness of this design methodology. It was also the first time that this class of controller was used for controlling a Sepic converter, that is a fourth order system and so is more complicated than the switched control design of second order Buck, Boost and Buck-Boost converters (Deaecto et al., 2010).

6. Acknowledgement

The authors gratefully acknowledge the financial support by CAPES, FAPESP and CNPq from Brazil.

7. References

Batlle, C., Fossas, E. & Olivar, G. (1996). Stabilization of periodic orbits in variable structure systems. Application to DC-DC power converters, *International Journal of Bifurcation and Chaos*, v. 6, n. 12B, p. 2635–2643.

Bolzern, P. & Spinelli, W. (2004). Quadratic stabilization of a switched affine system about a nonequilibrium point, *American Control Conference, 2004. Proceedings of the 2004*, v. 5, p. 3890–3895.

Boyd, S., El Ghaoui, L., Feron, E. & Balakrishnan, V. (1994). *Linear Matrix Inequalities in Systems and Control Theory*, Studies in Applied Mathematics, v. 15, 2 nd., SIAM Studies in Applied Mathematics.

Cardim, R., Teixeira, M. C. M., Assunção, E. & Covacic, M. R. (2009). Variable-structure control design of switched systems with an application to a DC-DC power converter, *IEEE Trans. Ind. Electronics*, v. 56, n. 9, p. 3505 –3513.

Cardim, R., Teixeira, M. C. M., Assunção, E., Covacic, M. R., Faria, F. A., Seixas, F. J. M. & Mainardi Júnior, E. I. (2011). Design and implementation of a DC-DC converter based on variable structure control of switched systems, *18th IFAC World Congress, 2011. Proceedings of the 2011*, v. 18, p. 11048–11054.

Deaecto, G. S., Geromel, J. C., Garcia, F. S. & Pomilio, J. A. (2010). Switched affine systems control design with application to DC-DC converters, *IET Control Theory & Appl.*, v. 4, n. 7, p. 1201–1210.

Decarlo, R. A., Branicky, M. S., Pettersson, S. & Lennartson, B. (2000). Perspectives and results on the stability and stabilizability of hybrid systems, *Proc. of the IEEE*, v. 88, n. 7, p. 1069 –1082.

Faria, F. A., Assunção, E., Teixeira, M. C. M., Cardim, R. & da Silva, N. A. P. (2009). Robust state-derivative pole placement LMI-based designs for linear systems, *International Journal of Control*, v. 82, n. 1, p. 1–12.

Feron, E. (1996). Quadratic stabilizability of switched systems via state and output feedback,*Technical report CICSP- 468* (MIT)..

Gahinet, P., Nemirovski, A., Laub, A. J. & Chilali, M. (1995). *LMI control toolbox - for use with MATLAB*.

He, Y., Xu, W. & Cheng, Y. (2010). A novel scheme for sliding-mode control of DC-DC converters with a constant frequency based on the averaging model, *Journal of Power Electronics*, v. 10, n.1, p. 1–8.

Ji, Z., Wang, L. & Xie, G. (2005). *Quadratic stabilization of switched systems*, v. 36, 7 edn, Taylor & Francis.

Liberzon, D. (2003). *Switching in Systems and Control*, Systems & Control, Birkhuser.

Liberzon, D. & Morse, A. S. (1999). *Basic problems in stability and design of switched systems*, v. 19, 5 edn, IEEE Constr. Syst. Mag.

Lofberg, J. (2004). Yalmip : a toolbox for modeling and optimization in MATLAB, *Computer Aided Control Systems Design, 2004 IEEE International Symposium on*, p. 284 –289.

Mazumder, S. K., Nayfeh, A. H. & Borojevic, D. (2002). Robust control of parallel DC-DC buck converters by combining integral-variable-structure and multiple-sliding-surface control schemes, *IEEE Trans. on Power Electron.*, v. 17, n. 3, p. 428–437.

Peaucelle, D., Henrion, D., Labit, Y. & Taitz, K. (2002). User's guide for sedumi interface 1.04.

Skafidas, E., Evans, R. J., Savkin, A. V. & Petersen, I. R. (1999). Stability results for switched controller systems, *Automatica*, v. 35, p. 553–564.

Sun, Z. & Ge, S. S. (2005). *Switched Linear Systems: Control and Design*, Springer, London.

Teixeira, M. C. M., Assunção, E. & Avellar, R. G. (2003). On relaxed LMI-based designs for fuzzy regulators and fuzzy observers, *IEEE Trans. Fuzzy Syst.* v. 11, n. 5, p. 613 – 623.

Teixeira, M. C. M., Covacic, M. R. & Assuncao, E. (2006). Design of SPR systems with dynamic compensators and output variable structure control, *Int. Workshop Var. Structure Syst.*, p. 328 –333.

Xu, X., Zhai, G. & He, S. (2008). On practical asymptotic stabilizability of switched affine systems, *Nonlinear Analysis: Hybrid Systems* v. 2, p. 196–208.

Yoshimura, V. L., Assunção, E., Teixeira, M. C. M. & Mainardi Júnior, E. I. (2011). A comparison of performance indexes in DC-DC converters under different stabilizing state-dependent switching laws, *Power Electronics Conference (COBEP), 2011 Brazilian*, p. 1069 –1075.

4

Neural and Genetic Control Approaches in Process Engineering

Javier Fernandez de Canete, Pablo del Saz-Orozco,
Alfonso Garcia-Cerezo and Inmaculada Garcia-Moral
University of Malaga,
Spain

1. Introduction

Nowadays, advanced control systems are playing a fundamental role in plant operations because they allow for effective plant management. Typically, advanced control systems rely heavily on real-time process modeling, and this puts strong demands on developing effective process models that, as a prime requirement, have to exhibit real-time responses. Because in many instances detailed process modeling is not viable, efforts have been devoted towards the development of approximate dynamic models.

Approximate process models are based either on first principles, and thus require good understanding of the process physics, or on some sort of black-box modeling. Neural network modeling represents an effective framework to develop models when relying on an incomplete knowledge of the process under examination (Haykin, 2008). Because of the simplicity of neural models, they exhibit great potentials in all those model-based control applications that require real-time solutions of dynamic process models. The better understanding acquired on neural network modeling has driven its exploitation in many process engineering applications (Hussain, 1999).

Genetic algorithms (GA) are model machine learning methodologies, which derive their behavior from a metaphor of the processes of evolution in nature and are able to overcome complex non-linear optimization tasks like non-convex problems, non-continuous objective functions, etc. (Michalewitz, 1992). They are based on an initial random population of solutions and an iterative procedure, which improves the characteristics of the population and produces solutions that are closer to the global optimum. This is achieved by applying a number of genetic operators to the population, in order to produce the next generation of solutions. GAs have been used successfully in combinations with neural and fuzzy systems (Fleming & Purhouse, 2002).

Distillation remains the most important separation technique in chemical process industries around the world. Therefore, improved distillation control can have a significant impact on reducing energy consumption, improving product quality and protecting environmental resources. However, both distillation modeling and control are difficult tasks because it is usually a nonlinear, non-stationary, interactive, and subject to constraints and disturbances process.

In this scenario, most of the contributions that have appeared in literature about advanced control schemes have been tested for nonlinear simulation models (Himmelblau, 2008), while applications with advanced control algorithms over industrial or pilot plants (Frattini et al, 2000) (Varshney and Panigrahi, 2005) (Escano et al, 2009) or even with classical control (Noorai et al, 1999) (Tellez-Anguiano et al, 2009) are hardly found.

Composition monitoring and composition control play an essential role in distillation control (Skogestad, 1997). In practice, on- line analyzer for composition is rarely used due to its costs and measurement delay. Therefore composition is often regulated indirectly using tray temperature close to product withdrawal location. In order to achieve the control purpose, many control strategies with different combination of manipulated variables configurations have been proposed (Skogestad, 2004). If a first-principles model describes the dynamics with sufficient accurately, a model-based soft sensor can be derived, such an extended Kalman filter or its adaptive versions (Venkateswarlu & Avantika, 2001), while inferential models can also be used when process data are available by developing heuristic models (Zamprogna et al, 2005). Artificial neural networks can be considered from an engineering viewpoint, as a nonlinear heuristic model useful to make predictions and data classifications, and have been also used as a soft sensors for process control (Bahar et al, 2004).

Nevertheless, few results are reported when is considered the composition control of experimental distillation columns, and some results are found either by applying direct temperature control (Marchetti et al, 1985) or by using the vapor-liquid equilibrium to estimate composition from temperature (Fileti et al, 2007), or even by using chromatographs (Fieg, 2002).

In this chapter we describe the application of adaptive neural networks to the estimation of the product compositions in a binary methanol-water continuous distillation column from available temperature measurements. This software sensor is then applied to train a neural network model so that a GA performs the searching for the optimal dual control law applied to the distillation column. The performance of the developed neural network estimator is further tested by observing the performance of the neural network control system designed for both set point tracking and disturbance rejection cases.

2. Neural networks and genetic algorithms for control

2.1 Neural networks for identification

Neural networks offer an alternative approach to modelling process behaviour as they do not require a priori knowledge of the process phenomena. They learn by extracting pre-existing patterns from a data set that describe the relationship between the inputs and the outputs in any given process phenomenon. When appropriate inputs are applied to the network, the network acquires knowledge from the environment in a process known as learning. As a result, the network assimilates information that can be recalled later. Neural networks are capable of handling complex and nonlinear problems, process information rapidly and can reduce the engineering effort required in controller model development (Basheer & Hajmeer, 2000).

Neural networks come in a variety of types, and each has their distinct architectural differences and reasons for their usage. The type of neural network used in this work is known as a feedforward network (Fig. 1) and has been found effective in many applications.

It has been shown that a continuous-valued neural network with a continuous differentiable nonlinear transfer function can approximate any continuous function arbitrarily well in a compact set (Cybenko, 1989).

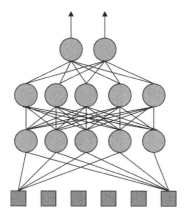

Fig. 1. Feedforward neural network architecture

There are several different approaches to neural network training, the process of determining an appropriate set of weights. Historically, training is developed with the backpropagation algorithm, but in practice quite a few simple improvements have been used to speed up convergence and to improve the robustness of the backpropagation algorithm (Hagan & Menhaj, 1994). The learning rule used here is common to a standard nonlinear optimization or least-squares technique. The entire set of weights is adjusted at once instead of adjusting them sequentially from the output layer to the input layer. The weight adjustment is done at the end of each epoch and the sum of squares of all errors for all patterns is used as the objective function for the optimization problem.

In particular we have employed the Levenberg-Marquardt algorithm to train the neural network used (Singh et al, 2007), which is a variation of the Newton's method, designed for minimizing functions that are sums of squares of other nonlinear functions. Newton's method for optimizing a performance index $F(x)$ is given by

$$x_{k+1} = x_k - A_k^{-1} g_k \qquad (1)$$

where $A_k = \nabla^2 F(x)|_{x=x_k}$ and $g_k = \nabla F(x)|_{x=x_k}$ are the hessian and the gradient of $F(x)$, respectively, and where x_k is the set of net parameters at time k. In cases where $F(x)$ is the sum of the square of errors $e(x)$ over the Q targets in the training set

$$F(x) = \sum_{i=1}^{Q} e_i^2(x) = e^T(x)e(x) \qquad (2)$$

then the gradient would be given by

$$\nabla F(x) = 2J^T(x)e(x) \qquad (3)$$

where $J(x)$ is the Jacobian matrix formed by elements $\frac{\partial e_i(x)}{\partial x_j}$. On the other hand, the hessian would be approximated by

$$\nabla^2 F(x) \cong 2 J^T(x) \cdot J(x) \tag{4}$$

Then, substituting (3) and (4) into (1), it results in the Gauss-Newton method

$$x_{k+1} = x_k - [J^T(x_k)J(x_k)]^{-1} J^T(x_k)e(x_k) \tag{5}$$

Adding a constant term $\mu_k I$ to $J^T(x_k)J(x_k)$, this lead to the Levenberg-Marquardt training rule so that

$$x_{k+1} = x_k - [J^T(x_k)J(x_k) + \mu_k I]^{-1} J^T(x_k)e(x_k) \tag{6}$$

where μ_k is the learning coefficient, which is set at a small value in the beginning of the training procedure (μ_k = 1e-03) and is increased (decreased) by a factor $\vartheta > 1$ (i.e. $\vartheta = 10$) according to the increase (decrease) of $F(x)$ in order to provide faster convergence. In fact, when μ_k is set to a small value the Levenberg-Marquardt algorithm approaches that of Gauss-Newton, otherwise it behaves as a gradient descent technique. The neural network was configured to stop training after the mean squared error went below 0.05, the minimum gradient went below 1e-10 or the maximum number of epochs was reached (normally a high number is selected so that this is a non-limiting condition).

The identification of the neural network model occurred via a dynamic structure constituted by a feedforward neural network representing the nonlinear relationship between input and output signals of the system to be modelled. The application of feedforward networks to dynamic systems modelling requires the use of external delay lines involving both input and output signals (Norgaard et al, 2000).

The network input vector dimension was associated with the time window length selected for each input variable, which was dependent on distillation column dynamics and is usually chosen according to the expertise of process engineers (Basheer & Hajmeer, 2000). The hidden layer dimension was defined by using a trial and error procedure after selecting the input vector, while the net's output vector dimension directly resulted from the selected controlled variables.

Therefore, the neural network identification model NN_I after selecting the optimal input vector was given by

$$\hat{x}(t + 1) = NN_I(x(t), u(t)) \tag{7}$$

where $\hat{x}(t + 1)$ stands for the predicted value of the neural network corresponding to the actual net input vector $u(t)$ and the state vector $x(t)$.

The resulting identification model was obtained after selecting the best neural network structure among the possible ones, after a training process. Finally, a neural network validation process was performed by comparing the network output with additional data that were not included in the training data (validation set).

2.2 Genetic algorithms for optimization and control

Genetic Algorithms are adaptive methods which can be used to solve optimization problems. They are based on genetic processes of biological organisms. Over many generations, natural populations evolve according to the principles of natural selection and survival of the fittest. In nature, individuals with the highest survival rate have relatively a large number of offspring, that is, the genes from the highly adapted or fit individuals spread to an increasing number of individuals in each successive generation. The strong characteristics from different ancestors can sometimes produce super-fit offspring, whose fitness is greater than that of either parent. In this way, species evolve to become better suited to their environment in an iterative way by following selection, recombination and mutation processes starting from an initial population.

The control scheme here proposed is based on the different strengths that neural network and genetic algorithms present. One of the most profitable characteristic of the neural networks is its capability of identification and generalization while genetic algorithms are used for optimizing functions.

If an accurate identification model is available, the controller can use the information provided by selecting the optimum input that makes the system as near as possible to the goal to achieve. So one of the main differences between this controller and the rest is the way it selects the inputs to the system.

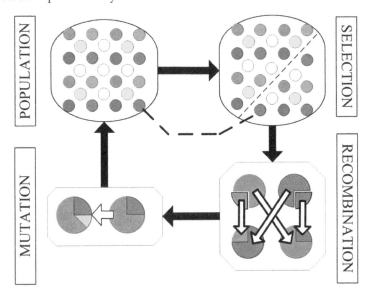

Fig. 2. Genetic Algorithm Structure

In this way, the function to minimize in each step is the absolute value of the difference between the predicted output (by means of the neural identification network) and the reference. This difference depends, usually, on known variables as past states of the system and past inputs and on unknown variables as are the current inputs to apply. Those inputs will be obtained from the genetic algorithm.

2.3 Neural networks for estimation

Most popular sensors used in process control are the ones that measure temperature, pressure and fluid level, due to the high accuracy, fast response properties and their cheapness. On the other hand, some of the most controlled variables, such as composition, present great difficulties in the measurement phase because it should be done off-line in laboratory, by involving both a high delay time and an extra cost due to the use of expensive equipment requiring both initial high investment and maintenance, such as occurs with chromatography.

The composition control is crucial in order to achieve the final product specifications during the distillation process. The use of sensors able to infer composition values from secondary variables (values easier to be measured) could be a solution to overcome the referred drawbacks, being this approach defined as a software sensor (Brosilow & Joseph, 2002).

In this way, an inferential system has been developed for achieving an on-line composition control. As the value of the controlled variable is inferred from other secondary variables, the model should be very accurate mainly in the operating region. The inferential system based on the first principles model approach presents the drawback of increasing computing time as the number of variables increase.

A black-box model approach relating the plant outputs with the corresponding sampled inputs has been used instead. Neural networks have proven to be universal approximators (Haykin, 2008), so they will be used to infer the composition from other secondary variables, defining thus the neural soft estimator.

One of the main difficulties in determining the complete structure of the neural estimator is the choice of the secondary variables to be used (both the nature and the location), selected among the ones provided by the set of sensors installed on the experimental pilot plant. In the literature there are several papers dedicated to the selection of variables for composition estimation and no consensus is reached in terms of number or position of the secondary sensors (here position is understood as the stage or plate where the variable is measured). In (Quintero-Marmol et al, 1991), the number that assures robust performance is $N_c + 2$, where N_c is the number of components. With respect to the location of the most sensitive trays, (Luyben, 2006) develops a very exhaustive study and concludes that the optimal position depends heavily on the plant and on the feed tray. In this way, the neural estimator should have as an input the optimum combination of selected secondary variables to determine accurately the product composition.

In order to select the most suitable secondary variables for our control purposes, a multivariate statistical technique based on the principal component analysis (PCA) methodology (Jackson,1991) has been used, following the same approach described by (Zamprogna et al,2005). The resulting neural network estimator NN_E is given by

$$\hat{x}_p(t) = NN_E(x_s(t)) \tag{8}$$

where $\hat{x}_p(t)$ and $x_s(t)$ stands for the primary and secondary selected variables.

2.4 Neurogenetic control structure

As an accurate neural network model that relates the past states, current states, and the current control inputs with the future outputs is available, the future output of the system can be predicted depending on the control inputs through a non linear function. In this way, the function to be minimized in each step is a cost function that is related to the absolute value of the difference between the predicted output and the desired reference to follow. This difference depends, usually, on known variables such as past inputs and past states of the system and on unknown variables such as the current control inputs to apply, which will be obtained from the genetic algorithm.

In this way, the optimization problem for controlling the distillation plant can be stated as the problem of finding the input that minimizes the norm of the difference, multiplied by a weighting matrix between the reference command to follow and the neural network model output, considering the input and the past and current states of the system. This procedure can be stated as $min\|K_w \cdot (x_r - NN_l(x, u))\|$, with x_r representing the reference command to follow, NN_l is the neural network model output, x represents the past and current states of the system, $u \in U$ is the control action and U is the universe of possible control actions and K_w is a weighting matrix.

In the present case, the reference command x_r will be given by the desired composition variables together with the desired level variables, while $u \in U$ represents the optimum neurogenetic control action, and the weighting matrix penalizes the errors in composition twice the errors in level, since composition control is more difficult to achieve than level control. In Fig. 3 the neurogenetic control strategy that is used here is shown, together with the neural composition estimator.

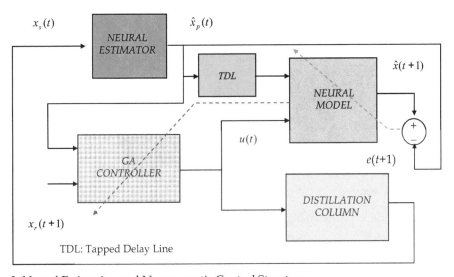

Fig. 3. Neural Estimation and Neurogenetic Control Structure

3. Application to a pilot distillation column

3.1 Description of the pilot distillation column

The pilot distillation column DELTALAB is composed of 9 plates, one condenser, and one boiler (Fig. 4). The instrumentation equipment consists of 12 Pt 100 RTD temperature sensors (T1-T12), 3 flow meters (FI1-FI3), 2 level sensors (LT1-LT2) and 1 differential pressure meter (PD), together with 3 pneumatic valves (LIC1-LIC2-TIC2) and a heating thermo-coil (TIC1), with up to four control loops for plant operation. Additionally, feed temperature and coolant flow control are included with corresponding valve (FIC1) and heating resistance (PDC1), being both variables considered as disturbances.

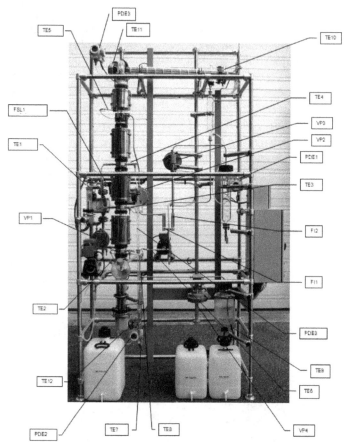

Fig. 4. Pilot distillation plant configuration

The condenser provides the necessary cooling to condense the distilled product. The condenser contains the cooling water provided by an external pump. The flow of the cooling liquid is regulated through a pneumatic valve with one flow controller, which as a last resort depends on the variable water flow supply. Two temperature sensors measure the temperature of the inlet and outlet flows.

Once the top stream is condensed, the liquid is stored in an intermediate reflux drum, endowed with level meter, temperature sensor and recirculation pump for reflux stream. The reflux to distillate ratio is controlled by 2 proportional pneumatic valves for reflux and distillate respectively, each flow measured through the corresponding flow meter with display.

The main body of the distillation column is composed of 9 bubble cap plates distributed into 3 sections. Two of them are connected to the feeding device, and can either function like feeding or normal plates, selecting each one through a manual valve. Four temperature sensors measure the temperature in each section junction.

The boiler provides the required heat to the distillation column by actuating on an electric heating thermo-coil located inside the boiler. A temperature sensor is located inside the boiler and a level meter measures the liquid stored in an intermediate bottom drum. A differential-pressure sensor indicates the pressure changes throughout the column which is operated at atmospheric pressure. The bottom flow is controlled by a proportional pneumatic valve and two temperature sensors measure the temperature of the inlet and outlet flows before cooling, with corresponding flow meters with display.

The feeding ethanol-water mixture is stored in a deposit, whose temperature is controlled by a pre-heating electric thermo-coil. The mixture to be distilled is fed into the column in small doses by a feeding pump with temperature controller (TIC3) and sensors installed to measure the temperature of the inlet and outlet feed flows.

The whole instrumentation of the distillation pilot plant is monitored under LabVIEW platform and is connected to the neural based controller designed under MATLAB platform, through a communication system based both on PCI and USB buses, with up to four control loops. In this experimental set-up, boiler heat flow Q_B, reflux valve opening V_R, distillate valve opening V_D and bottom valve opening V_B constitute the set of manipulated variables, while light composition C_D, bottom composition C_B, light product level L_D and heavy product level L_B define the corresponding set of controlled variables (Fig. 5), while the feed flow temperature T_F is considered as a disturbance.

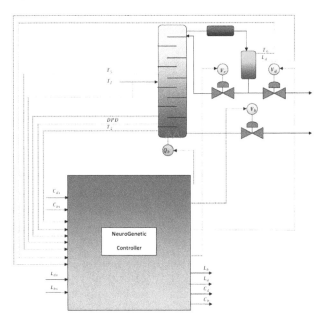

Fig. 5. Pilot distillation plant configuration

It is important to highlight that a dynamical model has not been derived to represent the pilot column behavior, instead of this we have made use of an approximate neural network model to identify the plant dynamics starting from selected I/O plant data operation.

3.2 Monitoring and control interface system

The monitoring and control interface system requires a communication system between the sensors and actuators on the one hand and the computer on the other hand throughout I/O modules, whose specifications are settled by the instrumentation characteristics utilized (Table 1 and 2).

In order to manage the I/O signals, USB and PCI buses have been chosen. On the one hand, the PCI bus enables the dynamic configuration of peripheral equipments, since during the operating system startup, the devices connected to PCI buses communicate with the BIOS and calculate the required resources for each one. On the other hand, the USB bus entails a substantial improvement regarding the 'plug and play' technology, having as main objective to suppress the necessity of acquiring different boards for computer ports. Besides this, an optimal performance is achieved for the set of different devices integrated into the instrumentation system, connectable without the needing to open the system.

Sensor	Variable	Physical Range	Magnitude	Signal Range	Measuring Accuracy
T1-T12	Temperature	-200-119 ° C	Resistance	18.5-145.7 Ω	0.01 °C
FI1-FI3	Flowrate	0-5 l/h	Current	4-20 mA	± 2.5 %
LT1	Level	0-495 mm	Current	4-20 mA	± 0.075 %
LT2	Level	0-950 mm	Current	4-20 mA	± 0.075 %
PD	Diff. Pressure	0-25 mbar	Current	4-20 mA	± 0.075 %

Table 1. Sensors characteristics for the pilot distillation column

The acquisition system configuration for the monitoring and control of the pilot plant is constituted by the next set of DAQ (Data Acquisition) boards: NI PCI-6220, NI-PCI-6722, NI-USB-6009, NI-USB-6210 for analog voltage signal acquisition and NI-PCI-6704 for analog current signal acquisition, all supplied by National Instruments (NI). Measurements obtained from the sensors have been conditioned to operate into the standard operational range, and signal averaging for noise cancelation has been applied using specific LabVIEW toolkits (Bishop, 2004).

The monitoring and control interface system developed for the pilot plant is configured throughout the interconnection of the NI Data acquisition system with both the LabVIEW monitoring subsystem and the neurogenetic controller implemented in MATLAB (Fig. 6), both environments linked together through the Mathscripts and running under a Intel core duo with 2.49 GHZ and 3 GB of RAM.

Control Loop	Actuator	Actuation Type	Magnitude	Signal Range
PDC1	Resistance	On/Off	Voltage	0-5 V
TIC1	Resistance	On/Off	Voltage	0-5 V
TIC2	Valve	Proportional	Current	4-20 mA
LIC1	Valve	Proportional	Current	4-20 mA
LIC2	Valve	Proportional	Current	4-20 mA
FIC1	Valve	Proportional	Current	4-20 mA

Table 2. Actuators characteristics for the pilot distillation column

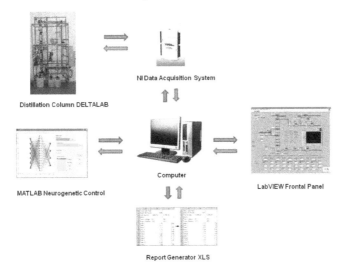

Fig. 6. Monitoring and control interface for pilot distillation plant

The process control scheme developed in each operation cycle implies the execution of five different actions: system initializing, buttons control reading from VI (Virtual Instruments), reading plant data from instruments, control action calculation and writing control data to instruments.

3.3 Neural composition estimator and neurogenetic controller

The complete controlled system is composed of a neural network model of the process and a control scheme based on a genetic algorithm which utilizes both the composition and the level variables to get the quasi-optimal control law, by using the neural composition estimator (Fig. 3) for both determining and monitoring the composition of light and heavy components from secondary variable measurements.

After applying the selection method, the inputs to the neural estimation network turned out to be four secondary variables, namely, three temperatures T_6, T_5, T_2, each corresponding to reflux, top and bottom temperatures, and differential pressure drop DPD, while C_D and C_B

compositions were the net outputs. This structure is in line with what the literature suggests (Quintero-Marmol et al, 1991) (Zamprogna et al, 2005) in terms of both the number of the selected measurements and its distribution. This fact contrasts with the standard approach consisting in selecting two temperatures for a two composition estimation (Medjell and Skogestad,1991) (Strandberg and Skogestad,2006). However, this assumption is not possible when the vapor-liquid equilibrium has a strong nonlinear behavior (Baratti et al.,1998) (Oisiovici and Cruz, 2001), so that holding the temperature constant does not imply that composition will also be constant (Rueda et al, 2006).

The final network structure selected for the neural composition estimator was a 4-25-2 net, trained using the Levenberg-Marquardt algorithm (Hagan et al, 2002), with a hidden layer configuration selected after a trial and error process and input layer determined by the PCA based algorithm for selection of the secondary variables previously exposed.

The training data set used herein consisted of 700 points collected randomly from a whole data set of more than 27000 acquired points, all obtained from several experiments carried out with the pilot distillation column by covering the whole range of operation. A different subset of 700 points has been also used for validation. For this purpose we have analyzed several samples of an ethanol-water mixture during the separation process by using a flash chromatograph VARIANT, and the composition error mean obtained was lower than 1.5%.

The final network structure selected for the neural plant model was a 22-25-6 neural feedforward architecture trained by using the Levenberg-Marquardt algorithm and validated throughout the set of I/O experimental data. The hidden layer configuration was selected after the algorithm as it was stated in the previous section, using this time V_R, V_D, V_B, Q_B, T_2, T_5, T_6, T_F, L_D, L_B, and DPD delayed values as inputs, while T_2, T_5, T_6, DPD, L_D, L_B were the estimated outputs. The neural net was trained with a different subset of 750 points selected randomly from the whole data set of 27000 acquired points with sampling T = 2 s, both by using a PID analog control module, by changing set-points for each of the controlled variables into its operating range and by working on open loop conditions. The neural net was also validated with another subset of 750 points comparing its outputs to the real system's outputs in independent experiments.

The neurogenetic controller is characterized by a population of 75 inhabitants, 50 generations and a codification of 8 bits. The maximum is accepted if it is invariant in 5 iterations. All these parameters were estimated for achieving a time response lower than 1.3 seconds for the computational system used for controlling the experimental distillation plant.

3.4 Results

In order to test the validity of the proposed control scheme, the performance of the neurogenetic control strategy is compared against a PID control strategy by using four decoupled PID controllers relating V_R, Q_B, V_D and V_B manipulated variables with the corresponding controlled variables C_D, C_B, L_D and L_B. Obviously in order to compare properly both strategies, the PID approach should control the same variables, in a way the composition is indirectly controlled, by following the standard LV configuration (Skogestad, 1997). The PID parameters set selected for each controlled variable has been heuristically tuned according to the analog PID values set by the DELTALAB field expert when the pilot column is supplied.

Several changes in composition set points on top and bottom purity have been made to test the neurogenetic controller performance (Fig. 7). As it is shown, the system is able to reach the required references in composition but is a bit slow in its response. The response obtained with the PID approach presents a bigger settling time and overshoot and a poorer response to changes in the targets in the coupled variables. In fact, the ISE (integral square error) which characterizes the accuracy of both control schemes during tracking of reference commands, is significantly lower for the neurogenetic control as compared to the PID control both controlled variables, with a $ISE_{d\_PID} = 4719.9\ (\%)^2 \cdot s$, $ISE_{d\_Neuro\_GA} = 3687.2\ (\%)^2 \cdot s$ for top composition and $ISE_{b\_PID} = 2427.6\ (\%)^2 \cdot s$, $ISE_{b\_Neuro\_GA} = 2071.8\ (\%)^2 \cdot s$ for bottom composition respectively. These facts imply a better performance even when changing conditions are present (variable feed changes), due to the adaptive nature of the neurogenetic controller.

In Fig. 8 are displayed the changes in control actions V_R, V_P, V_H (in % of opening) and Q_B (in % of maximum power) corresponding to the set point changes on top and bottom composition as described formerly for the neurogenetic control scheme. It must be emphasized that all control signal are within the operating range with minimum saturation effects, mainly due to mild conditions imposed to the time response profile during the neurogenetic design.

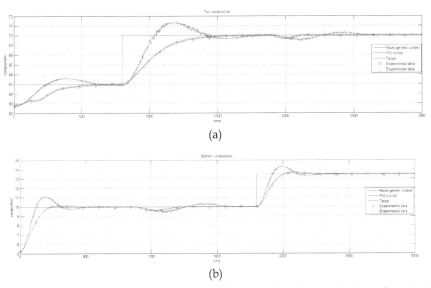

(a)

(b)

Fig. 7. Response of top and bottom composition for set point changes in ethanol purity in (a) 60-70 % range on top (b) 5-12 % range on bottom for pilot distillation column under decoupled PID and neurogenetic control

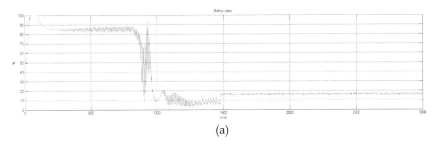

(a)

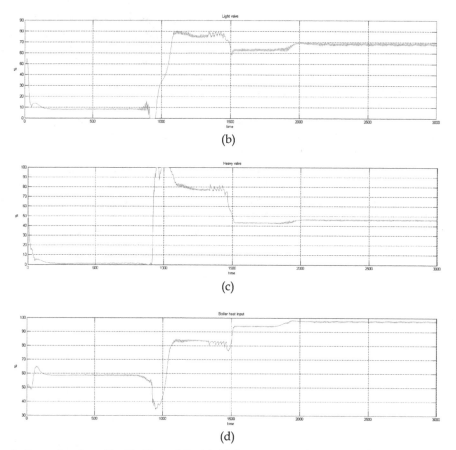

(b)

(c)

(d)

Fig. 8. Control actions V_R, V_P, V_H and Q_B (a)-(d) for set point changes in ethanol purity in 60-70 % range on top and 5-12 % range on bottom for pilot distillation column under neurogenetic control.

4. Conclusions

Adaptive neural networks have been applied to the estimation of product composition starting from on-line secondary variables measurements, by selecting the optimal net input vector for estimator by using PCA based algorithm. Genetic algorithms have been used to derive the optimum control law under MATLAB, based both on the neural network model of the pilot column and the estimation of composition. This neurogenetic approach has been applied to the dual control of distillate and bottom composition for a continuous ethanol water nonlinear pilot distillation column monitored under LabVIEW.

The proposed method gives better or equal performances over other methods such as fuzzy or adaptive control by using a simpler design based exclusively on the knowledge about the pilot distillation column in form of I/O operational data. It is also necessary to highlight the potential benefits of artificial neural networks combined with GA when are applied to the multivariable control of nonlinear plants, with unknown first-principles model and under an experimental set-up as was demonstrated with the distillation pilot plant.

Future work is directed toward the application of this methodology to industrial plants and also towards the stability and robustness analysis due to uncertainty generated by the neural network identification errors when the plant is approximated.

5. References

Bahar, A.; Ozgen, C.; Leblebicioglu, K. & Halici, U. (2004). Artificial neural network estimator design for the inferential model predictive control of an industrial distillation column. *Industrial Engineering Chemical Research*, Vol. 43 (2004), pp. 6102-6111.

Baratti, R.; Bertucco, A.; Da Rold, A. & Morbidelli, M. (1998). A composition estimator for multicomponent distillation columns-development and experimental tests on ternary mixture. *Chemical Engineering Science*, Vol. 53 (1998), pp. 3601–3612.

Basheer, I.A. & Hajmeer, M. (2000). Artificial neural networks: fundamentals, computing, design, and application, *Journal of Microbiology Methods*, Vol. 43 (2000), pp. 3–31.

Bishop, R. (2004). *Learning with LabVIEW 7 Express*. Prentice Hall, New Jersey.

Brosilow, C., & Joseph, B. (2002). *Techniques of Model Based Control*. Prentice Hall, New York, USA.

Cybenko, G. (1989). Approximation by superposition of sigmoidal functions. *Mathematics of Control, Signals and Systems*, Vol. 2 (1989), pp. 303–314.

Escano, J.M.; Bordons, C.; Vilas, C.; Garcia, M.R. & Alonso, A. A. (2009). Neurofuzzy model based predictive control for thermal batch processes. *Journal of Process Control*, Vol. 18 (200), pp. 1566-1575.

Fieg, G. (2002). Composition control of distillation columns with a sidestream by using gas chromatographs, Chemical Engineering Processing, Vol. 41, No. 2), pp. 123-133.

Fileti, A.M.; Antunes, A.J.; Silva, F.V.; Silveira, V. & Pereira. J.A. (2007). Experimental investigations on fuzzy logic for process control. *Control Engineering Practice*, Vol. 15 (2007), pp. 1149-1160.

Fleming, P. & Purshouse, R. (2002). Evolutionary algorithms in control systems engineering: a survey. *Control Engineering Practice*, Vol. 10 (2002), pp. 1223-1241.

Frattini, A.M; Cruz, S.L. & Pereira, J. (2000). Control strategies for a batch distillation column with experimental testing. *Chemical Engineering Processing*, Vol. 39 (2000), pp. 121-128.

Hagan, M.T. & Menhaj, M. (1994). Training feed-forward networks with the Marquardt algorithm. *IEEE Transactions on Neural Networks*, Vol. 5 (1994), pp. 989–993.

Hagan, M.T; Demuth, H.B. & De Jesus, O. (2002). An introduction to the use of neural networks in control systems. *International Journal of Robust and Nonlinear Control*, Vol. 12, No. 11 (2002), pp. 959-985.

Haykin, S. (2008). *Neural networks and learning machines*, Prentice Hall.

Himmelblau, D.M. (2008). Accounts of experiences in the application of artificial neural networks in chemical engineering. *Industrial Engineering Chemical Research*, Vol. 47, No. 16 (2008), pp. 5782–5796.

Hussain, M.A. (1999). Review of the applications of neural networks in chemical process control. Simulation and on-line implementations. *Artificial Intelligence in Engineering*, Vol. 13 (1999), pp. 55-68.

Jackson, J. E. (1991). *A user's guide to principal components*. John Wiley & Sons, New York, USA.

Luyben, W.L. (2006). Evaluation of criteria for selecting temperature control trays in distillation columns. *Journal of Process Control*, Vol. 16 (2006), pp. 115-134.

Marchetti, J.L.; Benallou, A.; Seborg D.E. & Mellichamp, D.A. (1985). A pilot-scale distillation facility for digital computer control research. *Computers and Chemical Engineering*, Vol. 9, No. 3 (1985), pp. 301-309.

Mejdell, T. & Skogestad, S. (1991). Composition estimator in a pilot plant distillation column using multiple temperatures. *Industrial Engineering Chemical Research*, Vol. 30 (1991), pp. 2555-2564.

Michalewitz, Z. (1992). *Genetic Algorithms + Data Structures = Evolution Programs*. Springer, Berlin, Germany.

Nooraii, A.; Romagnoli, J.A & Figueroa, J. (1999). Process identification, uncertainty characterization and robustness analysis of a pilot scale distillation column. *Journal of Process Control*, Vol. 9 (1999), pp. 247-264.

Norgaard, O.; Ravn, N.; Poulsen, K. & Hansen, L.K. (2000). *Neural networks for modeling and control of dynamic systems*, Springer Verlag.

Osiovici, R. & Cruz, S.L. (2001). Inferential control of high-purity multicomponent batch distillation columns using an extended Kalman filter. *Industrial Engineering Chemical Research*, Vol. 40,(2001), pp. 2628-2639.

Quintero-Marmol, E.; Luyben, W.L & Georgakis, C. (1991). Application of an extended Luenberger observer to the control of multicomponent batch distillation. *Industrial Engineering Chemical Research*, Vol. 30 (1991), pp 1870-1880.

Rueda, L.M.; Edgar, T.F. & Eldridge, R.B. (2006). A novel control methodology for a pilot plant azeotropic distillation column, *Industrial Engineering Chemical Research*, Vol. 45 (2006), pp. 8361-8372.

Singh, V.; Gupta, I. and Gupta, H.O. (2007). ANN-based estimator for distillation using Levenberg–Marquardt approach. *Engineering Applications of Artificial Intelligence*, Vol. 20 (2007), pp. 249-259.

Skogestad, S. (1997). Dynamics and control of distillation columns. A tutorial introduction. *Transactions on Industrial Chemical Engineering*, Vol. 75(A) (1997), pp. 539-562.

Skogestad, S. (2004). Control structure design for complete chemical plants. *Computers and Chemical Engineering*, Vol. 28 (2004), pp. 219-234.

Strandberg, J. & Skogestad, S. (2006). Stabilizing operation of a 4-product integrated Kaibel column. *Proceedings of IFAC Symposium on Advanced Control of Chemical Processes*,pp.623-628, Gramado, Brazil, 2-5 April 2006.

Tellez-Anguiano, A.; Rivas-Cruz, F.; Astorga-Zaragoza, C.M.; Alcorta-Garcia, E. & Juarez-Romero, D. (2009). Process control interface system for a distillation plant. *Computer Standards and Interfaces*, Vol. 31 (2009), pp. 471-479.

Varshney, K. & Panigrahi, P.K. (2005). Artificial neural network control of a heat exchanger in a closed flow air circuit. *Applied Soft Computing*, Vol. 5 (2005), pp. 441-465.

Venkateswarlu, C. & Avantika, S. (2001). Optimal state estimation of multicomponent batch distillation. *Chemical Engineering Science*, Vol. 56 (2001), pp. 5771-5786.

Zamprogna, E; Barolo, M. & Seborg, D.E. (2005). Optimal selection of soft sensor inputs for batch distillation columns using principal component analysis. *Journal of Process Control*, Vol. 15 (2005), pp. 39-52.

5

Online Adaptive Learning Solution of Multi-Agent Differential Graphical Games

Kyriakos G. Vamvoudakis[1] and Frank L. Lewis[2]
[1]*Center for Control, Dynamical-Systems,
and Computation (CCDC),
University of California, Santa Barbara,*
[2]*Automation and Robotics Research Institute,
The University of Texas at Arlington,
USA*

1. Introduction

Distributed networks have received much attention in the last year because of their flexibility and computational performance. The ability to coordinate agents is important in many real-world tasks where it is necessary for agents to exchange information with each other. Synchronization behavior among agents is found in flocking of birds, schooling of fish, and other natural systems. Work has been done to develop cooperative control methods for consensus and synchronization (Fax and Murray, 2004; Jadbabaie, Lin and Morse, 2003; Olfati-Saber, and Murray, 2004; Qu, 2009; Ren, Beard, and Atkins, 2005; Ren, and beard, 2005; Ren, and Beard, 2008; Tsitsiklis, 1984). See (Olfati-Saber, Fax, and Murray, 2007; Ren, Beard, and Atkins, 2005) for surveys. Leaderless consensus results in all nodes converging to common value that cannot generally be controlled. We call this the cooperative regulator problem. On the other hand the problem of cooperative tracking requires that all nodes synchronize to a leader or control node (Hong, Hu, and Gao, 2006; Li, Wang, and Chen, 2004; Ren, Moore, and Chen, 2007; Wang, and Chen, 2002). This has been called pinning control or control with a virtual leader. Consensus has been studied for systems on communication graphs with fixed or varying topologies and communication delays.

Game theory provides an ideal environment in which to study multi-player decision and control problems, and offers a wide range of challenging and engaging problems. Game theory (Tijs, 2003) has been successful in modeling strategic behavior, where the outcome for each player depends on the actions of himself and all the other players. Every player chooses a control to minimize independently from the others his own performance objective. Multi player cooperative games rely on solving coupled Hamilton-Jacobi (HJ) equations, which in the linear quadratic case reduce to the coupled algebraic Riccati equations (Basar, and Olsder, 1999; Freiling, Jank, and Abou-Kandil, 2002; Gajic, and Li, 1988). Solution methods are generally offline and generate fixed control policies that are then implemented in online controllers in real time. These coupled equations are difficult to solve.

Reinforcement learning (RL) is a sub-area of machine learning concerned with how to methodically modify the actions of an agent (player) based on observed responses from its environment (Sutton, and Barto, 1998). RL methods have allowed control systems researchers to develop algorithms to learn online in real time the solutions to optimal control problems for dynamic systems that are described by difference or ordinary differential equations. These involve a computational intelligence technique known as Policy Iteration (PI) (Bertsekas, and Tsitsiklis, 1996), which refers to a class of algorithms with two steps, *policy evaluation* and *policy improvement*. PI has primarily been developed for discrete-time systems, and online implementation for control systems has been developed through approximation of the value function (Bertsekas, and Tsitsiklis, 1996; Werbos, 1974; Werbos, 1992). PI provides effective means of learning solutions to HJ equations online. In control theoretic terms, the PI algorithm amounts to learning the solution to a nonlinear Lyapunov equation, and then updating the policy through minimizing a Hamiltonian function. Policy Iteration techniques have been developed for continuous-time systems in (Vrabie, Pastravanu, Lewis, and Abu-Khalaf, 2009).

RL methods have been used to solve multiplayer games for finite-state systems in (Busoniu, Babuska, and De Schutter, 2008; Littman, 2001). RL methods have been applied to learn online in real-time the solutions for optimal control problems for dynamic systems and differential games in (Dierks, and Jagannathan, 2010; Johnson, Hiramatsu, Fitz-Coy, and Dixon, 2010; Vamvoudakis 2010; Vamvoudakis 2011).

This book chapter brings together cooperative control, reinforcement learning, and game theory to solve multi-player differential games on communication graph topologies. There are four main contributions in this chapter. The first involves the formulation of a *graphical game* for dynamical systems networked by a communication graph. The dynamics and value function of each node depend only on the actions of that node and its neighbors. This graphical game allows for synchronization as well as Nash equilibrium solutions among neighbors. It is shown that standard definitions for Nash equilibrium are not sufficient for graphical games and a new definition of "Interactive Nash Equilibrium" is given. The second contribution is the derivation of coupled Riccati equations for solution of graphical games. The third contribution is a Policy Iteration algorithm for solution of graphical games that relies only on local information from neighbor nodes. It is shown that this algorithm converges to the best response policy of a node if its neighbors have fixed policies, and to the Nash solution if all nodes update their policies. The last contribution is the development of an online adaptive learning algorithm for computing the Nash equilibrium solutions of graphical games.

The book chapter is organized as follows. Section 2 reviews synchronization in graphs and derives an error dynamics for each node that is influenced by its own actions and those of its neighbors. Section 3 introduces differential graphical games cooperative Nash equilibrium. Coupled Riccati equations are developed and stability and solution for Nash equilibrium are proven. Section 4 proposes a policy iteration algorithm for the solution of graphical games and gives proofs of convergence. Section 5 presents an online adaptive learning solution based on the structure of the policy iteration algorithm of Section 4. Finally Section 6 presents a simulation example that shows the effectiveness of the proposed algorithms in learning in real-time the solutions of graphical games.

2. Synchronization and node error dynamics

2.1 Graphs

Consider a graph $G = (V, E)$ with a nonempty finite set of N nodes $V = \{v_1, \cdots, v_N\}$ and a set of edges or arcs $E \subseteq V \times V$. We assume the graph is simple, e.g. no repeated edges and $(v_i, v_i) \notin E, \forall i$ no self loops. Denote the connectivity matrix as $E = [e_{ij}]$ with $e_{ij} > 0$ if $(v_j, v_i) \in E$ and $e_{ij} = 0$ otherwise. Note $e_{ii} = 0$. The set of neighbors of a node v_i is $N_i = \{v_j : (v_j, v_i) \in E\}$, i.e. the set of nodes with arcs incoming to v_i. Define the in-degree matrix as a diagonal matrix $D = diag(d_i)$ with $d_i = \sum_{j \in N_i} e_{ij}$ the weighted in-degree of node i (i.e. i-th row sum of E). Define the graph Laplacian matrix as $L = D - E$, which has all row sums equal to zero.

A directed path is a sequence of nodes $v_0, v_1, \cdots, v_r$ such that $(v_i, v_{i+1}) \in E, i \in \{0,1,\cdots, r-1\}$. A directed graph is strongly connected if there is a directed path from v_i to v_j for all distinct nodes $v_i, v_j \in V$. A (directed) tree is a connected digraph where every node except one, called the root, has in-degree equal to one. A graph is said to have a spanning tree if a subset of the edges forms a directed tree. A strongly connected digraph contains a spanning tree.

General directed graphs with fixed topology are considered in this chapter.

2.2 Synchronization and node error dynamics

Consider the N systems or agents distributed on communication graph G with node dynamics

$$\dot{x}_i = A x_i + B_i u_i \tag{1}$$

where $x_i(t) \in \mathbb{R}^n$ is the state of node i, $u_i(t) \in \mathbb{R}^{m_i}$ its control input. Cooperative team objectives may be prescribed in terms of the *local neighborhood tracking error* $\delta_i \in \mathbb{R}^n$ (Khoo, Xie, and Man, 2009) as

$$\delta_i = \sum_{j \in N_i} e_{ij}(x_i - x_j) + g_i(x_i - x_0) \tag{2}$$

The pinning gain $g_i \geq 0$ is nonzero for a small number of nodes i that are coupled directly to the leader or control node x_0, and $g_i > 0$ for at least one i (Li, Wang, and Chen, 2004). We refer to the nodes i for which $g_i \neq 0$ as the pinned or controlled nodes. Note that δ_i represents the information available to node i for state feedback purposes as dictated by the graph structure.

The state of the control or target node is $x_0(t) \in \mathbb{R}^n$ which satisfies the dynamics

$$\dot{x}_0 = A x_0 \tag{3}$$

Note that this is in fact a *command generator* (Lewis, 1992) and we seek to design a cooperative control command generator tracker. Note that the trajectory generator A may not be stable.

The Synchronization control design problem is to design local control protocols for all the nodes in G to synchronize to the state of the control node, i.e. one requires $x_i(t) \to x_0(t)$, $\forall i$.

From (2), the overall error vector for network Gr is given by

$$\delta = \big((L+G) \otimes I_n \big)(x - \underline{x}_0) = \big((L+G) \otimes I_n \big) \zeta \tag{4}$$

where the global vectors are

$$x = \begin{bmatrix} x_1^T & x_2^T & \cdots & x_N^T \end{bmatrix}^T \in \mathbb{R}^{nN} \quad \delta = \begin{bmatrix} \delta_1^T & \delta_2^T & \cdots & \delta_N^T \end{bmatrix}^T \in \mathbb{R}^{nN} \quad \text{and} \quad \underline{x}_0 = \underline{I} x_0 \in \mathbb{R}^{nN}, \quad \text{with}$$

$\underline{I} = \underline{1} \otimes I_n \in R^{nN \times n}$ and $\underline{1}$ the N-vector of ones. The Kronecker product is $\otimes$ (Brewer, 1978). $G \in R^{N \times N}$ is a diagonal matrix with diagonal entries equal to the pinning gains g_i. The (global) consensus or synchronization error (e.g. the disagreement vector in (Olfati-Saber, and Murray, 2004)) is

$$\zeta = (x - \underline{x}_0) \in \mathbb{R}^{nN} \tag{5}$$

The communication digraph is assumed to be strongly connected. Then, if $g_i \neq 0$ for at least one i, $(L+G)$ is nonsingular with all eigenvalues having positive real parts (Khoo, Xie, and Man, 2009). The next result therefore follows from (4) and the Cauchy Schwartz inequality and the properties of the Kronecker product (Brewer, 1978).

Lemma 1. Let the graph be strongly connected and $G \neq 0$. Then the synchronization error is bounded by

$$\| \zeta \| \leq \| \delta \| / \underline{\sigma}(L+G) \tag{6}$$

with $\underline{\sigma}(L+G)$ the minimum singular value of $(L+G)$, and $\delta(t) \equiv 0$ if and only if the nodes synchronize, that is

$$x(t) = \underline{I} x_0(t) \tag{7}$$

∎

Our objective now shall be to make small the local neighborhood tracking errors $\delta_i(t)$, which in view of Lemma 1 will guarantee synchronization.

To find the dynamics of the local neighborhood tracking error, write

$$\dot{\delta}_i = A \delta_i + (d_i + g_i) B_i u_i - \sum_{j \in N_i} e_{ij} B_j u_j \tag{8}$$

with $\delta_i \in \mathbb{R}^n, u_i \in \mathbb{R}^{m_i}, \forall i$.

This is a dynamical system with multiple control inputs, from node i and all of its neighbors.

3. Cooperative multi-player games on graphs

We wish to achieve synchronization while simultaneously optimizing some performance specifications on the agents. To capture this, we intend to use the machinery of multi-player games (Basar, Olsder, 1999). Define $u_{G-i} = \{u_j : j \in N, j \neq i\}$ as the set of policies of all other nodes in the graph other than node i. Define $u_{-i}(t)$ as the vector of the control inputs $\{u_j : j \in N_i\}$ of the neighbors of node i.

3.1 Cooperative performance index

Define the local performance indices

$$J_i(\delta_i(0), u_i, u_{-i}) = \frac{1}{2}\int_0^\infty (\delta_i^T Q_{ii}\delta_i + u_i^T R_{ii}u_i + \sum_{j \in N_i} u_j^T R_{ij}u_j) \, dt \equiv \frac{1}{2}\int_0^\infty L_i(\delta_i(t), u_i(t), u_{-i}(t)) \, dt \quad (9)$$

where all weighting matrices are constant and symmetric with $Q_{ii} > 0, R_{ii} > 0, R_{ij} \geq 0$. Note that the i-th performance index includes only information about the inputs of node i and its neighbors.

For dynamics (8) with performance objectives (9), introduce the associated Hamiltonians

$$H_i(\delta_i, p_i, u_i, u_{-i}) \equiv p_i^T\left(A\delta_i + (d_i + g_i)B_i u_i - \sum_{j \in N_i} e_{ij}B_j u_j\right) + \frac{1}{2}\delta_i^T Q_{ii}\delta_i + \frac{1}{2}u_i^T R_{ii}u_i + \frac{1}{2}\sum_{j \in N_i} u_j^T R_{ij}u_j = 0 \quad (10)$$

where p_i is the costate variable. Necessary conditions (Lewis, and Syrmos, 1995) for a minimum of (9) are (1) and

$$-\dot{p}_i = \frac{\partial H_i}{\partial \delta_i} \equiv A^T p_i + Q_{ii}\delta_i \quad (11)$$

$$0 = \frac{\partial H_i}{\partial u_i} \Rightarrow u_i = -(d_i + g_i)R_{ii}^{-1}B_i^T p_i \quad (12)$$

3.2 Graphical games

Interpreting the control inputs u_i, u_j as state dependent policies or strategies, the value function for node i corresponding to those policies is

$$V_i(\delta_i(t)) = \frac{1}{2}\int_t^\infty (\delta_i^T Q_{ii}\delta_i + u_i^T R_{ii}u_i + \sum_{j \in N_i} u_j^T R_{ij}u_j) \, dt \quad (13)$$

Definition 1. Control policies u_i, $\forall i$ are defined as admissible if u_i are continuous, $u_i(0) = 0$, u_i stabilize systems (8) locally, and values (13) are finite.

When V_i is finite, using Leibniz' formula, a differential equivalent to (13) is given in terms of the Hamiltonian function by the Bellman equation

$$H_i(\delta_i, \frac{\partial V_i}{\partial \delta_i}, u_i, u_{-i}) \equiv \frac{\partial V_i}{\partial \delta_i}^T \left(A\delta_i + (d_i + g_i)B_i u_i - \sum_{j \in N_i} e_{ij}B_j u_j \right) + \frac{1}{2}\delta_i^T Q_{ii}\delta_i + \frac{1}{2}u_i^T R_{ii}u_i + \frac{1}{2}\sum_{j \in N_i} u_j^T R_{ij}u_j = 0 \quad (14)$$

with boundary condition $V_i(0) = 0$. (The gradient is disabused here as a column vector.) That is, solution of equation (14) serves as an alternative to evaluating the infinite integral (13) for finding the value associated to the current feedback policies. It is shown in the Proof of Theorem 2 that (14) is a Lyapunov equation. According to (13) and (10) one equates $p_i = \partial V_i / \partial \delta_i$.

The local dynamics (8) and performance indices (9) only depend for each node i on its own control actions and those of its neighbors. We call this a *graphical game*. It depends on the topology of the communication graph $G = (V,E)$. We assume throughout the chapter that the game is well-formed in the following sense.

Definition 2. The graphical game with local dynamics (8) and performance indices (9) is well-formed if $B_j \neq 0 \rightleftharpoons e_{ij} \in E$, $R_{ij} \neq 0 \rightleftharpoons e_{ij} \in E$.

The control objective of agent i in the graphical game is to determine

$$V_i^*(\delta_i(t)) = \min_{u_i} \int_t^\infty \frac{1}{2}(\delta_i^T Q_{ii}\delta_i + u_i^T R_{ii}u_i + \sum_{j \in N_i} u_j^T R_{ij}u_j)\, dt \quad (15)$$

Employing the stationarity condition (12) (Lewis, and Syrmos, 1995) one obtains the control policies

$$u_i = u_i(V_i) \equiv -(d_i + g_i)R_{ii}^{-1}B_i^T \frac{\partial V_i}{\partial \delta_i} \equiv -h_i(p_i) \quad (16)$$

The game defined in (15) corresponds to Nash equilibrium.

Definition 3. (Basar, and Olsder, 1999) (Global Nash equilibrium) An *N-tuple* of policies $\{u_1^*, u_2^*, ..., u_N^*\}$ is said to constitute a global Nash equilibrium solution for an N player game if for all $i \in N$

$$J_i^* \triangleq J_i(u_i^*, u_{G-i}^*) \leq J_i(u_i, u_{G-i}^*) \quad (17)$$

The *N-tuple* of game values $\{J_1^*, J_2^*, ..., J_N^*\}$ is known as a Nash equilibrium outcome of the N-player game.

The distributed multiplayer graphical game with local dynamics (8) and local performance indices (9) should be contrasted with standard multiplayer games (Abou-Kandil, Freiling, Ionescu, and Jank, 2003; Basar, and Olsder 1999) which have centralized dynamics

$$\dot{z} = Az + \sum_{i=1}^{N} B_i u_i \qquad (18)$$

where $z \in \mathbb{R}^n$ is the state, $u_i(t) \in \mathbb{R}^{m_i}$ is the control input for every player, and where the performance index of each player depends on the control inputs of all other players. In the graphical games, by contrast, each node's dynamics and performance index only depends on its own state, its control, and the controls of its immediate neighbors.

It is desired to study the distributed game on a graph defined by (15) with distributed dynamics (8). It is not clear in this scenario how global Nash equilibrium is to be achieved.

Graphical games have been studied in the computational intelligence community (Kakade, Kearns, Langford, and Ortiz, 2003; Kearns, Littman, and Singh, 2001; Shoham, and Leyton-Brown, 2009). A (nondynamic) graphical game has been defined there as a tuple (G, U, v) with $G = (V, E)$ a graph with N nodes, action set $U = U_1 \times \cdots \times U_N$ with U_i the set of actions available to node i, and $v = \begin{bmatrix} v_1 & \cdots & v_N \end{bmatrix}^T$ a payoff vector, with $v_i(U_i, \{U_j : j \in N_i\}) \in R$ the payoff function of node i. It is important to note that *the payoff of node i only depends on its own action and those of its immediate neighbors*. The work on graphical games has focused on developing algorithms to find standard Nash equilibria for payoffs generally given in terms of matrices. Such algorithms are simplified in that they only have complexity on the order of the maximum node degree in the graph, not on the order of the number of players N. Undirected graphs are studied, and it is assumed that the graph is connected.

The intention in this chapter is to provide online real-time adaptive methods for solving differential graphical games that are distributed in nature. That is, the control protocols and adaptive algorithms of each node are allowed to depend only information about itself and its neighbors. Moreover, as the game solution is being learned, all node dynamics are required to be stable, until finally all the nodes synchronize to the state of the control node. These online methods are discussed in Section V.

The following notions are needed in the study of differential graphical games.

Definition 4. (Shoham, and Leyton-Brown, 2009) Agent i's *best response* to fixed policies u_{-i} of his neighbors is the policy u_i^* such that

$$J_i(u_i^*, u_{-i}) \le J_i(u_i, u_{-i}) \qquad (19)$$

for all policies u_i of agent i.

For centralized multi-agent games, where the dynamics is given by (18) and the performance of each agent depends on the actions of all other agents, an equivalent definition of Nash equilibrium is that each agent is in best response to all other agents. In graphical games, if all agents are in best response to their neighbors, then all agents are in Nash equilibrium, as seen in the proof of Theorem 1.

However, a counterexample shows the problems with the definition of Nash equilibrium in graphical games. Consider the completely disconnected graph with empty edge set where each node has no neighbors. Then Definition 4 holds if each agent simply chooses his single-player optimal control solution $J_i^* = J_i(u_i^*)$, since, for the disconnected graph case one has

$$J_i(u_i) = J_i(u_i, u_{G-i}) = J_i(u_i, u'_{G-i}), \ \forall i \tag{20}$$

for any choices of the two sets u_{G-i}, u'_{G-i} of the policies of all the other nodes. That is, the value function of each node does not depend on the policies of any other nodes.

Note, however, that Definition 3 also holds, that is, the nodes are in a global Nash equilibrium. Pathological cases such as this counterexample cannot occur in the standard games with centralized dynamics (18), particularly because stabilizability conditions are usually assumed.

3.3 Interactive Nash equilibrium

The counterexample in the previous section shows that in pathological cases when the graph is disconnected, agents can be in Nash equilibrium, yet have no influence on each others' games. In such situations, the definition of coalition-proof Nash equilibrium (Shinohara, 2010) may also hold, that is, no set of agents has an incentive to break away from the Nash equilibrium and seek a new Nash solution among themselves.

To rule out such undesirable situations and guarantee that all agents in a graph are involved in the same game, we make the following stronger definition of global Nash equilibrium.

Definition 5. (Interactive Global Nash equilibrium) An *N-tuple* of policies $\{u_1^*, u_2^*, ..., u_N^*\}$ is said to constitute an interactive global Nash equilibrium solution for an N player game if, for all $i \in N$, the Nash condition (17) holds and in addition there exists a policy u'_k such that

$$J_i(u_k^*, u_{G-k}^*) \neq J_i(u'_k, u_{G-k}^*) \tag{21}$$

for all $i, k \in N$. That is, at equilibrium there exists a policy of every player k that influences the performance of all other players i.

If the systems are in Interactive Nash equilibrium, the graphical game is well-defined in the sense that all players are in a single Nash equilibrium with each player affecting the decisions of all other players. Condition (21) means that the reaction curve (Basar, and Olsder, 1999) of any player i is not constant with respect to all variations in the policy of any other player k.

The next results give conditions under which the local best responses in Definition 4 imply the interactive global Nash of Definition 5.

Consider the systems (8) in closed-loop with admissible feedbacks (12), (16) denoted by $u_k = K_k p_k - v_k$ for a single node k and $u_j = K_j p_j, \forall j \neq k$. Then

$$\dot{\delta}_i = A\delta_i + (d_i + g_i)B_i K_i p_i - \sum_{j \in N_i} e_{ij} B_j K_j p_j + e_{ik} B_k v_k, \quad k \neq i \tag{22}$$

The global closed-loop dynamics are

$$\begin{bmatrix} \dot{\delta} \\ \dot{p} \end{bmatrix} = \begin{bmatrix} (I_N \otimes A) & ((L+G) \otimes I_n) diag(B_i K_i) \\ -diag(Q_{ii}) & -(I_N \otimes A^T) \end{bmatrix} \begin{bmatrix} \delta \\ p \end{bmatrix} + \begin{bmatrix} ((L+G) \otimes I_n)\underline{B}_k \\ 0 \end{bmatrix} \overline{v}_k \equiv \overline{A} \begin{bmatrix} \delta \\ p \end{bmatrix} + \overline{B}\overline{v}_k \tag{23}$$

with $\underline{B}_k = diag(B_i)$ and $\overline{v}_k = \begin{bmatrix} 0 & \cdots & v_k^T & \cdots & 0 \end{bmatrix}^T$ has all block entries zero with v_k in block k. Consider node i and let $M > 0$ be the first integer such that $[(L+G)^M]_{ik} \neq 0$, where $[.]_{ik}$ denotes the element (i,k) of a matrix. That is, M is the length of the shortest directed path from k to i. Denote the nodes along this path by $k = k_0, k_1, \cdots, k_{M-1}, k_M = i$. Denote element (i,k) of $L+G$ by ℓ_{ik}. Then the $n \times m$ block element in block row i and block column k of matrix $\overline{A}^{2(M-1)}\overline{B}$ is equal to

$$\left[\overline{A}^{2(M-1)}\overline{B} \right]^{ik} = \sum_{k_{M-1}, \cdots, k_1} \ell_{i,k_{M-1}} \cdots \ell_{k_1,k} B_{k_{M-1}} K_{k_{M-1}} Q_{k_{M-1}} B_{k_{M-2}} \cdots B_{k_1} K_{k_1} Q_{k_1} B_k \equiv \sum_{k_{M-1}} B_{k_{M-1}} \overline{B}_{k_{M-1},k} \tag{24}$$

where $\overline{B}_{k_{M-1},k} \in R^{m_{k_{M-1}} \times m_k}$ and $[\quad]^{ik}$ denotes the position of the block element in the block matrix.

Assumption 1.

a. $\overline{B}_{k_{M-1},k} \in R^{m_{k_{M-1}} \times m_k}$ has rank $m_{k_{M-1}}$.

All shortest paths to node i from node k pass through a single neighbor $k_M - 1$ of i.

An example case where Assumption 1a holds is when there is a single shortest path from k to i, $m_i = m, \forall i$, $rank(B_i) = m, \forall i$.

Lemma 2. Let (A, B_j) be reachable for all $j \in N$ and let Assumption 1 hold. Then the i-th closed-loop system (22) is reachable from input v_k if and only if there exists a directed path from node k to node i.

Proof:

Sufficiency. If $k = i$ the result is obvious. Otherwise, the reachability matrix from node k to node i has the $n \times m$ block element in block row i and block column k given as

$$\left[\overline{A}^{2(M-1)}\overline{B} \quad \overline{A}^{2(M-1)+1}\overline{B} \quad \overline{A}^{2(M-1)+2}\overline{B} \quad \cdots \right]^{ik} = \left[\sum_{k_{M-1}} B_{k_{M-1}} \quad \sum_{k_{M-1}} AB_{k_{M-1}} \quad \sum_{k_{M-1}} A^2 B_{k_{M-1}} \quad \cdots \right]$$

$$\times \begin{bmatrix} \overline{B}_{k_{M-1},k} & * & * & \\ 0 & \overline{B}_{k_{M-1},k} & * & \\ \vdots & 0 & \overline{B}_{k_{M-1},k} & \\ 0 & \cdots & 0 & \ddots \end{bmatrix}$$

where * denotes nonzero entries. Under the assumptions, the matrix on the right has full row rank and the matrix on the left is written as $\left[B_{k_{M-1}} \quad AB_{k_{M-1}} \quad A^2 B_{k_{M-1}} \quad \cdots \right]$.

However, $(A, B_{k_{M-1}})$ is reachable.

Necessity. If there is no path from node k to node i, then the control input of node k cannot influence the state or value of node i.

∎

Theorem 1. Let (A, B_i) be reachable for all $i \in N$. Let every node i be in best response to all his neighbors $j \in N_i$. Let Assumption 1 hold. Then all nodes in the graph are in interactive global Nash equilibrium if and only if the graph is strongly connected.

Proof:

Let every node i be in best response to all his neighbors $j \in N_i$. Then $J_i(u_i^*, u_{-i}) \le J_i(u_i, u_{-i})$, $\forall i$. Hence $u_j = u_j^*$, $\forall u_j \in u_{-i}$ and $J_i(u_i^*, u_{-i}^*) \le J_i(u_i, u_{-i}^*)$, $\forall i$. However, according to (9) $J_i(u_i^*, u_{-i}^*, u_k) = J_i(u_i^*, u_{-i}^*, u_k)$, $\forall k \notin \{i\} \cup N_i$ so that $J_i(u_i^*, u_{G-i}^*) \le J_i(u_i, u_{G-i}^*)$, $\forall i$ and the nodes are in Nash equilibrium.

Necessity. If the graph is not strongly connected, then there exist nodes k and i such that there is no path from node k to node i. Then, the control input of node k cannot influence the state or the value of node i. Therefore, the Nash equilibrium is not interactive.

Sufficiency. Let (A, B_i) be reachable for all $i \in N$. Then if there is a path from node k to node i, the state δ_i is reachable from u_k, and from (9) input u_k can change the value J_i. Strong connectivity means there is a path from every node k to every node i and condition (21) holds for all $i, k \in N$.

∎

The reachability condition is sufficient but not necessary for Interactive Nash equilibrium.

According to the results just established, the following assumptions are made.

Assumptions 2.

a. (A, B_i) is reachable for all $i \in N$.

b. The graph is strongly connected and at least one pinning gain g_i is nonzero. Then $(L + G)$ is nonsingular.

3.4 Stability and solution of graphical games

Substituting control policies (16) into (14) yields the coupled cooperative game Hamilton-Jacobi (HJ) equations

$$\frac{\partial V_i}{\partial \delta_i}^T A_i^c + \frac{1}{2}\delta_i^T Q_{ii}\delta_i + \frac{1}{2}(d_i + g_i)^2 \frac{\partial V_i}{\partial \delta_i}^T B_i R_{ii}^{-1}B_i^T \frac{\partial V_i}{\partial \delta_i} + \frac{1}{2}\sum_{j\in N_i}(d_j + g_j)^2 \frac{\partial V_j}{\partial \delta_j}^T B_j R_{jj}^{-1}R_{ij}R_{jj}^{-1}B_j^T \frac{\partial V_j}{\partial \delta_j} = 0, i \in N \quad (25)$$

where the closed-loop matrix is

$$A_i^c = A\delta_i - (d_i + g_i)^2 B_i R_{ii}^{-1}B_i^T \frac{\partial V_i}{\partial \delta_i} + \sum_{j\in N_i} e_{ij}(d_j + g_j)B_j R_{jj}^{-1}B_j^T \frac{\partial V_j}{\partial \delta_j}, i \in N \quad (26)$$

For a given V_i, define $u_i^* = u_i(V_i)$ as (16) given in terms of V_i. Then HJ equations (25) can be written as

$$H_i(\delta_i, \frac{\partial V_i}{\partial \delta_i}, u_i^*, u_{-i}^*) = 0 \quad (27)$$

There is one coupled HJ equation corresponding to each node, so solution of this N-player game problem is blocked by requiring a solution to N coupled partial differential equations. In the next sections we show how to solve this N-player cooperative game online in a distributed fashion at each node, requiring only measurements from neighbor nodes, by using techniques from reinforcement learning.

It is now shown that the coupled HJ equations (25) can be written as coupled Riccati equations. For the global state δ given in (4) we can write the dynamics as

$$\dot{\delta} = (I_N \otimes A)\delta + (L + G) \otimes I_n diag(B_i)u \quad (28)$$

where u is the control given by

$$u = -diag(R_{ii}^{-1}B_i^T)((D + G) \otimes I_n p) \quad (29)$$

where $diag(.)$ denotes diagonal matrix of appropriate dimensions. Furthermore the global costate dynamics are

$$-\dot{p} = \frac{\partial H}{\partial \delta} \equiv (I_N \otimes A)^T p + diag(Q_{ii})\delta \quad (30)$$

This is a set of coupled dynamic equations reminiscent of standard multi-player games (Basar, and Olsder, 1999) or single agent optimal control (Lewis, and Syrmos, 1995). Therefore the solution can be written without any loss of generality as

$$p = \overline{P}\delta \quad (31)$$

for some matrix $\overline{P} > 0 \in \mathbb{R}^{nNxnN}$.

Lemma 3. HJ equations (25) are equivalent to the coupled Riccati equations

$$\delta^T \overline{P}^T \overline{A}_i \delta - \delta^T \overline{P}^T \overline{B}_i \overline{P} \delta + \tfrac{1}{2}\delta^T \overline{Q}_i \delta + \tfrac{1}{2}\delta^T \overline{P}^T \overline{R}_i \overline{P}\delta = 0 \tag{32}$$

or equivalently, in closed-loop form,

$$(\overline{P}^T \overline{A}_{ic} + \overline{A}_{ic}{}^T \overline{P} + \overline{Q}_i + \overline{P}^T \overline{R}_i \overline{P}) = 0 \tag{33}$$

where $\overline{P}$ is defined by (31), and

$$\overline{A}_i = \begin{bmatrix} 0 & & & \\ & 0 & & \\ & & [A]^{ii} & \\ & & & 0 \end{bmatrix}, \overline{B}_i = \begin{bmatrix} 0 & & & \\ & \left[(d_i + g_i)I_n\right]^{ii} & \left[-a_{ij}I_n\right]^{ij} & diag((d_i + g_i)B_iR_{ii}^{-1}B_i^{T}) \\ & & 0 & \end{bmatrix}$$

$$\overline{A}_{ic} = \overline{A}_i - \overline{B}_i \overline{P}$$

$$\overline{Q}_i = \begin{bmatrix} 0 & & & \\ & 0 & & \\ & & [Q_{ii}]^{ii} & \\ & & & 0 \end{bmatrix}, \overline{R}_i = diag((d_i + g_i)B_iR_{ii}^{-1}) \begin{bmatrix} R_{i1} & & & & \\ & \ddots & & & \\ & & R_{ij} & & \\ & & & \ddots & \\ & & & & R_{ii} \\ & & & & & R_{iN} \end{bmatrix} diag((d_i + g_i)R_{ii}^{-1}B_i^{T})$$

Proof:

Take (14) and write it with respect to the global state and costate as

$$H_i \equiv \begin{bmatrix} \dfrac{\partial V_1}{\partial \delta_1} \\ \vdots \\ \vdots \\ \dfrac{\partial V_N}{\partial \delta_N} \end{bmatrix}^T \begin{bmatrix} 0 & & & \\ & 0 & & \\ & & [A]^{ii} & \\ & & & 0 \end{bmatrix} \delta$$

$$+ \begin{bmatrix} \dfrac{\partial V_1}{\partial \delta_1} \\ \vdots \\ \vdots \\ \dfrac{\partial V_N}{\partial \delta_N} \end{bmatrix}^T \begin{bmatrix} 0 & \cdots & 0 & & 0 \\ \vdots & 0 & \vdots & & \vdots \\ \vdots & \vdots & \left[(d_i + g_i)I_n\right]^{ii} & \left[-a_{ij}I_n\right]^{ij} & \\ 0 & \cdots & 0 & & 0 \end{bmatrix} \begin{bmatrix} B_1 & & & \\ & \ddots & & \\ & & B_i & \\ & & & B_N \end{bmatrix} \begin{bmatrix} u_1 \\ \vdots \\ u_i \\ \vdots \\ u_N \end{bmatrix} \tag{34}$$

$$
+\frac{1}{2}\delta^T
\begin{bmatrix}
0 & & & \\
& 0 & & \\
& & [Q_{ii}]^{ii} & \\
& & & 0
\end{bmatrix}
\delta + \frac{1}{2}
\begin{bmatrix}
u_1 \\ \vdots \\ u_i \\ u_N
\end{bmatrix}^T
\begin{bmatrix}
R_{i1} & & & \\
& R_{ij} & & \\
& & R_{ii} & \\
R_{iN} & & & R_{iN}
\end{bmatrix}
\begin{bmatrix}
u_1 \\ \vdots \\ u_i \\ u_N
\end{bmatrix} = 0
$$

By definition of the costate one has

$$
p \equiv \begin{bmatrix} \dfrac{\partial V_1}{\partial \delta_1} & \cdots & \cdots & \dfrac{\partial V_N}{\partial \delta_N} \end{bmatrix}^T = \overline{P}\delta \tag{35}
$$

∎

From the control policies (16), (34) becomes (32).

It is now shown that if solutions can be found for the coupled design equations (25), they provide the solution to the graphical game problem.

Theorem 2. Stability and Solution for Cooperative Nash Equilibrium.

Let Assumptions 1 and 2a hold. Let $V_i > 0 \in C^1$, $i \in N$ be smooth solutions to HJ equations (25) and control policies u_i^*, $i \in N$ be given by (16) in terms of these solutions V_i. Then

a.   Systems (8) are asymptotically stable so all agents synchronize.

$\{u_1^*, u_2^*, ..., u_N^*\}$ are in global Nash equilibrium and the corresponding game values are

$$
J_i^*(\delta_i(0)) = V_i \ , i \in N \tag{36}
$$

Proof:

If $V_i > 0$ satisfies (25) then it also satisfies (14). Take the time derivative to obtain

$$
\dot{V}_i = \frac{\partial V_i}{\partial \delta_i}^T \dot{\delta}_i = \frac{\partial V_i}{\partial \delta_i}^T \left(A\delta_i + (d_i + g_i)B_i u_i - \sum_{j \in N_i} e_{ij} B_j u_j \right) = -\frac{1}{2}\left(\delta_i^T Q_{ii} \delta_i + u_i^T R_{ii} u_i + \sum_{j \in N_i} u_j^T R_{ij} u_j \right) \tag{37}
$$

which is negative definite since $Q_{ii} > 0$. Therefore V_i is a Lyapunov function for δ_i and systems (8) are asymptotically stable.

According to part a, $\delta_i(t) \to 0$ for the selected control policies. For any smooth functions $V_i(\delta_i)$, $i \in N$, such that $V_i(0) = 0$, setting $V_i(\delta_i(\infty)) = 0$ one can write (9) as

$$
J_i(\delta_i(0), u_i, u_{-i}) = \frac{1}{2}\int_0^\infty (\delta_i^T Q_{ii} \delta_i + u_i^T R_{ii} u_i + \sum_{j \in N_i} u_j^T R_{ij} u_j) \, dt + V_i(\delta_i(0))
$$

$$
+ \int_0^\infty \frac{\partial V_i}{\partial \delta_i}^T (A\delta_i + (d_i + g_i)B_i u_i - \sum_{j \in N_i} e_{ij} B_j u_j) dt
$$

Now let V_i satisfy (25) and u_i^*, u_{-i}^* be the optimal controls given by (16). By completing the squares one has

$$J_i(\delta_i(0), u_i, u_{-i}) = V_i\,(\delta_i(0)) + \int_0^\infty (\tfrac{1}{2} \sum_{j \in N_i} (u_j - u_j^*)^T R_{ij}(u_j - u_j^*) + \tfrac{1}{2}(u_i - u_i^*)^T R_{ii}(u_i - u_i^*)$$

$$-\frac{\partial V_i}{\partial \delta_i}^T \sum_{j \in N_i} e_{ij} B_j (u_j - u_j^*) + \sum_{j \in N_i} u_j^{*T} R_{ij}(u_j - u_j^*))dt$$

At the equilibrium point $u_i = u_i^*$ and $u_j = u_j^*$ so

$$J_i^*(\delta_i(0), u_i^*, u_{-i}^*) = V_i\,(\delta_i(0))$$

Define

$$J_i(u_i, u_{-i}^*) = V_i\,(\delta_i(0)) + \frac{1}{2}\int_0^\infty (u_i - u_i^*)^T R_{ii}(u_i - u_i^*)dt$$

and $J_i^* = V_i\,(\delta_i(0))$. Then clearly J_i^* and $J_i(u_i, u_{-i}^*)$ satisfy (19). Since this is true for all i, Nash condition (17) is satisfied.

∎

The next result shows when the systems are in Interactive Nash equilibrium. This means that the graphical game is well defined in the sense that all players are in a single Nash equilibrium with each player affecting the decisions of all other players.

Corollary 1. Let the hypotheses of Theorem 2 hold. Let Assumptions 1 and 2 hold so that the graph is strongly connected. Then $\{u_1^*, u_2^*, ..., u_N^*\}$ are in interactive Nash equilibrium and all agents synchronize.

Proof:

From Theorems 1 and 2.

∎

3.5 Global and local performance objectives: Cooperation and competition

The overall objective of all the nodes is to ensure synchronization of all the states $x_i(t)$ to $x_0(t)$. The multi player game formulation allows for considerable freedom of each agent while achieving this objective. Each agent has a performance objective that can embody team objectives as well as individual node objectives.

The performance objective of each node can be written as

$$J_i = \frac{1}{N_i} \sum_{j \in N_i} J_j + \frac{1}{N_i} \sum_{j \in N_i} (J_i - J_j) \equiv J_{team} + J_i^{conflict}$$

where J_{team} is the overall ('center of gravity') performance objective of the networked team and $J_i^{conflict}$ is the conflict of interest or competitive objective. J_{team} measures how much the players are vested in common goals, and $J_i^{conflict}$ expresses to what extent their objectives differ. The objective functions can be chosen by the individual players, or they may be assigned to yield some desired team behavior.

4. Policy iteration algorithms for cooperative multi-player games

Reinforcement learning (RL) techniques have been used to solve the single-player optimal control problem online using adaptive learning techniques to determine the optimal value function. Especially effective are the approximate dynamic programming (ADP) methods (Werbos, 1974; Werbos, 1992). RL techniques have also been applied for multiplayer games with centralized dynamics (18). See for example (Busoniu, Babuska, and De Schutter, 2008; Vrancx, Verbeeck, and Nowe, 2008). Most applications of RL for solving optimal control problems or games online have been to finite-state systems or discrete-time dynamical systems. In this section is given a policy iteration algorithm for solving continuous-time differential games on graphs. The structure of this algorithm is used in the next section to provide online adaptive solutions for graphical games.

4.1 Best response

Theorem 2 and Corollary 1 reveal that, under assumptions 1 and 2, the systems are in interactive Nash equilibrium if, for all $i \in N$ node i selects his best response policy to his neighbors policies and the graph is strongly connected. Define the best response HJ equation as the Bellman equation (14) with control $u_i = u_i^*$ given by (16) and arbitrary policies $u_{-i} = \{u_j : j \in N_i\}$

$$0 = H_i(\delta_i, \frac{\partial V_i}{\partial \delta_i}, u_i^*, u_{-i}) \equiv \frac{\partial V_i}{\partial \delta_i}^T A_i^c + \frac{1}{2}\delta_i^T Q_{ii}\delta_i + \frac{1}{2}(d_i + g_i)^2 \frac{\partial V_i}{\partial \delta_i}^T B_i R_{ii}^{-1} B_i^T \frac{\partial V_i}{\partial \delta_i} + \frac{1}{2}\sum_{j \in N_i} u_j^T R_{ij}u_j \quad (38)$$

where the closed-loop matrix is

$$A_i^c = A\delta_i - (d_i + g_i)^2 B_i R_{ii}^{-1} B_i^T \frac{\partial V_i}{\partial \delta_i} - \sum_{j \in N_i} e_{ij}B_j u_j \quad (39)$$

Theorem 3. Solution for Best Response Policy

Given fixed neighbor policies $u_{-i} = \{u_j : j \in N_i\}$, assume there is an admissible policy u_i . Let $V_i > 0 \in C^1$ be a smooth solution to the best response HJ equation (38) and let control policy u_i^* be given by (16) in terms of this solution V_i . Then

a. Systems (8) are asymptotically stable so that all agents synchronize.
b. u_i^* is the best response to the fixed policies u_{-i} of its neighbors.

Proof:

a. $V_i > 0$ satisfies (38). Proof follows Theorem 2, part a.

b.   According to part a, $\delta_i(t) \to 0$ for the selected control policies. For any smooth functions $V_i(\delta_i), i \in N$, such that $V_i(0) = 0$, setting $V_i(\delta_i(\infty)) = 0$ one can write (9) as

$$J_i(\delta_i(0), u_i, u_{-i}) = \tfrac{1}{2} \int_0^\infty (\delta_i^T Q_{ii} \delta_i + u_i^T R_{ii} u_i + \sum_{j \in N_i} u_j^T R_{ij} u_j) \, dt + V_i(\delta_i(0))$$

$$+ \int_0^\infty \frac{\partial V_i}{\partial \delta_i}^T (A\delta_i + (d_i + g_i) B_i u_i - \sum_{j \in N_i} e_{ij} B_j u_j) dt$$

Now let V_i satisfy (38), u_i^* be the optimal controls given by (16), and u_{-i} be arbitrary policies. By completing the squares one has

$$J_i(\delta_i(0), u_i, u_{-i}) = V_i \, (\delta_i(0)) + \int_0^\infty \tfrac{1}{2} (u_i - u_i^*)^T R_{ii} (u_i - u_i^*) dt$$

The agents are in best response to fixed policies u_{-i} when $u_i = u_i^*$ so

$$J_i(\delta_i(0), u_i^*, u_{-i}) = V_i \, (\delta_i(0))$$

Then clearly $J_i(\delta_i(0), u_i, u_{-i})$ and $J_i(\delta_i(0), u_i^*, u_{-i})$ satisfy (19).

■

4.2 Policy iteration solution for graphical games

The following algorithm for the N-player distributed games is motivated by the structure of policy iteration algorithms in reinforcement learning (Bertsekas, and Tsitsiklis, 1996; Sutton, and Barto, 1998) which rely on repeated policy evaluation (e.g. solution of (14)) and policy improvement (solution of (16)). These two steps are repeated until the policy improvement step no longer changes the present policy. If the algorithm converges for every i, then it converges to the solution to HJ equations (25), and hence provides the distributed Nash equilibrium. One must note that the costs can be evaluated only in the case of admissible control policies, admissibility being a condition for the control policy which initializes the algorithm.

Algorithm 1. Policy Iteration (PI) Solution for N-player distributed games.

Step 0: Start with admissible initial policies u_i^0, $\forall i$.

Step 1: (Policy Evaluation) Solve for V_i^k using (14)

$$H_i(\delta_i, \frac{\partial V_i^k}{\partial \delta_i}, u_i^k, u_{-i}^k) = 0, \forall i = 1, \ldots, N \tag{40}$$

Step 2: (Policy Improvement) Update the N-tuple of control policies using

$$u_i^{k+1} = \arg\min_{u_i} H_i(\delta_i, \frac{\partial V_i}{\partial \delta_i}^k, u_i, u_{-i}^k), \forall i = 1,\ldots,N$$

which explicitly is

$$u_i^{k+1} = -(d_i + g_i)R_{ii}^{-1}B_i^T \frac{\partial V_i}{\partial \delta_i}^k, \forall i = 1,\ldots,N. \tag{41}$$

Go to step 1.

On convergence- End

■

The following two theorems prove convergence of the policy iteration algorithm for distributed games for two different cases. The two cases considered are the following, i) *only* agent i updates its policy and ii) all the agents update their policies.

Theorem 4. Convergence of Policy Iteration algorithm when only i^{th} agent updates its policy and all players u_{-i} in its neighborhood do not change. Given fixed neighbors policies u_{-i}, assume there exists an admissible policy u_i. Assume that agent i performs Algorithm 1 and the its neighbors do not update their control policies. Then the algorithm converges to the best response u_i to policies u_{-i} of the neighbors and to the solution V_i to the best response HJ equation (38).

Proof:

It is clear that

$$H_i^o(\delta_i, \frac{\partial V_i^k}{\partial \delta_i}, u_{-i}^k) \equiv \min_{u_i} H_i(\delta_i, \frac{\partial V_i^k}{\partial \delta_i}, u_i^k, u_{-i}^k) = H_i(\delta_i, \frac{\partial V_i^k}{\partial \delta_i}, u_i^{k+1}, u_{-i}^k) \tag{42}$$

Let $H_i(\delta_i, \frac{\partial V_i}{\partial \delta_i}^k, u_i^k, u_{-i}^k) = 0$ from (40) then according to (42) it is clear that

$$H_i^o(\delta_i, \frac{\partial V_i^k}{\partial \delta_i}, u_{-i}^k) \leq 0 \tag{43}$$

Using the next control policy u_i^{k+1} and the current policies u_{-i}^k one has the orbital derivative (Leake, Wen Liu, 1967)

$$\dot{V}_i^k = H_i(\delta_i, \frac{\partial V_i^k}{\partial \delta_i}, u_i^{k+1}, u_{-i}^k) - L_i(\delta_i, u_i^{k+1}, u_{-i}^k)$$

From (42) and (43) one has

$$\dot{V}_i^k = H_i^0(\delta_i, \frac{\partial V_i}{\partial \delta_i}^k, u_{-i}^k) - L_i(\delta_i, u_i^{k+1}, u_{-i}^k) \le -L_i(\delta_i, u_i^{k+1}, u_{-i}^k) \tag{44}$$

Because only agent i update its control it is true that $u_{-i}^{k+1} = u_{-i}^k$ and

$$H_i(\delta_i, \frac{\partial V_i^{k+1}}{\partial \delta_i}, u_i^{k+1}, u_{-i}^k) = 0 .$$

But since $\dot{V}_i^{k+1} = -L_i(\delta_i, u_i^{k+1}, u_{-i}^{k+1})$, from (44) one has

$$\dot{V}_i^k = H_i^0(\delta_i, \frac{\partial V_i}{\partial \delta_i}^k, u_{-i}^k) - L_i(\delta_i, u_i^{k+1}, u_{-i}^k) \le -L_i(\delta_i, u_i^{k+1}, u_{-i}^k) = \dot{V}_i^{k+1} \tag{45}$$

So that $\dot{V}_i^k \le \dot{V}_i^{k+1}$ and by integration it follows that

$$V_i^{k+1} \le V_i^k \tag{46}$$

Since $V_i^* \le V_i^k$, the algorithm converges, to V_i^*, to the best response HJ equation (38).

∎

The next result concerns the case where all nodes update their policies at each step of the algorithm. Define the relative control weighting as $\rho_{ij} = \bar{\sigma}(R_{jj}^{-1}R_{ij})$, where $\bar{\sigma}(R_{jj}^{-1}R_{ij})$ is the maximum singular value of $R_{jj}^{-1}R_{ij}$.

Theorem 5. Convergence of Policy Iteration algorithm when all agents update their policies. Assume all nodes i update their policies at each iteration of PI. Then for small enough edge weights e_{ij} and ρ_{ij}, u_i converges to the global Nash equilibrium and for all i, and the values converge to the optimal game values $V_i^k \to V_i^*$.

Proof:

It is clear that

$$H_i(\delta_i, \frac{\partial V_i}{\partial \delta_i}^{k+1}, u_i^{k+1}, u_{-i}^{k+1}) \equiv H_i^0(\delta_i, \frac{\partial V_i}{\partial \delta_i}^{k+1}, u_{-i}^k) + \frac{1}{2}\sum_{j \in N_i}(u_j^{k+1} - u_j^k)^T R_{ij}(u_j^{k+1} - u_j^k)$$

$$+ \sum_{j \in N_i} u_j^{kT} R_{ij}(u_j^{k+1} - u_j^k) + \frac{\partial V_i}{\partial \delta_i}^{k+1T} \sum_{j \in N_i} e_{ij}B_j(u_j^k - u_j^{k+1})$$

and so

$$\dot{V}_i^{k+1} = -L_i(\delta_i, u_i^{k+1}, u_{-i}^{k+1}) = -L_i(\delta_i, u_i^{k+1}, u_{-i}^k) + \frac{1}{2}\sum_{j \in N_i}(u_j^{k+1} - u_j^k)^T R_{ij}(u_j^{k+1} - u_j^k)$$

$$+ \frac{\partial V_i}{\partial \delta_i}^{k+1T} \sum_{j \in N_i} e_{ij}B_j(u_j^k - u_j^{k+1}) + \sum_{j \in N_i} u_j^{kT} R_{ij}(u_j^{k+1} - u_j^k)$$

Therefore,

$$\dot{V}_i^k \le \dot{V}_i^{k+1} - \frac{1}{2} \sum_{j \in N_i} (u_j^{k+1} - u_j^{k})^T R_{ij} (u_j^{k+1} - u_j^{k})$$

$$+ \frac{\partial V_i}{\partial \delta_i}^{k+1T} \sum_{j \in N_i} e_{ij} B_j (u_j^{k+1} - u_j^{k}) - \sum_{j \in N_i} u_j^{kT} R_{ij} (u_j^{k+1} - u_j^{k})$$

A sufficient condition for $\dot{V}_i^k \le \dot{V}_i^{k+1}$ is

$$\tfrac{1}{2} \Delta u_j^T R_{ij} \Delta u_j - e_{ij} (p_i^{k+1})^T B_j \Delta u_j - (d_j + g_j)(p_j^{k-1}) B_j^T R_{jj}^{-1} R_{ij} \Delta u_j > 0$$

$\tfrac{1}{2} \underline{\sigma}(R_{ij}) \|\Delta u_j\| > e_{ij} \|p_i^{k+1}\| \cdot \|B_j\| + (d_j + g_j) \rho_{ij} \|p_j^{k-1}\| \cdot \|B_j\|$ where $\Delta u_j = (u_j^{k+1} - u_j^{k})$, p_i the costate and $\underline{\sigma}(R_{ij})$ is the minimum singular value of R_{ij}.

This holds if $e_{ij} = 0$, $\rho_{ij} = 0$. By continuity, it holds for small values of e_{ij}, ρ_{ij}.

∎

This proof indicates that for the PI algorithm to converge, the neighbors' controls should not unduly influence the i-th node dynamics (8), and the j-th node should weight its own control u_j in its performance index J_j relatively more than node i weights u_j in J_i. These requirements are consistent with selecting the weighting matrices to obtain proper performance in the simulation examples. An alternative condition for convergence in Theorem 5 is that the norm $\|B_j\|$ should be small. This is similar to the case of weakly coupled dynamics in multi-player games in (Basar, and Olsder, 1999).

5. Online solution of multi-agent cooperative games using neural networks

In this section an online algorithm for solving cooperative Hamilton-Jacobi equations (25) based on (Vamvoudakis, Lewis 2011) is presented. This algorithm uses the structure in the PI Algorithm 1 to develop an actor/critic adaptive control architecture for approximate online solution of (25). Approximate solutions of (40), (41) are obtained using value function approximation (VFA). The algorithm uses two approximator structures at each node, which are taken here as neural networks (NN) (Abu-Khalaf, and Lewis, 2005; Bertsekas, and Tsitsiklis, 1996; Vamvoudakis, Lewis 2010; Werbos, 1974; Werbos, 1992). One critic NN is used at each node for value function approximation, and one actor NN at each node to approximate the control policy (41). The critic NN seeks to solve Bellman equation (40). We give tuning laws for the actor NN and the critic NN such that equations (40) and (41) are solved simultaneously online for each node. Then, the solutions to the coupled HJ equations (25) are determined. Though these coupled HJ equations are difficult to solve, and may not even have analytic solutions, we show how to tune the NN so that the approximate solutions are learned online. The next assumption is made.

Assumption 2. For each admissible control policy the nonlinear Bellman equations (14), (40) have smooth solutions $V_i \ge 0$.

In fact, only local smooth solutions are needed. To solve the Bellman equations (40), approximation is required of both the value functions V_i and their gradients $\partial V_i / \partial \delta_i$. This requires approximation in Sobolev space (Abu-Khalaf, and Lewis, 2005).

5.1 Critic neural network

According to the Weierstrass higher-order approximation Theorem (Abou-Khalaf, and Lewis, 2005) there are NN weights W_i such that the smooth value functions V_i are approximated using a critic NN as

$$V_i(\delta_i) = W_i^T \phi_i(z_i) + \varepsilon_i \tag{47}$$

where $z_i(t)$ is an information vector constructed at node i using locally available measurements, e.g. $\delta_i(t), \{\delta_j(t) : j \in N_i\}$. Vectors $\phi_i(z_i) \in \mathbb{R}^h$ are the critic NN activation function vectors, with h the number of neurons in the critic NN hidden layer. According to the Weierstrass Theorem, the NN approximation error ε_i converges to zero uniformly as $h \to \infty$. Assuming current weight estimates $\hat{W}_i$, the outputs of the critic NN are given by

$$\hat{V}_i = \hat{W}_i^T \phi_i \tag{48}$$

Then, the Bellman equation (40) can be approximated at each step k as

$$H_i(\delta_i, \hat{W}_i, u_i, u_{-i}) = \delta_i^T Q_{ii} \delta_i + u_i^T R_{ii} u_i + \sum_{j \in N_i} u_j^T R_{ij} u_j + \hat{W}_i^T \frac{\partial \phi_i}{\partial \delta_i}(A\delta_i + (d_i + g_i)B_i u_i - \sum_{j \in N_i} e_{ij} B_j u_j) = e_{H_i} \tag{49}$$

It is desired to select $\hat{W}_i$ to minimize the square residual error

$$E_1 = \tfrac{1}{2} e_{H_i}^T e_{H_i} \tag{50}$$

Then $\hat{W}_i \to W_i$ which solves (49) in a least-squares sense and e_{H_i} becomes small. Theorem 6 gives a tuning law for the critic weights that achieves this.

5.2 Action neural network and online learning

Define the control policy in the form of an action neural network which computes the control input (41) in the structured form

$$\hat{u}_i \equiv \hat{u}_{i+N} = -\tfrac{1}{2}(d_i + g_i)R_{ii}^{-1}B_i^T \frac{\partial \phi_i}{\partial \delta_i}^T \hat{W}_{i+N} \tag{51}$$

where $\hat{W}_{i+N}$ denotes the current estimated values of the ideal actor NN weights W_i. The notation $\hat{u}_{i+N}$ is used to keep indices straight in the proof. Define the critic and actor NN estimation errors as $\tilde{W}_i = W_i - \hat{W}_i$ and $\tilde{W}_{i+N} = W_i - \hat{W}_{i+N}$.

The next results show how to tune the critic NN and actor NN in real time at each node so that equations (40) and (41) are simultaneously solved, while closed-loop system stability is also guaranteed. Simultaneous solution of (40) and (41) guarantees that the coupled HJ equations (25) are solved for each node i. System (8) is said to be uniformly ultimately bounded (UUB) if there exists a compact set $S \subset \mathbb{R}^n$ so that for all $\delta_i(0) \in S$ there exists a bound B and a time $T(B, \delta_i(0))$ such that $\|\delta_i(t)\| \le B$ for all $t \ge t_0 + T$.

Select the tuning law for the ith critic NN as

$$\dot{\hat{W}}_i = -a_i \frac{\partial E_1}{\partial \hat{W}_i} = -a_i \frac{\sigma_{i+N}}{(1 + \sigma_{i+N}^T \sigma_{i+N})^2} [\sigma_{i+N}^T \hat{W}_i + \delta_i^T Q_{ii} \delta_i + \tfrac{1}{4} \hat{W}_{i+N}^T \bar{D}_i \hat{W}_{i+N}$$

$$+ \tfrac{1}{4} \sum_{j \in N_i} (d_j + g_j)^2 \hat{W}_{j+N}^T \frac{\partial \phi_j}{\partial \delta_j} B_j R_{jj}^{-T} R_{ij} R_{jj}^{-1} B_j^T \frac{\partial \phi_j^T}{\partial \delta_j} \hat{W}_{j+N}]$$

(52)

where $\sigma_{i+N} = \dfrac{\partial \phi_i}{\partial \delta_i} (A\delta_i + (d_i + g_i) B_i \hat{u}_{i+N} - \sum_{j \in N_i} e_{ij} B_j \hat{u}_{j+N})$, and the tuning law for the ith actor NN as

$$\dot{\hat{W}}_{i+N} = -a_{i+N} \{ (S_i \hat{W}_{i+N} - F_i \bar{\sigma}_{i+N}^T \hat{W}_i) - \frac{1}{4} \bar{D}_i \hat{W}_{i+N} \frac{\bar{\sigma}_{i+N}}{m_{si}}^T \hat{W}_i$$

$$- \frac{1}{4} \sum_{j \in N_i} (d_j + g_j)^2 \frac{\partial \phi_j}{\partial \delta_j} B_j R_{jj}^{-T} R_{ij} R_{jj}^{-1} B_j^T \frac{\partial \phi_j^T}{\partial \delta_j} \hat{W}_j \frac{\bar{\sigma}_{i+N}}{m_{s_i}}^T \hat{W}_{i+N} \}$$

(53)

where

$$\bar{D}_i(x) \equiv \frac{\partial \phi_i}{\partial \delta_i} B_i R_{ii}^{-1} B_i^T \frac{\partial \phi_i^T}{\partial \delta_i}, \quad m_{s_i} \equiv (\sigma_{i+N}^T \sigma_{i+N} + 1), \quad \bar{\sigma}_{i+N} = \sigma_{i+N} / (\sigma_{i+N}^T \sigma_{i+N} + 1), \quad \text{and}$$

$a_i > 0, \dots a_{i+N} > 0$ and $F_i > 0, G_i > 0, \quad i \in N$ are tuning parameters.

Theorem 6. Online Cooperative Games.

Let the error dynamics be given by (8), and consider the cooperative game formulation in (15). Let the critic NN at each node be given by (48) and the control input be given for each node by actor NN (51). Let the tuning law for the ith critic NN be provided by (52) and the tuning law for the ith actor NN be provided by (53). Assume $\bar{\sigma}_{i+N} = \sigma_{i+N} / (\sigma_{i+N}^T \sigma_{i+N} + 1)$ is persistently exciting. Then the closed-loop system states $\delta_i(t)$, the critic NN errors $\tilde{W}_i$, and the actor NN errors $\tilde{W}_{i+N}$ are uniformly ultimately bounded.

Proof:

The proof is similar to (Vamvoudakis, 2011).

∎

Remark 1. Theorem 6 provides algorithms for tuning the actor/critic networks of the N agents at the same time to guarantee stability and make the system errors $\delta_i(t)$ small and the NN approximation errors bounded. Small errors guarantee synchronization of all the node trajectories.

Remark 2. Persistence of excitation is needed for proper identification of the value functions by the critic NNs, and nonstandard tuning algorithms are required for the actor NNs to guarantee stability. It is important to notice that the actor NN tuning law of every agent needs information of the critic weights of all his neighbors, while the critic NN tuning law of every agent needs information of the actor weights of all his neighbors,

Remark 3. NN usage suggests starting with random, nonzero control NN weights in (51) in order to converge to the coupled HJ equation solutions. However, extensive simulations show that convergence is more sensitive to the persistence of excitation in the control inputs than to the NN weight initialization. If the proper persistence of excitation is not selected, the control weights may not converge to the correct values.

Remark 4. The issue of which inputs $z_i(t)$ to use for the critic and actor NNs needs to be addressed. According to the dynamics (8), the value functions (13), and the control inputs (16), the NN inputs at node i should consist of its own state, the states of its neighbors, and the costates of its neighbors. However, in view of (31) the costates are functions of the states. In view of the approximation capabilities of NN, it is found in simulations that it is suitable to take as the NN inputs at node i its own state and the states of its neighbors.

The next result shows that the tuning laws given in Theorem 6 guarantee approximate solution to the coupled HJ equations (25) and convergence to the Nash equilibrium.

Theorem 7. Convergence to Cooperative Nash Equilibrium.

Suppose the hypotheses of Theorem 6 hold. Then:

a. $H_i(\delta_i, \hat{W}_i, \hat{u}_i, \hat{u}_{-i})$, $\forall i \in N$ are uniformly ultimately bounded, where

$\hat{u}_i = -\frac{1}{2}(d_i + g_i)R_{ii}^{-1}B_i^T \frac{\partial \phi_i}{\partial \delta_i}^T \hat{W}_i$. That is, $\hat{W}_i$ converge to the approximate cooperative

coupled HJ-solution.
b. $\hat{u}_{i+N}$ converge to the approximate cooperative Nash equilibrium (Definition 2) for every i .

Proof:

The proof is similar to (Vamvoudakis, 2011) but is done only with respect to the neighbors (local information) of each agent and not with respect to all agents.

Consider the weights $\hat{W}_i, \hat{W}_{i+N}$ to be UUB as proved in Theorem 6.

a. The approximate coupled HJ equations are $H_i(\delta_i, \hat{W}_i, \hat{u}_i, \hat{u}_{-i})$, $\forall i \in N$.

$$H_i(\delta_i, \hat{W}_i, \hat{u}_i, \hat{u}_{-i}) \equiv H_i(\delta_i, \hat{W}_i \, \hat{W}_{-i}) = \delta_i^T Q_{ii} \delta_i + \hat{W}_i^T \frac{\partial \phi_i}{\partial \delta_i} A \delta_i - \frac{1}{4}(d_i + g_i)^2 \hat{W}_i^T \frac{\partial \phi_i}{\partial \delta_i} B_i R_{ii}^{-1} B_i^T \frac{\partial \phi_i}{\partial \delta_i}^T \hat{W}_i$$

$$+\frac{1}{4}\sum_{j\in N_i}(d_j+g_j)^2\hat{W}_j^T\frac{\partial\phi_j}{\partial\delta_j}B_jR_{jj}^{-1}R_{ij}R_{jj}^{-1}B_j^T\frac{\partial\phi_j}{\partial\delta_j}^T\hat{W}_j+\frac{1}{2}\hat{W}_i^T\frac{\partial\phi_i}{\partial\delta_i}\sum_{j\in N_i}e_{ij}B_jR_{jj}^{-1}B_j^T\frac{\partial\phi_j}{\partial\delta_j}^T\hat{W}_j-\varepsilon_{HJ_i}$$

where ε_{HJ_i}, $\forall i$ are the residual errors due to approximation.

After adding zero we have

$$H_i(\delta_i,\hat{W}_i\;\hat{W}_{-i})=-\tilde{W}_i^T\frac{\partial\phi_i}{\partial\delta_i}A\delta_i-\frac{1}{4}(d_i+g_i)^2\tilde{W}_i^T\frac{\partial\phi_i}{\partial\delta_i}B_iR_{ii}^{-1}B_i^T\frac{\partial\phi_i}{\partial\delta_i}^T\tilde{W}_i$$

$$+\frac{1}{2}(d_i+g_i)^2W_i^T\frac{\partial\phi_i}{\partial\delta_i}B_iR_{ii}^{-1}B_i^T\frac{\partial\phi_i}{\partial\delta_i}^TW_i+\frac{1}{2}(d_i+g_i)^2W_i^T\frac{\partial\phi_i}{\partial\delta_i}B_iR_{ii}^{-1}B_i^T\frac{\partial\phi_i}{\partial\delta_i}^T\hat{W}_i$$

$$-\frac{1}{4}\sum_{j\in N_i}(d_j+g_j)^2\tilde{W}_j^T\frac{\partial\phi_j}{\partial\delta_j}B_jR_{jj}^{-1}R_{ij}R_{jj}^{-1}B_j^T\frac{\partial\phi_j}{\partial\delta_j}^T\tilde{W}_j+\frac{1}{2}\sum_{j\in N_i}(d_j+g_j)^2W_j^T\frac{\partial\phi_j}{\partial\delta_j}B_jR_{jj}^{-1}R_{ij}R_{jj}^{-1}B_j^T\frac{\partial\phi_j}{\partial\delta_j}^TW_j$$

$$+\frac{1}{2}\sum_{j\in N_i}(d_j+g_j)^2W_j^T\frac{\partial\phi_j}{\partial\delta_j}B_jR_{jj}^{-1}R_{ij}R_{jj}^{-1}B_j^T\frac{\partial\phi_j}{\partial\delta_j}^T\hat{W}_j+\hat{W}_i^T\frac{\partial\phi_i}{\partial\delta_i}\sum_{j\in N_i}e_{ij}B_jR_{jj}^{-1}B_j^T\frac{\partial\phi_j}{\partial\delta_j}^T\hat{W}_j-\varepsilon_{HJ_i}$$

$$-\frac{1}{2}\tilde{W}_i^T\frac{\partial\phi_i}{\partial\delta_i}\sum_{j\in N_i}e_{ij}B_jR_{jj}^{-1}B_j^T\frac{\partial\phi_j}{\partial\delta_j}^T\tilde{W}_j-\frac{1}{2}W_i^T\frac{\partial\phi_i}{\partial\delta_i}\sum_{j\in N_i}e_{ij}B_jR_{jj}^{-1}B_j^T\frac{\partial\phi_j}{\partial\delta_j}^T\hat{W}_j$$

$$-\frac{1}{2}\hat{W}_i^T\frac{\partial\phi_i}{\partial\delta_i}\sum_{j\in N_i}e_{ij}B_jR_{jj}^{-1}B_j^T\frac{\partial\phi_j}{\partial\delta_j}^TW_j \tag{54}$$

But

$$\hat{W}_i=-\tilde{W}_i+W_i,\quad\forall i. \tag{55}$$

After taking norms in (55) and letting $\|W_i\|<W_{i\max}$ one has

$$\left\|\hat{W}_i\right\|=\left\|-\tilde{W}_i+W_i\right\|\le\left\|\tilde{W}_i\right\|+\left\|W_i\right\|\le\left\|\tilde{W}_i\right\|+W_{i\max}$$

Now (54) with $\sup\left\|\varepsilon_{HJ_i}\right\|<\bar{\varepsilon}_i$ becomes

$$\left\|H_i(\delta_i,\hat{W}_i\;\hat{W}_{-i})\right\|\le\left\|\tilde{W}_i\right\|\left\|\frac{\partial\phi_i}{\partial\delta_i}\right\|\left\|A\delta_i\right\|+\frac{1}{4}(d_i+g_i)^2\left\|\tilde{W}_i\right\|^2\left\|\frac{\partial\phi_i}{\partial\delta_i}\right\|^2\left\|B_i\right\|^2\left\|R_{ii}^{-1}\right\|$$

$$+\frac{1}{2}(d_i+g_i)^2\left\|W_{i\max}\right\|^2\left\|\frac{\partial\phi_i}{\partial\delta_i}\right\|^2\left\|B_i\right\|^2\left\|R_{ii}^{-1}\right\|+\frac{1}{2}(d_i+g_i)^2\left\|W_{i\max}\right\|\left\|\frac{\partial\phi_i}{\partial\delta_i}\right\|^2\left\|B_i\right\|^2\left\|R_{ii}^{-1}\right\|\left(\left\|\tilde{W}_i\right\|+W_{i\max}\right)$$

$$+\frac{1}{4}\sum_{j\in N_i}(d_j+g_j)^2\left\|\tilde{W}_j\right\|^2\left\|\frac{\partial\phi_j}{\partial\delta_j}\right\|^2\left\|B_j\right\|^2\left\|R_{jj}^{-1}R_{ij}R_{jj}^{-1}\right\|+\frac{1}{2}\sum_{j\in N_i}(d_j+g_j)^2\left\|W_{j\max}\right\|^2\left\|\frac{\partial\phi_j}{\partial\delta_j}\right\|^2\left\|B_j\right\|^2\left\|R_{jj}^{-1}R_{ij}R_{jj}^{-1}\right\|$$

$$+\left(\left\|\tilde{W}_i\right\|+W_{i\max}\right)\left\|\frac{\partial\phi_i}{\partial\delta_i}\right\|\sum_{j\in N_i}e_{ij}\left\|B_j\right\|^2\left\|R_{jj}^{-1}\right\|\left\|\frac{\partial\varphi_j}{\partial\delta_j}\right\|\left(\left\|\tilde{W}_j\right\|+W_{j\max}\right)+\frac{1}{2}\left\|\tilde{W}_i\right\|\left\|\frac{\partial\phi_i}{\partial\delta_i}\right\|\sum_{j\in N_i}e_{ij}\left\|B_j\right\|^2\left\|R_{jj}^{-1}\right\|\left\|\frac{\partial\phi_j}{\partial\delta_j}\right\|\left\|\tilde{W}_j\right\|$$

$$+\frac{1}{2}\left\|W_{i\max}\right\|\left\|\frac{\partial\phi_i}{\partial\delta_i}\right\|\sum_{j\in N_i}e_{ij}\left\|B_j\right\|^2\left\|R_{jj}^{-1}\right\|\left\|\frac{\partial\phi_j}{\partial\delta_j}\right\|\left(\left\|\tilde{W}_j\right\|+W_{j\max}\right)$$

$$+\frac{1}{2}\left(\left\|\tilde{W}_i\right\|+W_{i\max}\right)\left\|\frac{\partial\phi_i}{\partial\delta_i}\right\|\sum_{j\in N_i}e_{ij}\left\|B_j\right\|^2\left\|R_{jj}^{-1}\right\|\left\|\frac{\partial\phi_j}{\partial\delta_j}\right\|W_{j\max}+\bar{\varepsilon}_2 \tag{56}$$

All the signals on the right hand side of (56) are UUB and convergence to the approximate coupled HJ solution is obtained for every agent.

b.   According to Theorem 6, $\left\|\hat{W}_{i+N}-W_i\right\|$, $\forall i$ are UUB. Then it is obvious that $\hat{u}_{i+N}$, $\forall i$ give the approximate cooperative Nash equilibrium (Definition 2).

∎

6. Simulation results

This section shows the effectiveness of the online approach described in Theorem 6 for two different cases.

Consider the three-node strongly connected digraph structure shown in Figure 1 with a leader node connected to node 3. The edge weights and the pinning gains are taken equal to 1 so that $d_1=d_2=1, d_3=2$.

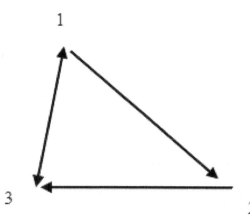

Fig. 1. Three agent communication graph showing the interactions.

Select the weight matrices in (9) as

$$Q_{11} = Q_{22} = Q_{33} = \begin{bmatrix} 1 & 0 \\ 0 & 1 \end{bmatrix}, R_{11} = 4, R_{12} = 1, R_{13} = -1,$$

$$R_{31} = -4, R_{22} = 9, R_{23} = 1, R_{33} = 9, R_{32} = 1, R_{21} = 1$$

In the examples below, every node is a second-order system. Then, for every agent $\delta_i = \begin{bmatrix} \delta_{i1} & \delta_{i2} \end{bmatrix}^T$.

According to the graph structure, the information vector at each node is

$$z_1 = \begin{bmatrix} \delta_1^T & \delta_3^T \end{bmatrix}^T, z_2 = \begin{bmatrix} \delta_1^T & \delta_2^T \end{bmatrix}^T, z_3 = \begin{bmatrix} \delta_1^T & \delta_2^T & \delta_3^T \end{bmatrix}^T$$

Since the value is quadratic, the critic NNs basis sets were selected as the quadratic vector in the agent's components and its neighbors' components. Thus the NN activation functions are

$$\phi_1(\delta_1, 0, \delta_3) = \begin{bmatrix} \delta_{11}^2 & \delta_{11}\delta_{12} & \delta_{12}^2 & 0 & 0 & 0 & \delta_{31}^2 & \delta_{31}\delta_{32} & \delta_{32}^2 \end{bmatrix}^T$$

$$\phi_1(\delta_1, \delta_2, 0) = \begin{bmatrix} \delta_{11}^2 & \delta_{11}\delta_{12} & \delta_{12}^2 & \delta_{21}^2 & \delta_{21}\delta_{22} & \delta_{21}^2 & 0 & 0 & 0 \end{bmatrix}^T$$

$$\phi_3(\delta_1, \delta_2, \delta_3) = \begin{bmatrix} \delta_{11}^2 & \delta_{11}\delta_{12} & \delta_{12}^2 & \delta_{21}^2 & \delta_{21}\delta_{22} & \delta_{22}^2 & \delta_{31}^2 & \delta_{31}\delta_{32} & \delta_{32}^2 \end{bmatrix}^T$$

6.1 Position and velocity regulated to zero

For the graph structure shown, consider the node dynamics

$$\dot{x}_1 = \begin{bmatrix} -2 & 1 \\ -4 & -1 \end{bmatrix} x_1 + \begin{bmatrix} 2 \\ 1 \end{bmatrix} u_1, \dot{x}_2 = \begin{bmatrix} -2 & 1 \\ -4 & -1 \end{bmatrix} x_2 + \begin{bmatrix} 2 \\ 3 \end{bmatrix} u_2, \dot{x}_3 = \begin{bmatrix} -2 & 1 \\ -4 & -1 \end{bmatrix} x_3 + \begin{bmatrix} 2 \\ 2 \end{bmatrix} u_3$$

and the command generator $\dot{x}_0 = \begin{bmatrix} -2 & 1 \\ -4 & -1 \end{bmatrix} x_0$.

The graphical game is implemented as in Theorem 6. Persistence of excitation was ensured by adding a small exponentially decreasing probing noise to the control inputs. Figure 2 shows the convergence of the critic parameters for every agent. Figure 3 shows the evolution of the states for the duration of the experiment.

6.2 All the nodes synchronize to the curve behavior of the leader node

For the graph structure shown above consider the following node dynamics

$$\dot{x}_1 = \begin{bmatrix} 0 & 1 \\ -1 & 0 \end{bmatrix} x_1 + \begin{bmatrix} 2 \\ 1 \end{bmatrix} u_1, \dot{x}_2 = \begin{bmatrix} 0 & 1 \\ -1 & 0 \end{bmatrix} x_2 + \begin{bmatrix} 2 \\ 3 \end{bmatrix} u_2, \dot{x}_3 = \begin{bmatrix} 0 & 1 \\ -1 & 0 \end{bmatrix} x_3 + \begin{bmatrix} 2 \\ 2 \end{bmatrix} u_3$$

with target generator $\dot{x}_0 = \begin{bmatrix} 0 & 1 \\ -1 & 0 \end{bmatrix} x_0$.

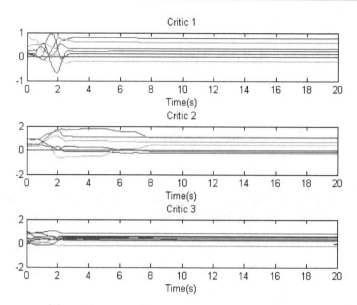

Fig. 2. Convergence of the critic parameters.

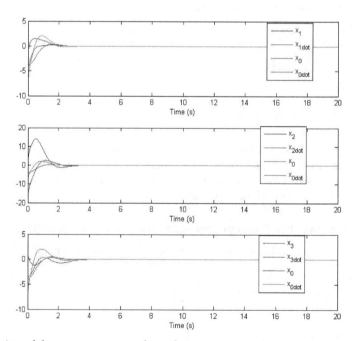

Fig. 3. Evolution of the system states and regulation.

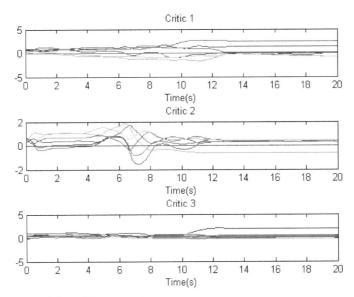

Fig. 4. Convergence of the critic parameters.

The command generator is marginally stable with poles at $s = \pm j$, so it generates a sinusoidal reference trajectory.

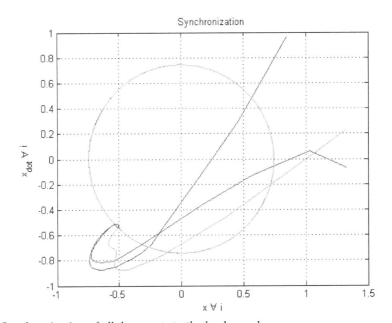

Fig. 5. Synchronization of all the agents to the leader node.

The graphical game is implemented as in Theorem 6. Persistence of excitation was ensured by adding a small exponential decreasing probing noise to the control inputs. Figure 4 shows the critic parameters converging for every agent. Figure 5 shows the synchronization of all the agents to the leader's behavior as given by the circular Lissajous plot.

7. Conclusion

This chapter brings together cooperative control, reinforcement learning, and game theory to solve multi-player differential games on communication graph topologies. It formulates graphical games for dynamic systems and provides policy iteration and online learning algorithms along with proof of convergence to the Nash equilibrium or best response. Simulation results show the effectiveness of the proposed algorithms.

8. References

Abou-Kandil H., Freiling G., Ionescu V., & Jank G., (2003). *Matrix Riccati Equations in Control and Systems Theory*, Birkhäuser.

Abu-Khalaf M., & Lewis F. L., (2005). Nearly Optimal Control Laws for Nonlinear Systems with Saturating Actuators Using a Neural Network HJB Approach, *Automatica*, 41(5), 779-791.

Başar T., & Olsder G. J.,(1999). *Dynamic Noncooperative Game Theory*, 2nd ed. Philadelphia, PA: SIAM.

Bertsekas D. P., & Tsitsiklis J. N. (1996). *Neuro-Dynamic Programming*, Athena Scientific, MA.

Brewer J.W., (1978). Kronecker products and matrix calculus in system theory, *IEEE Transactions Circuits and Systems*, 25, 772-781.

Busoniu L., Babuska R., & De Schutter B., (2008). A Comprehensive Survey of Multi-Agent Reinforcement Learning, *IEEE Transactions on Systems, Man, and Cybernetics – Part C: Applications and Reviews*, 38(2), 156–172.

Dierks T. & Jagannathan S., (2010). Optimal Control of Affine Nonlinear Continuous-time Systems Using an Online Hamilton-Jacobi-Isaacs Formulation1, *Proc. IEEE Conf Decision and Control*, Atlanta, 3048-3053.

Fax J. & Murray R., (2004). Information flow and cooperative control of vehicle formations, *IEEE Trans. Autom. Control*, 49(9), 1465–1476.

Freiling G., Jank G., & Abou-Kandil H., (2002). On global existence of Solutions to Coupled Matrix Riccati equations in closed loop Nash Games, *IEEE Transactions on Automatic Control*, 41(2), 264- 269.

Hong Y., Hu J., & Gao L., (2006). Tracking control for multi-agent consensus with an active leader and variable topology, *Automatica*, 42 (7), 1177–1182.

Gajic Z., & Li T-Y., (1988). Simulation results for two new algorithms for solving coupled algebraic Riccati equations, *Third Int. Symp. On Differential Games*, Sophia, Antipolis, France.

Jadbabaie A., Lin J., & Morse A., (2003). Coordination of groups of mobile autonomous agents using nearest neighbor rules, *IEEE Trans. Autom. Control*, 48(6), 988–1001.

Johnson M., Hiramatsu T., Fitz-Coy N., & Dixon W. E., (2010). Asymptotic Stackelberg Optimal Control Design for an Uncertain Euler Lagrange System, *IEEE Conference on Decision and Control*, 6686-6691.

Kakade S., Kearns M., Langford J., & Ortiz L., (2003). Correlated equilibria in graphical games, *Proc. 4th ACM Conference on Electronic Commerce, 42–47.*

Kearns M., Littman M., & Singh S., (2001) Graphical models for game theory, *Proc. 17th Annual Conference on Uncertainty in Artificial Intelligence, 253–260.*

Khoo S., Xie L., & Man Z., (2009) Robust Finite-Time Consensus Tracking Algorithm for Multirobot Systems, *IEEE Transactions on Mechatronics, 14, 219-228.*

Leake R. J., Liu Ruey-Wen, (1967). Construction of Suboptimal Control Sequences, *J. SIAM Control, 5 (1), 54-63.*

Lewis F. L., Syrmos V. L. (1995). *Optimal Control,* John Wiley.

Lewis F. (1992). *Applied Optimal Control and Estimation: Digital Design and Implementation,* New Jersey: Prentice-Hall.

Li X., Wang X., & Chen G., (2004). Pinning a complex dynamical network to its equilibrium, *IEEE Trans. Circuits Syst. I, Reg. Papers, 51(10), 2074–2087.*

Littman M.L., (2001). Value-function reinforcement learning in Markov games, *Journal of Cognitive Systems Research 1.*

Olfati-Saber R., Fax J., & Murray R., (2007). Consensus and cooperation in networked multi-agent systems, *Proc. IEEE, vol. 95(1), 215–233.*

Olfati-Saber R., & Murray R.M., (2004). Consensus Problems in Networks of Agents with Switching Topology and Time-Delays, *IEEE Transaction of Automatic Control, 49, 1520-1533.*

Qu Z., (2009). *Cooperative Control of Dynamical Systems: Applications to Autonomous Vehicles,* New York: Springer-Verlag.

Ren W., Beard R., & Atkins E., (2005). A survey of consensus problems in multi-agent coordination, in *Proc. Amer. Control Conf., 1859–1864.*

Ren W. & Beard R., (2005). Consensus seeking in multiagent systems under dynamically changing interaction topologies, *IEEE Trans. Autom. Control, 50(5), 655–661.*

Ren W. & Beard R.W., (2008) *Distributed Consensus in Multi-vehicle Cooperative Control,* Springer, Berlin.

Ren W., Moore K., & Chen Y., (2007). High-order and model reference consensus algorithms in cooperative control of multivehicle systems, *J. Dynam. Syst., Meas., Control, 129(5), 678–688.*

R. Shinohara, "Coalition proof equilibria in a voluntary participation game," *International Journal of Game Theory,* vol. 39, no. 4, pp. 603-615, 2010.

Shoham Y., Leyton-Brown K., (2009). *Multiagent Systems: Algorithmic, Game-Theoretic, and Logical Foundations,* Cambridge University Press.

Sutton R. S., Barto A. G. (1998) *Reinforcement Learning – An Introduction,* MIT Press, Cambridge, Massachusetts.

Tijs S (2003), *Introduction to Game Theory,* Hindustan Book Agency, India.

Tsitsiklis J., (1984). Problems in Decentralized Decision Making and Computation, Ph.D. dissertation, Dept. Elect. Eng. and Comput. Sci., MIT, Cambridge, MA.

Vamvoudakis Kyriakos G., & Lewis F. L., (2010). Online Actor-Critic Algorithm to Solve the Continuous-Time Infinite Horizon Optimal Control Problem, *Automatica,46(5),* 878-888.

Vamvoudakis K.G., & Lewis F. L., (2011). Multi-Player Non-Zero Sum Games: Online Adaptive Learning Solution of Coupled Hamilton-Jacobi Equations, to appear in *Automatica.*

Vrabie, D., Pastravanu, O., Lewis, F. L., & Abu-Khalaf, M. (2009). Adaptive Optimal Control for continuous-time linear systems based on policy iteration. Automatica, 45(2), 477-484.

Vrancx P., Verbeeck K., & Nowe A., (2008). Decentralized learning in markov games, *IEEE Transactions on Systems, Man and Cybernetics, 38(4), 976-981.*

Wang X. & Chen G., (2002). Pinning control of scale-free dynamical networks, Physica A, *310(3-4), 521-531.*

Werbos P. J. (1974). *Beyond Regression: New Tools for Prediction and Analysis in the Behavior Sciences,* Ph.D. Thesis.

Werbos P. J. (1992). Approximate dynamic programming for real-time control and neural modeling, *Handbook of Intelligent Control,* ed. D.A. White and D.A. Sofge, New York: Van Nostrand Reinhold.

6

A Comparative Study Using Bio-Inspired Optimization Methods Applied to Controllers Tuning

Davi Leonardo de Souza[1], Fran Sérgio Lobato[2] and Rubens Gedraite[2]
[1]*Department of Chemical Engineering and Statistics*
Universidade Federal de São João Del-Rei,
[2]*School of Chemical Engineering*
Universidade Federal de Uberlândia
Brazil

1. Introduction

The evolution of process control techniques have increased in a significant way during last years. Even so, in the industry, the Proportional-Integral-Derivative (PID) controller is frequently used in closed loops due to its simplicity, applicability, and easy implementation (Astrom & Hagglund, 1995); Shinskey, 1998; (Desbourough & Miller, 2002). An extensive research concerning regulatory control of loops used in refinery, chemical, and pulp and paper processes reveals that 97% of the applications make use of classical PID structure even though sophisticated control techniques, like advanced control strategies, are also based on PID algorithms with lower hierarchy level (Desbourough & Miller, 2002).

Traditionally, the controllers tuning is obtained using classical methods, such as Ziegler-Nichols (ZN), Cohen-Coon (CC) and hybridization. However, these methodologies are found to present quite satisfactory results for first-order processes, but they usually fail to provide acceptable performance for higher-order processes and especially for nonlinear ones due to large overshoots and poor regulation on loading (Hang et al., 1991; Mudi et al., 2008). In addition, it has been quite difficult to tune properly the PID parameters, during typical operation plant, due to difficulties related to production goals (Coelho & Pessoa, 2011).

Recently, optimization methods through use of information about real or synthetic data, has been used as alternative to controllers tuning (Lobato & Souza, 2008). Among these strategies, one can cite the based on evolutionary optimization techniques to controllers tuning, such as fuzzy logic (Hamid et al., 2010), genetic algorithms (Bandyopadhyay et al., 2001; Pan et al., 2011), augmented Lagrangian particle swarm optimization algorithm (Sedlaczek & Eberhard, 2006), particle swarm optimization (Kim et al., 2008; Solihin et al., 2011); differential evolution algorithm (Lobato & Souza, 2008); and differential evolution combined with chaotic Zaslavskii map (Coelho & Pessoa, 2011). Basically, the interest in evolutionary approach is due to following characteristics: easy code building and implementation, no usage of information about gradients and, capacity to escape from local optimal (Lobato & Souza, 2008; Souza, 2007).

According to this search area, biological systems have contributed significantly to the development of new optimization techniques. These methodologies - known as Bio-inspired Optimization Methods (BiOM) - are based on usage of strategies that seek to mimic the behavior observed in species of nature to update a population of candidates to solve optimization problems (Lobato et al., 2010). These systems have the capacity to notice and to modify its "atmosphere" in order to seek diversity and convergence. In addition, this capacity turns possible the communication among the agents (individuals of population) that capture the changes in "atmosphere" generated by local interactions (Parrich et al., 2002).

Among the most recent bio-inspired strategies, one can cite the Bees Colony Algorithm - BCA (Pham et al., 2006), the Fish Swarm Algorithm - FSA (Li et al., 2002) and the Firefly Colony Algorithm - FCA (Yang, 2008). The classical form of BCA is based on the behavior of bees' colonies in their search of raw materials for honey production. In each hive, groups of bees (called scouts) are recruited to explore new areas in search for pollen and nectar. These bees, returning to the hive, share the acquired information so that new bees are indicated to explore the best regions visited in an amount proportional to the previously passed assessment. Thus, the most promising regions are best explored and eventually the worst ones end up being discarded. This cycle repeats itself, with new areas being visited by scouts at each iteration (Pham et al., 2006). The FSA is a random search algorithm based on the behaviour of fish swarm which contains searching, swarming and chasing behaviour. It constructs the simple behaviours of artificial fish firstly, and then makes the global optimum appear finally based on animal individuals' local searching behaviours (Li et al., 2002). Finally, the FCA is inspired in social behaviour of fireflies and their communication through the phenomenon of bioluminescence. This optimization technique admits that the solution of an optimization problem can be perceived as an agent (firefly) which "glows" proportionally to its quality in a considered problem setting. Consequently each brighter firefly attracts its partners (regardless of their sex), which makes the search space being explored more efficiently (Yang, 2008).

In the present contribution, BiOM are used for the controllers tuning in chemical engineering problems.   For this finality, three problems are studied, with emphasis on a realistic application: the control design of heat exchangers on pilot scale. The results obtained with the methodology proposed are compared with those from the classical methods. This chapter is organized as follows. Classical methods to controllers tuning are reviewed in Section 2. In Section 3 the main characteristics of BiOM are briefly presented. The results and discussion are described in Section 4. Finally, the conclusions and suggestions for future work complete the chapter.

2. Controllers tuning using classical methods

As mentioned earlier, about 97% of industrial controllers are of PID type, and implement them in practice, or even during maintenance of same, there are several technical adjustments of its parameters. In literature, there are several classical methods for controllers tuning, such as strategies based on minimization of integral error, and correlation-based methods such as ZN and CC, among others.

The majority of works involving the controllers design use the ZN and CC methods (Conner & Seborg, 2005; Lobato & Souza, 2008; Solihin et al., 2011; Xi et al., 2007). In this context, the ZN and CC methods are brief described.

2.1 Reaction curve method

The principle of this method is the correlation between controller parameters (K_c, τ_I e τ_D) with model parameters (K, τ and θ) through the temporal response of open-loop system (called the process reaction curve), compared to a step input. In open loop, leads to a unit step of input variable to obtain the reaction curve as in Fig. 1. With the parameters θ and τ, we can obtain

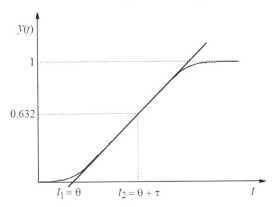

Fig. 1. Time response to open-loop system with step input $y(t)$.

the controller parameters according to Tab. 1.

Controller Parameter	ZN		CC	
P	$KK_c = \dfrac{\tau}{\theta}$		$KK_c = \dfrac{\tau}{\theta} + \dfrac{1}{3}$	
PI	$KK_c = 0.9\dfrac{\tau}{\theta}$ $\dfrac{\tau_I}{\tau}=3.33\dfrac{\theta}{\tau}$	$KK_c = 0.9\dfrac{\tau}{\theta}+0.083$	$\dfrac{\tau_I}{\tau} = \dfrac{\theta(3.33 + 0.33(\theta/\tau))}{1 + 2.2(\theta/\tau)}$	
PID	$KK_c = 1.2\dfrac{\tau}{\theta}$ $\dfrac{\tau_I}{\tau}=2\dfrac{\theta}{\tau}$ $\dfrac{\tau_D}{\tau} = 0.5\dfrac{\theta}{\tau}$	$KK_c = 1.35\dfrac{\tau}{\theta}+0.27$	$\dfrac{\tau_I}{\tau} = \dfrac{\theta(32 + 6(\theta/\tau))}{13 + 8(\theta/\tau)}$ $\dfrac{\tau_D}{\tau} = \dfrac{0.37(\theta/\tau)}{1 + 0.2(\theta/\tau)}$	

Table 1. Controllers tuning with ZN and CC methods through reaction curve (Seborg et al., 1989).

2.2 Continuous cycling method

This classical method is based on sustained oscillation, known as Continuous Cycling Method (Seborg et al., 1989). This procedure is valid only for open-loop stable plants, and conducted with the following steps: (*i*) establishment of parameter proportional to a very small gain, (*ii*) increase the gain to obtain an oscillatory response with constant amplitude and period (*iii*) registration of critical value (K_u), critical period (P_u) and (*iv*) adjustment of the parameters, as shown in Tab. 2. Although the vast majority of PID controllers design is tuned by ZN and CC methods, some difficulties can be observed, such as the need for knowledge of process dynamics in open-loop, and in the Continuous Cycling method, the need to work near of system instability limit (Seborg et al., 1989).

Controller/Parameter	K_c	τ_I	τ_D
P	$0.5K_u$	-	-
PI	$0.45K_u$	$P_u/1.2$	-
PID	$0.6K_u$	$0.5P_u$	$P_u/8$

Table 2. Controllers tuning with Continuous Cycling (Seborg et al., 1989).

3. Bio-inspired optimization methods

In the last decades, nature has inspired the development of various optimization methods. These techniques try to imitate behaviors of species found in nature, such as ants, birds, bees, fireflies, bacteria, among others, to extract information that can be used to promote the development of simple and robust strategies.

This section presents briefly three bio-inspired algorithms in nature: the Bee Colony Algorithm, the Firefly Colony Algorithm and the Fish Swarm Algorithm.

3.1 Bee colony algorithm - BCA

The algorithm proposed by Pham et al. (2006) and described in this section is based on the following characteristics observed in nature (von Frisch, 1976): (*i*) a bees' colony can extend itself over long distances (more than 10 km) and in multiple directions simultaneously to exploit a large number of food sources, and (*ii*) capacity of memorization, learning and transmission of information in colony, so forming the swarm intelligence.

In a colony the foraging process begins by scout bees being sent to search randomly for promising flower patches. When they return to the hive, those scout bees that found a patch which is rated above a certain quality threshold (measured as a combination of some constituents, such as sugar content) deposit their nectar or pollen and go to the "waggle dance".

This dance is responsible by the transmission (colony communication) of information regarding a flower patch: the direction in which it will be found, its distance from the hive and its quality rating (or fitness) (von Frisch, 1976). This dance enables the colony to evaluate the relative merit of different patches according to both the quality of the food they provide and the amount of energy needed to harvest it (Camazine et al., 2003). Mathematically this dance can be represented by following expression:

$$x = x - ngh + 2ngh \times rand \tag{1}$$

where x is the new position, ngh is the patch radius for neighbourhood search and $rand$ is the random generator.

After waggle dancing on the dance floor, the dancer (scout bee) goes back to the flower patch with follower bees that were waiting inside the hive. More follower bees are sent to more promising patches. This allows the colony to gather food quickly and efficiently. While harvesting from a patch, the bees monitor its food level. This is necessary to decide upon the next waggle dance when they return to the hive (Camazine et al., 2003). If the patch is still good enough as a food source, then it will be advertised in the waggle dance and more bees will be recruited to that source.

In this context, Pham et al. (2006) proposed an optimization algorithm inspired by the natural foraging behavior of honey bees and presented in Tab. 3.

| 1. Initialise population with random solutions. |
| 2. Evaluate fitness of the population. |
| 3. While (stopping criterion not met) |
| 4. Select sites for neighborhood search. |
| 5. Recruit bees for selected sites (more bees for the best e sites) and evaluate fitnesses. |
| 6. Select the fittest bee from each site. |
| 7. Assign remaining bees to search randomly and evaluate their fitnesses. |
| 8. End While. |

Table 3. Bees Colony Algorithm (Pham et al., 2006).

The BCA requires a number of parameters to be set, namely, the number of scout bees (n), number of sites selected for neighborhood search (out of n visited sites) (m), number of top-rated (elite) sites among m selected sites (e), number of bees recruited for the best e sites (nep), number of bees recruited for the other (m-e) selected sites (ngh), and the stopping criterion.

The BCA starts with the n scout bees being placed randomly in the search space. The fitnesses of the sites visited by the scout bees are evaluated in step 2.

In step 4, bees that have the highest fitnesses are chosen as "selected bees" and sites visited by them are chosen for neighborhood search. Then, in steps 5 and 6, the algorithm conducts searches in the neighborhood of the selected sites, assigning more bees to search near to the best e sites. The bees can be chosen directly according to the fitnesses associated with the sites they are visiting.

Alternatively, the fitness values are used to determine the probability of the bees being selected. Searches in the neighborhood of the best e sites, which represent more promising solutions, are made more detailed by recruiting more bees to follow them than the other selected bees. Together with scouting, this differential recruitment is a key operation of the BCA.

However, in step 6, for each patch only the bee with the highest fitness will be selected to form the next bee population. In nature, there is no such a restriction. This restriction is introduced here to reduce the number of points to be explored. In step 7, the remaining bees in the population are assigned randomly around the search space scouting for new potential solutions.

In the literature, various applications using this bio-inspired approach can be found, such as: modeling combinatorial optimization transportation engineering problems (Lucic & Teodorovic, 2001), engineering system design (Lobato et al., 2010; Yang, 2005), transport problems (Teodorovic & Dell'Orco, 2005), mathematical function optimization (Pham et al., 2006), dynamic optimization (Chang, 2006), optimal control problems (Afshar et al., 2001), parameter estimation in control problems (Azeem & Saad, 2004), estimation of radiative properties in a one-dimensional participating medium (Ribeiro Neto et al., 2011), among other applications (http://www.bees-algorithm.com/).

3.2 Firefly colony algorithm - FCA

The FCA is based on the characteristics of fireflies' bioluminescence, insects notorious for their light emission. Although biology does not have a complete knowledge to determine all the utilities that firefly luminescence can bring to, at least three functions have been identified (Lukasik & Zak, 2009; Yang, 2008): (*i*) as a communication tool and appeal to potential partners in the reproduction, (*ii*) as a bait to lure prey for the firefly, (*iii*) as a warning mechanism for potential predators reminding them that fireflies have a bitter taste.

It were idealized some of the flashing characteristics of the fireflies so as to develop firefly-inspired algorithms. The following three idealized rules were used (Yang, 2008):

- all fireflies are unisex so that one firefly will be attracted to other fireflies regardless of their sex;
- attractiveness is proportional to their brightness, thus for any two flashing fireflies the less bright will move towards the brightest one. The attractiveness is proportional to the brightness and they both decrease as their distance increases. If there is no brightest one, than a particular firefly will move randomly;
- the brightness of a firefly is affected or determined by the landscape of the objective function. For a maximization problem, the brightness can simply be proportional to the value of the objective function.

According to Yang (2008), in the firefly algorithm there are two important issues: the variation of light intensity and the formulation of the attractiveness. For simplicity, it is always assumed that the attractiveness of a firefly is determined by its brightness, which in turn is associated with the encoded objective function.

This swarm intelligence optimization technique is based on the assumption that the solution of an optimization problem can be perceived as agent (firefly) which "glows" proportionally to its quality in a considered problem setting. Consequently, each brighter firefly attracts its partners (regardless of their sex) which make the search space being explored more efficiently.

The algorithm makes use of a synergic local search. Each member of the swarm explores the problem space taking into account results obtained by others, still applying its own randomized moves as well. The influence of other solutions is controlled by the value of attractiveness (Lukasik & Zak, 2009).

According to Lukasik & Zak (2009), the FA is presented as follows. Consider a continuous constrained optimization problem where the task is to minimize the cost function $f(x)$. Assume that there exists a swarm of N agents (fireflies) solving the above mentioned problem iteratively, and x_i represents a solution for a firefly i at the algorithm's iteration k, whereas $f(x_i)$ denotes its cost. Initially, all fireflies are dislocated in S (randomly or employing some deterministic strategy). Each firefly has its distinctive attractiveness β which implies how strong it attracts other members of the swarm. As the firefly attractiveness, one should select any monotonically decreasing function of the distance $r_j = d(x_i, x_j)$ to the chosen firefly j, e.g., the exponential function:

$$\beta = \beta_0 e^{-\gamma r_j} \tag{2}$$

where β_0 and γ are the following predetermined algorithm parameters: maximum attractiveness value and absorption coefficient, respectively. Furthermore, every member of the swarm is characterized by its light intensity, I_i, which can be directly expressed as the inverse of a cost function $f(x_i)$. To effectively explore the considered search space S, it is assumed that each firefly i changes its position iteratively by taking into account two factors: attractiveness of other swarm members with higher light intensity, e.g., $I_j > I_i$, $\forall j=1, ..., m,$

$j \neq i$, which is varying across the distance and a fixed random step vector u_i. It should be noted as well that if no brighter firefly can be found only such randomized step is being used.

Thus, moving at a given time step t of a firefly i toward a better firefly j is defined as:

$$x_i^t = x_i^{t-1} + \beta \left(x_j^{t-1} - x_i^{t-1} \right) + \alpha \left(rand - \frac{1}{2} \right) \tag{3}$$

where the second term on the right hand side of the equation inserts the attractiveness factor, β while the third term (governed by the parameter α) governs the insertion of certain randomness in the path followed by the firefly, $rand$ is a random number between 0 and 1.

In the literature, few works using the FCA can be found. In this context, the application of the technique is emphasized in continuous constrained optimization task (Lukasik & Zak, 2009), multimodal optimization (Yang, 2009), solution of singular optimal control problems (Pfeifer & Lobato, 2010) and load dispatch problem (Apostolopoulos & Vlachos, 2011).

3.3 Fish swarm algorithm - FSA

In the development of FSA, based on fish swarm and observed in nature, the following characteristics are considered (Madeiro, 2010): (*i*) each fish represents a candidate solution of optimization problem; (*ii*) food density is related to an objective function to be optimized (in an optimization problem, the amount of food in a region is inversely proportional to value of objective function); and (*iii*) the aquarium is the design space where the fish can be found.

As noted earlier, the fish weight at the swarm represents the accumulation of food (e.g., the objective function) received during the evolutionary process. In this case, the weight is an indicator of success (Madeiro, 2010).

Basically, the FSA presents four operators classified into two class: "food search" and "movement". Details on each of these operators are shown as follows.

3.3.1 Individual movement operator

This operator contributes for the movement individual and collective of fishes in swarm. Each fish updates its new position using the Eq. (4):

$$x_i (t + 1) = x_i (t) + rand \times s_{ind} \tag{4}$$

where x_i is the final position of fish i at current generation, $rand$ is a random generator and s_{ind} is a weighted parameter.

3.3.2 Food operator

The weight of each fish is a metaphor used to measure the success of food search. The higher the weight of a fish, the more likely this fish be in a potentially interesting region in design space.

According to Madeiro (2010), the amount of food that a fish eats depends on improvement in its objective function in current generation and the value of greatest value considering the swarm. The weight is updated according to Eq. (5):

$$W_i (t + 1) = W_i (t) + \frac{\Delta f_i}{\max (\Delta f)} \tag{5}$$

where $W_i(t)$ is the fish weight i at generation t and Δf_i is the difference of objective function between the current position and the new position of fish i. It is important to emphasize that $\Delta f_i=0$ for the fishes in same position.

3.3.3 Instinctive collective movement operator

This operator is important for the individual movement of fishes when $\Delta f_i \neq 0$. Thus, only the fishes whose individual execution of the movement resulted in improvement of their fitness will influence the direction of motion of the school, resulting in instinctive collective movement. In this case, the resulting direction (I_d), calculated using the contribution of the directions taken by the fish, and the new position of the ith fish are given by:

$$I_d(t) = \frac{\sum\limits_{i=1}^{N} \Delta x_i \Delta f_i}{\sum\limits_{i=1}^{N} \Delta f_i} \tag{6}$$

$$x_i(t+1) = x_i(t) + I_d(t) \tag{7}$$

It is important to emphasize that in the application of this operator, the direction chosen by a fish that located the largest portion of food to exert the greatest influence on the swarm. Therefore, the instinctive collective movement operator tends to guide the swarm in the direction of motion chosen by fish who found the largest portion of food in it individual movement.

3.3.4 Non-Instinctive collective movement operator

As noted earlier, the fish weight is a good indication of search success for food. In this way, the swarm weight is increasing, this means that the search process is successful. So, the "radius" of the swarm must decrease for that other regions can be explored. Otherwise, if the swarm weight remains constant, the radius should increase to allow the exploration of new regions.

For the swarm contraction, the centroid concept is used. This is obtained by means of an average position of all fish weighted with the respective fish weights, according to Eq. (8):

$$B(t) = \frac{\sum\limits_{i=1}^{N} x_i W_i(t)}{\sum\limits_{i=1}^{N} W_i(t)} \tag{8}$$

If the swarm weight remains constant in the current iteration, all fish must update their positions using the Eq. (9):

$$x(t+1) = x(t) - s_{vol} \times rand \times \frac{x(t) - B(t)}{d(x(t), B(t))} \tag{9}$$

where d is a function that calculates the Euclidean distance between the centroid and the current position of fish, and s_{vol} is the step size used to control the displacement of fish.

In the literature, few works using the FSA can be found. In this context, feed forward neural networks (Wang et al., 2005), parameter estimation in engineering systems (Li et al, 2004), combinatorial optimization problem (Cai, 2010), global optimization (Yang, 2010), Augmented Lagrangian fish swarm based method for global optimization (Rocha et al., 2011), forecasting stock indices using radial basis function neural networks optimized (Shen et al., 2011), and hybridization of the FSA with the Particle Swarm Algorithm to solve engineering systems (Tsai & Lin, 2011).

4. Applications

For evaluating the methodology proposed in this work for controllers tuning, some practical points should be emphasized:

- the objective function (Sum Quadratic Error - SQE) considered in all case studies is given by Eq. (10):

$$\min SQE = \sum_{k=1}^{np} Error = \sum_{k=1}^{np} \left(X^{setpoint} - X^{calculated} \right)^2 \tag{10}$$

where $X^{setpoint}$ and $X^{calculated}$ are the values of variables considered at *setpoint* and calculated using the mathematical model, respectively, and np is the points number used to formulate this objective function (np equals to 1000).

- in all case studies the following parameters used are presented in Tab. (Li et al., 2002; Pham et al., 2006; Yang, 2008).
- it should be emphasized that is necessary, with the parameters listed in this table, 1510 objective function evaluations in each algorithm.
- all case studies were run 20 times independently to obtain the values and standard deviations shown in the upcoming tables.
- the stopping criterion used was the maximum number of iterations (generations).
- to compare the results obtained by the BiOM, the following strategies were used: Ziegler-Nichols Sensibility-Limiar Method (ZN-SL), Ziegler-Nichols Reaction Curve Method (ZN-RC) and Cohen-Coon Reaction Curve Method (CC-RC).

4.1 Distillation column

This first study proposed by (Skogestad & Morari, 1987) considers a distillation column of high purity consisting of 25 plates, a condenser and a reboiler. The reflow ration and the composition of distillate are the input and output system, respectively. The dynamic model that describes this system is given by following transfer function (Skogestad & Morari, 1987):

$$G(z) = \frac{-0.75448z + 0.149199}{z - 0.6386913} \tag{11}$$

The objective is to maintain the composition of distillate in 0.99 by manipulating the reflow ratio, which has a nominal value of 1.477 Kmol/min. In this case, the following ranges to controllers tuning are considered: $0 \leq K_c \leq 150, 0 \leq \tau_I \leq 50$ and $0 \leq \tau_D \leq 50$.

Table 5 presents the best value and standard deviation for the distillation column case study.

In this table can be observed that both the algorithms presented good estimates for the unknown parameters. When the results are analyzed in terms of the objective function (OF),

BCA	
Number of scout bees	10
Number of bees recruited for the best e sites	5
Number of bees recruited for the other selected sites	5
Number of sites selected for neighbourhood search	5
Number of top-rated (elite) sites among m selected sites	5
Neighborhood search	10^{-6}
Generation Number	50
FCA	
Number of fireflies	15
Maximum attractiveness value	0.9
Absorption coefficient	0.9
Generation Number	50
FSA	
Number of fishes	15
Weighted parameter (s_{ind})	0.01
Weighted parameter (s_{vol})	1
Generation Number	50

Table 4. Parameters used by the BiOM.

Method	K_c	τ_I (min^{-1})	τ_D (min^{-1})	OF (Eq. 10)
ZN-SL	67.2000	12.500	3.1250	8.10×10^{-3}
ZN-RC	2.6578	2.0000	0.5000	1.24×10^{-2}
CC-RC	3.2890	2.1154	0.3364	1.09×10^{-2}
BCA	24.282	0.008	43.103	8.102×10^{-3}
	(36.265)	(12.12)	(12.710)	(3.026×10^{-6})
FCA	77.412	0.003	26.955	8.100×10^{-3}
	(37.324)	(0.037)	(15.926)	(6.834×10^{-8})
FSA	128.009	0.009	7.176	8.101×10^{-3}
	(28.260)	(0.237)	(18.779)	(3.145×10^{-8})

Table 5. Results obtained by the BiOM - distillation column case study.

is clear that the combination of control parameters lead us to very close values, also seen in the value of standard deviation presented.

Figure 2 present the distillation top and the control action (reflow profile), respectively, using the classical methods and the BiOM. The behaviour observed in this simple case study is practically the same for all strategies used.

4.2 Heat exchanger

Consider a heat exchanger type shell-tube counter-current as illustrated in Fig. 3 (Garcia, 2005). In this figure, $Q_{c,i}$ and $T_{c,i}$ represent, the flow rate and inlet temperature of the hot fluid, respectively, $Q_{t,e}$ and $T_{t,e}$, the flow rate and inlet temperature of cold fluid, respectively. T_c is the fluid temperature on the side of hull and T_t is the fluid temperature in the side of pipe. The objective of this system is to heat a water stream at 40 °C to 41 °C manipulating a hot water stream ($Q_{t,e}$) with nominal flow rate 0.0004 m^3/s. The thermal exchanges are

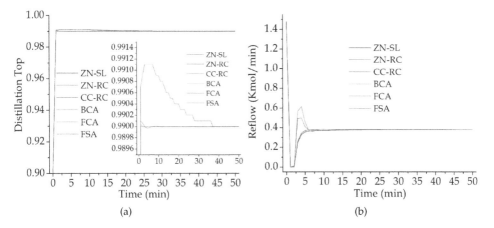

Fig. 2. Distillation top profile (a) and reflow profile (b) using classical and BiOM.

considered: heat transfer fluid circulating between the tubes and the hull, heat transfer fluid circulating between the hull and its walls, and transport of energy (enthalpy) due to fluid flow in pipes and shell. More information about the design and the considerations are in Garcia (2005).

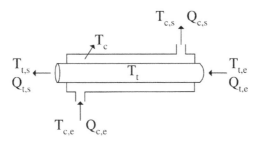

Fig. 3. Schematic heat exchanger.

The dynamic model that describes this system is given by transfer function:

$$G(s) = \frac{0.0189}{10s^3 + 5.114s^2 + 0.825s + 0.041} \tag{12}$$

In this case, the following ranges to controllers tuning are considered: $0 \leq K_c \leq 150, 0 \leq \tau_I \leq 50$ and $0 \leq D \leq 50$.

Table 6 presents the best value and standard deviation for the heat exchanger case study.

As observed in the previous case, that both the algorithms presented good estimates for the unknown parameters, but the best results were obtained by the BiOM. It is possible to observe the fluctuation of control parameters, found in the value of standard deviation.

Figure 4 present the temperature and flow profiles using the classical methods and the BiOM. It should be emphasized the oscillatory behaviour observed with the application of the BiOM, even for a short period of time (see Fig. 4(a)).

Method	K_c	τ_I (s^{-1})	τ_D (s^{-1})	OF (Eq. 10)
ZN-SL	5.8800	30.0000	7.5000	0.6206
ZN-RC	3.3600	27.0000	6.7500	2.1733
CC-RC	4.4300	26.0200	4.2500	0.9025
BCA	26.289 (10.949)	10.086 (5.845)	17.120 (4.467)	0.0219 (0.0202)
FCA	48.638 (13.425)	6.577 (5.146)	24.354 (4.898)	0.0228 (0.033)
FSA	47.274 (9.751)	9.012 (7.209)	17.508 (5.409)	0.0238 (0.0157)

Table 6. Results obtained by the BiOM - heat exchanger case study.

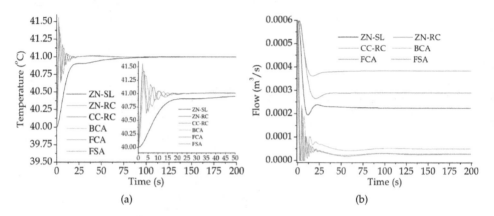

(a)

(b)

Fig. 4. Temperature and flow (control action) profiles.

4.3 Shell and tube heat exchanger

Finally, consider the real system shown in Fig. 5 for analysis and application of the previously studied concepts. This system consists essentially of: (*i*) the main tank (P5) in stainless steel with a capacity of approximately 0.250 m^3, (*ii*) stainless steel shell and tube heat exchanger (P1), (*iii*) positive displacement pump for movement of food products (P3), (*iv*) centrifugal pump for the heating agent movement (P2) and (*v*) vertical cylindrical storage tank water heater (P4) (Gedraite et al., 2011).

The Vettore-Manghi heat exchanger is responsible for heating the liquid food product that flows inside the tube bundle, considering four passes. The displacement of the process fluid inside the tubes of the heat exchanger is driven by a model RE50-110 Robuschi positive displacement pump (pump 2). The heating of the heat exchanger is done by hot water which flows through the shell side of the exchanger. Hot water is transported by Robuschi , model RE50-160 centrifugal pump (pump 3). Hot water is heated at the expense of saturated steam produced in the H. BREMER steam generator, installed in suitable and safe environment. The temperature of the process fluid is controlled by manipulating the flow of steam fed to the system, whose setting is done by the Fluxotrol model PK2117 control valve (P4), with reverse action. The hot water removed from the shell of heat exchanger returns to the vertical tank, that is equipped with a safety valve. For cooling the product, the procedure is reversed, ie, the

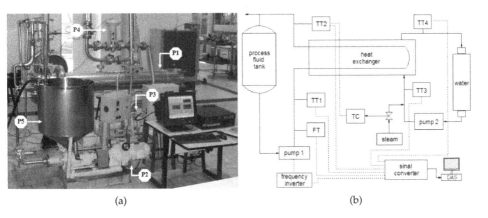

(a)   (b)

Fig. 5. Process and instrumentation diagram for the system studied (Gedraite et al., 2011).

flow control valve used for manipulate the value of heating steam flow is gradually closed. In this process, the response time is slower when compared with heating time. Cooling only occurs as a result of the heat exchange between the body of the heat exchanger, the process fluid and the environment.

The data acquisition and control system is composed by an PC based DAS (Data Acquisition System), working also as a computer control. This system consists of the following items: (*i*) PC microcomputer for the collection and storage of process data, (*ii*) LabVIEW® version 2009 application to perform monitoring, data acquisition and process control in real-time, (*iii*) data acquisition board, National Instruments (NI) PCI-6259 model, with 4 analog output channels and 32 analog input channels with both operating range of -10 V to +10 V and resolution of 16 bits, and 48 channel digital input/output programmable, (*iv*) set of cables to acquisition board NI model SHC68-68-EPM, (*v*) a connections terminal NI model CB-68LP, (*vi*) signal conditioners INCON model CS01-1360 to match the signals from the sensing elements of temperature, (*vii*) temperature sensors IOPE model 49312 type Pt 100, (*viii*) METROVAL flow meter model OI-2-SMRX/FS, (*ix*) ENGINSTREL model 621-IPB electrical current to pressure signal converter, (*x*) pneumatic control valve Fluxotrol model PK2117 and (x_i) Micronal model B474 pH meter.

4.3.1 Approximate model system

The non-parametric identification process employs basically the response curves of the system when excited by input signals like step, impulse or sinusoidal. From these curves, one can extract approximate models of low order, which describe the dynamic behavior of the process (Aguirre, 2007). These models are reasonably accurate and can be assumed to be good enough to represent the system studied. In this work, they were used to perform the pre-tuning PID controllers and to mathematically model the dynamic behavior of pH versus time.

The input most commonly used as non-parametric excitation to identify a process dynamics is the step (Aguirre, 2007). These tests usually can generate by means of graphical representation, empirical dynamic models that consists of low order transfer functions (1st or 2nd order, possibly including a dead time) with a maximum of four parameters to be determined experimentally.

Astrom & Hagglund (1995) state that many of the processes can be represented in an approximate way, by combining four elements typically found in industrial processes, namely: (*i*) gain, (*ii*) transport delay, (*iii*) transfer delay and (*iv*) integrating element. The approach of overdamped systems of order 2 or higher for transfer delay plus dead time (transport delay) can be represented by the transfer function shown in Eq. (13) (Aguirre, 2007):

$$G(s) = \frac{K.e^{-\theta s}}{1 + \tau s} \qquad (13)$$

where K is the gain, τ is the transfer delay and θ is the dead time (or transport delay).

4.4 Plant reaction curves

Tests were made to obtain the process parameters related to plant response to changes in flow and temperature. In this test, the equipment was put into operation with steady flow of 7 Lmin^{-1} and applied positive step 3 Lmin^{-1} at time 32 s, waiting for the system stabilization. In the sequence, a negative step of 3 Lmin^{-1} at time 203 s was applied. The first step (7 to 10 Lmin^{-1}) was adopted to obtain the process parameters, whose results are presented below. Figure 6 shows the system behavior to the situation examined.

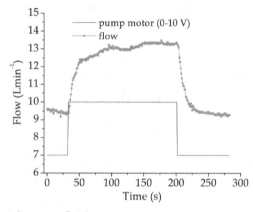

Fig. 6. Step test at flow of process fluid.

In the assay realized for the temperature, whose response time is illustrated in Fig. 7, a constant flow of 9 Lmin^{-1} was used. The outlet temperature of process fluid was adjusted equal to 60 °C and a step into the control valve installed at the steam line stem position was applied at the time 50 s, starting from the condition of fully closed until 50% opening. Following the instant 1430 s, we applied a second step of amplitude equal to 10%. The analysis to obtain the process parameters were calculated considering the first step (50%).

The process parameters K (process gain), τ (process time constant) and θ (process dead time) were calculated using the method proposed by Aguirre (2007). The transfer functions obtained are presented in Eqs. (14) and (15):

i) flow: K=1.3 Lmin^{-1}V^{-1}, τ=14 s and θ=2 s

$$G(s) = \frac{1.3 \exp(-2s)}{1 + 14s} \qquad (14)$$

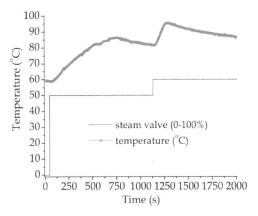

Fig. 7. Step test at temperature position at steam line.

ii) temperature: $K=7.22\,^\circ$C, $\tau=378$ s and $\theta=78$ s

$$G(s) = \frac{7.22\exp(-78s)}{1+378s} \tag{15}$$

In these simulations, the following ranges to design parameters are considered: $0 \le K_c \le 50$, $0 \le \tau_I \le 248$ and $0 \le \tau_D \le 50$.

Tables 7 and 8 present the average and standard deviation for the flow and the temperature case studies.

Method	K_c	τ_I (s^{-1})	τ_D (s^{-1})	OF (Eq. 10)
ZN-SL	6.9240	3.25	0.8125	0.0119
ZN-RC	3.1185	8.0000	2.0000	0.0629
CC-RC	3.7162	8.9937	1.3972	0.0366
BCA	9.4859	3.7011	0.6687	1.0707×10^{-5}
	(1.6760)	(0.5122)	(0.1531)	(0.0055)
FCA	9.1635	3.9305	0.7154	2.2288×10^{-7}
	(1.2686)	(0.7119)	(0.1616)	(0.0021)
FSA	9.3345	3.8780	0.7149	3.3536×10^{-7}
	(0.8472)	(0.7640)	(0.2455)	(0.0001)

Table 7. Results obtained by BiOM - Flow case study.

In these tables is possible to observe that both the algorithms presented good estimates for the controllers tuning, but the best results were obtained by the BiOM (this represent a reduction of approximately 97% in comparison to ZN-SL method). In addition, it is important to comment that if a larger range for the design variables was used, the value of the objective function would reduce. However, in spite of this reduction, the design found cannot be physically viable, e.g., can represent an infeasible condition in industrial context, as illustrated in Fig. 11(a) for the classical methods.

Figures 8 and 9 present the flow and temperature profiles using the classical methods and the BiOM. Also can be observed in these figures the control action (motor pump signal (8(a)) and valve steam signal (9(b)).

Method	K_c	τ_I (s^{-1})	τ_D (s^{-1})	OF (Eq. 10)
ZN-SL	0.8880	112.5	28.125	87303
ZN-RC	0.7453	124.0000	31.0000	64056
CC-RC	0.8758	142.8046	21.9605	77922
BCA	1.3713 (0.0010)	248.0000 (0)	38.1768 (0.0307)	1929.1287 (0.2111)
FCA	1.3708 (0.0018)	248.0000 (0.0008)	38.1905 (0.0518)	1929.1073 (0.1990)
FSA	1.3725 (0.0015)	248.0000 (0.0023)	38.1428 (0.0438)	1929.1481 (0.1501)

Table 8. Results obtained by BiOM - Temperature case study.

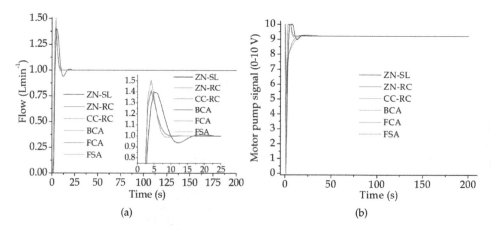

Fig. 8. Flow profile and control action (motor pump signal).

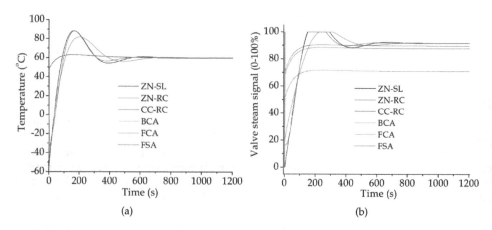

Fig. 9. Temperature profile and control action (valve steam signal).

5. Conclusions

In the present contribution, the effectiveness of using the BiOM for controllers tuning through formulation of an optimization problem was analyzed.

In this sense, three cases were studied and it was possible to conclude that both bio-inspired algorithms led to good results for an acceptable number of generations (1510) when compared to the classical methods. It should be pointed out that the quality of solution obtained is dependent of design space considered, e.g., if other ranges were used, other results can be found. Besides, also can be observed that the combination of control parameters, can take to values close, in terms of the objective function.

It is important to emphasize that the use of the BiOM not have the pretension of substituting the classical techniques for the controllers tuning, but to represent an interesting alternative for this purpose.

6. Acknowledgment

The authors acknowledge the financial support provided by CNPq, Conselho Nacional de Desenvolvimento Científico e Tecnológico and FAPEMIG, Fundação de Amparo à Pesquisa do Estado de Minas Gerais.

7. References

Aguirre, L. A. (2007). *Introduction to System Identification: Linear and Non-linear Techniques Applied to Real System*. Ed. UFMG, Minas Gerais - Brasil.

Apostolopoulos, T. & Vlachos, A. (2011). Application of the Firefly Algorithm for Solving the Economic Emissions Load Dispatch Problem. *International Journal of Combinatorics*, Vol. 3, 1–23.

Astrom, K J. & Hagglund, T. (1995). *PID Controller: Theory, Design and Tuning*. Triangle Park, NC: ISA Publication Research.

Afshar, A.; Haddad, O. B.; Mariño, M. A. & Adams, B. J. (2001). Honey-Bee Mating Optimization (HBMO) Algorithm for Optimal Reservoir Operation. *Journal of the Franklin Institute*, Vol. 344, 452–462.

Azeem, M. F. & Saad, A. M. (2004). Modified Queen Bee Evolution based Genetic Algorithm for Tuning of Scaling Factors of Fuzzy Knowledge Base Controller. *IEEE INDICON 2004 Proceedings of the India Annual Conference*, 299–303.

Bandyopadhyay, R.; Chakraborty, U. K. & Patranabis, D. (2001). Auto Tuning a PID Controller: A Fuzzy-Genetic Approach. *Journal of Systems Architecture*, Vol. 47 (7), 663–73.

Cai, Y. (2010). Artificial Fish School Algorithm Applied in a Combinatorial Optimization Problem. *Intelligent Systems and Applications*, Vol. 1, 37–43.

Camazine, S.; Deneubourg, J.; Franks, N. R.; Sneyd, J.; Theraula, G. & Bonabeau, E. (2003). *Self-Organization in Biological Systems*, Princeton University Press.

Chang, H. S. (2006). Converging Marriage in Honey-Bees Optimization and Application to Stochastic Dynamic Programming. *Journal of Global Optimization*, Vol. 35, 423–441.

Coelho, L. S. & Pessoa, M. W. (2011). A Tuning Strategy for Multivariable PI and PID Controllers using Differential Evolution Combined with Chaotic Zaslavskii Map, *Expert Systems with Applications*, Vol. 38, 13694–13701.

Conner, J. S. & Seborg, D. E. (2005). Assessing de Need for Process Re-identification. *Industrial Engineering Chemical Research*, Vol. 44, 2767–2775.

Desbourough, L. & Miller, R. (2002). Increasing Customer Value of Industrial Control Performance Monitoring-Honeywell's Experience. *Proc. Int. Conference on Chemical Process Control*. AIChE symposium series, N. 326, Vol. 98.

Garcia, C. (2005). *Modeling and Simulation*. EdUSP, São Paulo (in portuguese), 120 pages.

Gedraite, R.; Lobato, F. S.; Neiro, S. M. da S.; Melero Jr., V.; Augusto, S. R. & Kunigk, L. (2011). CIP System kinetics mathematical modeling: dynamic behavior of residuals removal kinetics. *XXXII Iberian Latin-American Congress on Computational Methods in Engineering*.

Hang, C. C.; Åström, K. J. & Ho, W. K. (1991). Refinements of Ziegler-Nichols Tuning Formula. *IEE Proc-D*, Vol. 138 (2), 111–118.

Hamid, B.; Mohamed, T.; Pierre-Yves, G. & Salim, L. (2010). Tuning Fuzzy PD and PI Controllers using Reinforcement Learning. *ISA Transactions*, Vol. 49 (4), 543–51.

Kim, T-H.; Maruta, I. & Sugie, T. (2008). Robust PID Controller Tuning based on the Constrained Particle Swarm Optimization, *Automatica*, Vol. 44, 1104–1110.

Li, X. L.; Shao, Z. J. & Qian, J. X. (2002). An Optimizing Method based on Autonomous Animate: Fish Swarm Algorithm, *System Engineering Theory and Practice*, Vol. 22 (11), 32–38.

Li, X. L.; Xue, Y. C.; Lu, F. & Tian, G. H. (2004). Parameter Estimation Method based on Artificial Fish School Algorithm, *Journal of Shan Dong University* (Engineering Science), Vol. 34 (3), 84–87.

Lobato, F. S. & Souza, D. L. (2008). Adaptive Differential Evolution Method Applied To Controllers Tuning. *7th Brazilian Conference on Dynamics, Control and Applications*.

Lobato, F. S.; Sousa, J. A.; Hori, C. E. & Steffen Jr, V. (2010), Improved Bees Colony Algorithm Applied to Chemical Engineering System Design. *International Review of Chemical Engineering* (Rapid Communications), Vol. 6, 1–7.

Lucic, P. & Teodorovic, D. (2001). Bee System: Modeling Combinatorial Optimization Transportation Engineering Problems by Swarm Intelligence. *TRISTAN IV Triennial Symposium on Transportation Analysis*, 441–445.

Lukasik, S. & Zak, S. (2009). *Firefly Algorithm for Continuous Constrained Optimization Task*, ICCCI 2009, Lecture Notes in Artificial Intelligence (Eds. N. T. Ngugen, R. Kowalczyk, S. M. Chen), Vol. 5796, 97–100.

Madeiro, S. S. (2010). *Modal Search for Swarm based on Density*, Dissertation, Universidade de Pernambuco (in portuguese).

Mudi, R. K.; Dey, C. & Lee, T. T. (2008). An Improved Auto-Tuning Scheme for PI Controllers. *ISA Trans*. Vol. 47, 45–52.

Pan, I.; Das, S. & Gupta, A. (2011). Tuning of an Optimal Fuzzy PID Controller with Stochastic Algorithms for Networked Control Systems with Random Time Delay, *ISA Transactions*, Vol. 50, 28–36.

Parrich, J.; Viscido, S. & Grunbaum, D. (2002). Self-organized Fish Schools: An Examination of Emergent Properties. *Biological Bulletin*, Vol. 202 (3), 296–305.

Pfeifer, A. A. & Lobato, F. S. (2010). Solution of Singular Optimal Control Problems using the Firefly Algorithm. *Proceedings of VI Congreso Argentino de Ingeniería Química - CAIQ2010*.

Pham, D. T.; Kog, E.; Ghanbarzadeh, A.; Otri, S.; Rahim, S. & Zaidi, M. (2006). The Bees Algorithm - A Novel Tool for Complex Optimisation Problems. *Proceedings of 2nd International Virtual Conference on Intelligent Production Machines and Systems*, Oxford, Elsevier.

Ribeiro Neto, A. C.; Lobato, F. S.; Steffen Jr, V. & Silva Neto, A. J. (2011). Solution of Inverse Radiative Transfer Problems with the Bees Colony Algorithm, *21st Brazilian Congress of Mechanical Engineering*, October 24-28, 2011, Natal, RN, Brazil.

Rocha, A. M. A. C.; Martins, T. F. M. C. & Fernandes, E. M. G. P. (2011). An Augmented Lagrangian Fish Swarm based Method for Global Optimization. *Journal of Computational and Applied Mathematics*, Vol. 235, 4611–4620.

Seborg, D. E.; Edgar, T. F. & Mellichamp, D. A. (1989). *Process Dynamics and Control*. Wiley Series in Chemical Engineering.

Sedlaczek, K. & Eberhard, P. (2006). Using Augmented Lagrangian Particle Swarm Optimization for Constrained Problems in Engineering. *Structural and Multidisciplinary Optimization*, Vol. 32(4), 277–286.

Shen, W.; Guo, X.; Wu, C. & Wu, D. (2011). Forecasting Stock Indices using Radial Basis Function Neural Networks Optimized by Artificial Fish Swarm Algorithm. *Knowledge-Based Systems*, Vol. 24, 378–385.

Solihin, M. I.; Tack, L. F. & Kean, M. L. (2011). Tuning of PID Controller Using Particle Swarm Optimization (PSO). *Proceeding of the International Conference on Advanced Science, Engineering and Information Technology*, ISBN 978-983-42366-4-9.

Souza, D. L. (2007). *Analysis and Performance of Control Systems*. Dissertação de Mestrado, Faculdade de Engenharia Química, Universidade Federal de Uberlândia, Brasil (in portuguese).

Skogestad, S. & Morari, M. (1987). The dominant time constant for distillation columns, *Computers and Chemical Engineering*, Vol. 1, 607–619.

Teodorovic, D. & Dell'Orco, M. (2005). Bee Colony Optimization - A Cooperative Learning Approach to Complex Transportation Problems. *Proceedings of the 10th EWGT Meeting and 16th Mini-EURO Conference*.

Tsai, H. C. & Lin, Y. H. (2011). Modification of the Fish Swarm Algorithm with Particle Swarm Optimization Formulation and Communication Behavior, *Applied Soft Computing Journal* (2011), Vol. 11, 5367–5374.

von Frisch, K. (1976). *Bees: Their Vision, Chemical Senses and Language*. Revised edn, Cornell University Press, N.Y., Ithaca.

Xi, M.; Sun, J. & Xu, W. (2007). Parameter Optimization of PID Controller Based on Quantum-behaved Particle Swarm Optimization Algorithm, Complex System and Application - Modeling, Control and Simulation, Vol. 14 (S2), 603–607.

Yang, X. S. (2005). *Engineering Optimizations via Nature-Inspired Virtual Bee Algorithms*. IWINAC 2005, Lecture Notes in Computer Science, 3562, Yang, J. M. and J.R. Alvarez (Eds.), Springer-Verlag, Berlin Heidelberg, 317–323.

Yang, X. S. (2008). *Nature-Inspired Metaheuristic Algorithms*, Luniver Press, Cambridge.

Yang, X. S. (2009). Firefly Algorithm for Multimodal Optimization. Stochastic Algorithms: Foundations and Applications, Vol. 5792, 169–178.

Yang, X. S. (2010). *Firefly algorithm, Lévy Flights and Global Optimization*, Research and Development in Intelligent Systems XXVI (Eds M. Bramer, R. Ellis, M. Petridis), Springer London, 209–218.

Wang, C. R.; Zhou, C. L. & Ma, J. W. (2005). An Improved Artificial Fish-Swarm Algorithm and Its Application in Feedforward Neural Networks. *Proc. of the Fourth Int. Conf. on Machine Learning and Cybernetics*, 2890–2894.

7

Adaptive Coordinated Cooperative Control of Multi-Mobile Manipulators

Víctor H. Andaluz[1], Paulo Leica[2],
Flavio Roberti[2], Marcos Toibero[2] and Ricardo Carelli[2]
[1]Universidad Técnica de Ambato,
Facultad de Ingeniería en Sistemas, Electrónica e Industrial
[2]Universidad Nacional de San Juan,
Instituto de Automática,
[1]Ecuador
[2]Argentina

1. Introduction

A coordinated group of robots can execute certain tasks, *e.g.* surveillance of large areas (Hougen et al., 2000), search and rescue (Jennings et al., 1997), and large objects-transportation (Stouten and De Graaf, 2004), more efficiently than a single specialized robot (Cao et al., 1997). Other tasks are simply not accomplishable by a single mobile robot, demanding a group of coordinated robots to perform it, like the problem of sensors and actuators positioning (Bicchi et al., 2008), and the entrapment/escorting mission (Antonelli et al., 2008). In such context, the term formation control arises, which can be defined as the problem of controlling the relative postures of the robots of a platoon that moves as a single structure (Consolini et al., 2007).

Mobile manipulator is nowadays a widespread term that refers to robots built by a robotic arm mounted on a mobile platform. This kind of system, which is usually characterized by a high degree of redundancy, combines the manipulability of a fixed-base manipulator with the mobility of a wheeled platform. Such systems allow the most usual missions of robotic systems which requiere both locomotion and manipulation abilities. Coordinated control of multiple mobile manipulators have attracted the attention of many researchers (Khatib et al., 1996; Fujii et al., 2007; Tanner et al., 2003; Yasuhisa et al., 2003). The interest in such systems stems from the capability for carrying out complex and dexterous tasks which cannot be simply made using a single robot. Moreover, multiple small mobile manipulators are also more appropriate for realizing several tasks in the human environments than a large and heavy mobile manipulator from a safety point of view.

Main coordination schemes for multiple mobile manipulators that can be found in the literature are:

1. Leader–follower control for mobile manipulator, where one or a group of mobile manipulators plays the role of a leader, which track a preplanned trajectory, and the

rest of the mobile manipulators form the follower group which moves in conjunction with the leader mobile manipulators (Fujii et al., 2007; Hirata et al., 2004; Thomas et al., 2002). In Xin and Yangmin, 2006, a leader-follower type formation control is designed for a group of mobile manipulators. To overcome parameter uncertainty in the model of the robot, a decentralized control law is applied to individual robots, in which an adaptive NN is used to model robot dynamics online.

2. Hybrid position–force control by decentralized/centralized scheme, where the position of the object is controlled in a certain direction of the workspace and the internal force of the object is controlled in a small range of the origin (Khatib et al., 1996; Tanner et al., 2003; Yamamoto et al., 2004). In Zhijun et al., 2008, robust adaptive controllers of multiple mobile manipulators carrying a common object in a cooperative manner have been investigated with unknown inertia parameters and disturbances. At first, a concise dynamics consisting of the dynamics of mobile manipulators and the geometrical constraints between the end-effectors and the object is developed for coordinated multiple mobile manipulators. In Zhijun et al., 2009 coupled dynamics are presented for two cooperating mobile manipulators manipulating an object with relative motion in the presence of uncertainties and external disturbances. Centralized robust adaptive controllers are introduced to guarantee the motion and force trajectories of the constrained object. A simulation study to the decentralized dynamic control for a robot collective consisting of nonholonomic wheeled mobile manipulators is performed in Hao and Venkat, 2008, by tracking the trajectories of the load, where two reference signals are used for each robot, one for the mobile platform and another for end-effector of the manipulating arm.

To reduce performance degradation, on-line parameter adaptation is relevant in applications where the mobile manipulator dynamic parameters may vary, such as load transportation. It is also useful when the knowledge of the dynamic parameters is limited. As an example, the trajectory tracking task can be severely affected by the change imposed to the robot dynamics when it is carrying an object, as shown in (Martins et al., 2008). Hence, some formation control architectures already proposed in the literature have considered the dynamics of the mobile robots (Zhijun et al., 2008; Zhijun et al., 2009).

In this Chapter, it is proposed a novel method for centralized-decentralized coordinated cooperative control of multiple wheeled mobile manipulators. Also, it is worth noting that, differently to the work in Hao and Venkat, 2008, we use a single reference for the end-effector of the robot mobile manipulator.

Although centralized control approaches present intrinsic problems, like the difficulty to sustain the communication between the robots and the limited scalability, they have technical advantages when applied to control a group of robots with defined geometric formations. Therefore, there still exists significant interest in their use. As an example, in Antonelli et al., 2008, a centralized multi-robot system is proposed for an entrapment/escorting mission, where the escorted agent is kept in the centroid of a polygon of n sides, surrounded by n robots positioned in the vertices of the polygon. Another task for which it is important to keep a formation during navigation is large-objects transportation, since the load has a fixed geometric form. Another recent work dealing with centralized formation control is Mas et al., 2008, where a control approach based on a virtual structure, called Cluster Space Control, is presented. There, the positioning control is carried out considering the centroid of a geometric structure corresponding to a three-robot formation.

In this Chapter, the proposed strategy conceptualizes the mobile manipulators system (with $n \geq 3$) as a single group, and the desired motions are specified as a function of cluster attributes, such as position, orientation, and geometry. These attributes guide the selection of a set of independent system state variables suitable for specification, control, and monitoring. The control is based on a virtual 3-dimensional structure, where the position control (or tracking control) is carried out considering the centroid of the upper side of a geometric structure (shaped as a prism) corresponding to a three-mobile manipulators formation. It is worth noting that in control problem formulating first it is considered three mobile manipulators robots, and then is generalized to mobile manipulators robots.

The proposed multi-layer control scheme is mainly divided in five modules: 1) the upper module is responsible for planning the trajectory to be followed by the team of mobile manipulators; 2) the next module controls the formation, whose shape is determined by the distance and angle between the end-effector of a mobile manipulator and the two other ones; 3) another module is responsible to generate the control signals to the end-effectors of the mobile manipulators, through the inverse kinematics of each robot. As a mobile manipulator is usually a redundant system, this redundancy can be used for the achievement of additional performances. In this layer two secondary objectives are considered: the avoidance of obstacles by the mobile platforms and the singular configuration prevention through the control of the system's manipulability; introduced by Yoshikawa (1985). 4) The adaptive dynamic compensation module compensates the dynamics of each mobile manipulator to reduce the velocity tracking error. It is worth noting that this controller has been designed based on a dynamic model having reference velocities as input signals. Also, it uses a robust updating law, which makes the dynamic compensation system robust to parameter variations and guarantees that no parameter drift will occur; 5) finally, the robots module represents the mobile manipulators.

It is worth noting that we propose a methodology to avoid obstacles in the trajectory of any mobile manipulator based on the concept of mechanical impedance of the interaction robots-environment, without deforming the virtual structure and maintaining its desired trajectory. It is considered that the obstacle is placed at a maximum height that it does not interfere with the workspace, so that the arm of the mobile manipulators can follow the desired trajectory even when the platform is avoiding the obstacle.

This Chapter is organized as follows. Section 2 shows the kinematic and dynamic models of the mobile manipulator. Section 3 presents the proposed multi-layer control scheme for the coordinated and cooperative control of mobile manipulators. While the forward and inverse kinematics transformations, necessary for the control scheme, are presented in Section 4. Section 5 describes the scalability for coordinated cooperative control of mobile manipulators. By its turn, Section 6 presents the design of the controller, and the analysis of the system's stability is developed. Next, simulation experiments results are presented and discussed in Section 7, and finally the Chapter conclusions are given in Section 8.

2. Mobile manipulator models

The mobile manipulator configuration is defined by a vector $\mathbf{q}$ of n independent coordinates, called *generalized coordinates of the mobile manipulator*, where $\mathbf{q} = [q_1 \quad q_2 \quad \cdots \quad q_n]^T = [\mathbf{q}_p^T \quad \mathbf{q}_a^T]^T$ where $\mathbf{q}_a$ represents the generalized coordinates of the arm, and $\mathbf{q}_p$ the generalized coordinates of the mobile platform. We notice that $n = n_a + n_p$, where n_a and n_p are respectively the dimensions of the generalized spaces associated to the robotic arm and to the mobile platform. The configuration $\mathbf{q}$ is an element of the mobile manipulator *configuration space;* denoted by $\mathcal{N}$. The location of the end-effector of the mobile manipulator is given by the m–dimensional vector $\mathbf{h} = [h_1 \quad h_2 \quad ... \quad h_m]^T$ which define the position and the orientation of the end-effector of the mobile manipulator in $\mathcal{R}$. Its m coordinates are the *operational coordinates of the mobile manipulator.* The set of all locations constitutes the *mobile manipulator operational space,* denoted by $\mathcal{M}$.

The location of the mobile manipulator end-effector can be defined in different ways according to the task, *i.e.,* it can be considered only the position of the end-effector or both its position and its orientation.

2.1 Mobile manipulator kinematic model

The *kinematic model of a mobile manipulator* gives the location of the end-effector $\mathbf{h}$ as a function of the robotic arm configuration and the platform location (or its operational coordinates as functions of the robotic arm generalized coordinates and the mobile platform operational coordinates).

$$f : \mathcal{N}_a \times \mathcal{M}_p \to \mathcal{M}$$

$$\left(\mathbf{q}_a, \mathbf{q}_p\right) \mapsto \mathbf{h} = f\left(\mathbf{q}_a, \mathbf{q}_p\right)$$

where, $\mathcal{N}_a$ is the *configuration space* of the robotic arm, $\mathcal{M}_p$ is the *operational space of the platform.*

The *instantaneous kinematic model of a mobile manipulator* gives the derivative of its end-effector location as a function of the derivatives of both the robotic arm configuration and the location of the mobile platform,

$$\dot{\mathbf{h}} = \frac{\partial f}{\partial \mathbf{q}}\left(\mathbf{q}_a, \mathbf{q}_p\right)\mathbf{v}$$

where, $\dot{\mathbf{h}} = [\dot{h}_1 \quad \dot{h}_2 \quad ... \quad \dot{h}_m]^T$ is the vector of the end-effector velocity, $\mathbf{v} = [v_1 \quad v_2 \quad ... \quad v_{\delta_n}]^T$ $= [\mathbf{v}_p^T \quad \mathbf{v}_a^T]^T$ is the control vector of mobility of the mobile manipulator. Its dimension is $\delta_n = \delta_{np} + \delta_{na}$, where δ_{np} and δ_{na} are respectively the dimensions of the control vector of mobility associated to the mobile platform and to the robotic arm. Now, after replacing $\mathbf{J}(\mathbf{q}) = \frac{\partial f}{\partial \mathbf{q}}\left(\mathbf{q}_a, \mathbf{q}_p\right)\mathbf{T}(\mathbf{q})$ in the above equation, we obtain

$$\dot{\mathbf{h}}(t) = \mathbf{J}(\mathbf{q})\mathbf{v}(t) \tag{1}$$

where, $J(q)$ is the Jacobian matrix that defines a linear mapping between the vector of the mobile manipulator velocities $v(t)$ and the vector of the end-effector velocity $\dot{h}(t)$, and $T(q)$ is the transformation matrix that relates joints velocities $\dot{q}(t)$ and mobile manipulator velocities $v(t)$ such that $\dot{q}(t) = T(q)v(t)$.

Remark 1: The transformation matrix $T(q)$ includes the non-holonomic constraints of the mobile platform.

The Jacobian matrix is, in general, a function of the configuration q; those configurations at which $J(q)$ is rank-deficient are termed *singular kinematic configurations*. It is fundamental to notice that, in general, the dimension of the operational space m is less than the degree of mobility of the mobile manipulator, therefore the system is redundant.

2.2 Mobile manipulator dynamic model

The mathematic model that represents the dynamics of a mobile manipulator can be obtained from Lagrange's dynamic equations, which are based on the difference between the kinetic and the potential energy of each of the joints of the robot (energy balance) (Spong and Vidyasagar, 1989; Yoshikawa, 1990; Sciavicco and Siciliano, 2000). The dynamic equation of the mobile manipulator can be represented as follows,

$$\bar{M}(q)\dot{v} + \bar{C}(q,v)v + \bar{G}(q) + \bar{d} = \bar{B}(q)\tau \tag{2}$$

where, $q = [q_1,...,q_n]^T \in \mathfrak{R}^n$ is the general coordinate system vector of the mobile manipulator, $v = [v_1,....,v_{\delta_n}]^T \in \mathfrak{R}^{\delta_n}$ is the velocity vector of the mobile manipulator, $\bar{M}(q) \in \mathfrak{R}^{\delta_n \times \delta_n}$ is a symmetrical positive definite matrix that represents the system's inertia, $\bar{C}(q,v)v \in \mathfrak{R}^{\delta_n}$ represents the components of the centripetal and Coriolis forces, $\bar{G}(q) \in \mathfrak{R}^{\delta_n}$ represents the gravitational forces, $\bar{d}$ denotes bounded unknown disturbances including the unmodeled dynamics, $\tau \in \mathfrak{R}^{\delta_n}$ is the torque input vector, $\bar{B}(q) \in \mathfrak{R}^{\delta_n \times \delta_n}$ is the transformation matrix of the control actions.

Most of the commercially available robots have low level PID controllers in order to follow the reference velocity inputs, thus not allowing controlling the voltages of the motors directly. Therefore, it becomes useful to express the dynamic model of the mobile manipulator in a more appropriate way, taking the rotational and longitudinal reference velocities as the control signals. To do so, the velocity servo controller dynamics are included in the model. The dynamic model of the mobile manipulator, having as control signals the reference velocities of the system, can be represented as follows,

$$M(q)\dot{v} + C(q,v)v + G(q) + d = v_{ref} \tag{3}$$

where $\mathbf{M}(\mathbf{q}) = \mathbf{H}^{-1}(\bar{\mathbf{M}} + \mathbf{D})$, $\mathbf{C}(\mathbf{q}, \mathbf{v}) = \mathbf{H}^{-1}(\bar{\mathbf{C}} + \mathbf{P})$, $\mathbf{G}(\mathbf{q}) = \mathbf{H}^{-1}\bar{\mathbf{G}}(\mathbf{q})$ and $\mathbf{d} = \mathbf{H}^{-1}\bar{\mathbf{d}}$. Thus, $\mathbf{M}(\mathbf{q}) \in \Re^{\delta_n \times \delta_n}$ is a positive definite matrix, $\mathbf{C}(\mathbf{q}, \mathbf{v})\mathbf{v} \in \Re^{\delta_n}$, $\mathbf{G}(\mathbf{q}) \in \Re^{\delta_n}$, $\mathbf{d} \in \Re^{\delta_n}$ and $\mathbf{v}_{ref} \in \Re^{\delta_n}$ is the vector of velocity control signals, $\mathbf{H} \in \Re^{\delta_n \times \delta_n}$, $\mathbf{D} \in \Re^{\delta_n \times \delta_n}$ and $\mathbf{P} \in \Re^{\delta_n \times \delta_n}$ are positive definite constant diagonal matrices containing the physical parameters of the mobile manipulator, motors, and velocity controllers of both the mobile platform and the manipulator. It is important to remark that $\mathbf{H}$, $\mathbf{D}$ and $\mathbf{P}$ are positive definite constant diagonal matrices, hence the properties for the dynamic model with reference velocities as control signals (3) were obtained on based of the properties of the dynamic model (2):

Property 1. Matrix $\mathbf{M}(\mathbf{q})$ is positive definite, additionally it is known that

$$\|\mathbf{M}(\mathbf{q})\| < k_M$$

Property 2. Furthermore, the following inequalities are also satisfied

$$\|\mathbf{C}(\mathbf{q}, \mathbf{v})\| < k_c \|\mathbf{v}\|$$

Property 3. Vector $\mathbf{G}(\mathbf{q})$ and $\mathbf{d}$ are bounded

$$\|\mathbf{G}(\mathbf{q})\| < k_G \; ; \qquad \|\mathbf{d}\| < k_d$$

where, k_c, k_M, k_G and k_d denote some positive constants.

Property 4. The dynamic model of the mobile manipulator can be represented by

$$\mathbf{M}(\mathbf{q})\dot{\mathbf{v}} + \mathbf{C}(\mathbf{q}, \mathbf{v})\mathbf{v} + \mathbf{G}(\mathbf{q}) + \mathbf{d} = \mathbf{\Phi}(\mathbf{q}, \mathbf{v})\chi$$

where, $\mathbf{\Phi}(\mathbf{q}, \mathbf{v}) \in \Re^{\delta_n \times l}$ and $\chi = [\chi_1 \quad \chi_2 \quad \cdots \quad \chi_l]^T$ is the vector of l unknown parameters of the mobile manipulator, *i.e.*, mass of the mobile robot, mass of the robotic arm, physical parameters of the mobile manipulator, motors, velocity, etc.

For the sake of simplicity, from now on it will be written $\mathbf{M} = \mathbf{M}(\mathbf{q})$, $\mathbf{C} = \mathbf{C}(\mathbf{q}, \mathbf{v})$ and $\mathbf{G} = \mathbf{G}(\mathbf{q})$.

Hence, the full mathematical model of the mobile manipulator robot is represented by (1), the *instantaneous kinematic model* and (3), the *dynamic model*, taking the reference velocities of the system as input signals.

3. Multi-layers control scheme

Figure 1, shows the Multi-layer control Scheme of the coordinated cooperative control of mobile manipulators which is taken into account in this Chapter.

Each layer works as an independent module, dealing with a specific part of the problem of coordinated cooperative control, and such control scheme includes a basic structure defined by the formation control layer, the kinematic control layer, the robots layer and the environment layer.

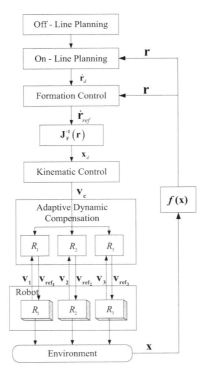

Fig. 1. Multi-layer control scheme

- *The Off-line Planning layer* is responsible for setting up the initial conditions, thus generating the trajectory of the object to be tracked, and for establishing the desired structure. The On-line Planning layer is capable of changing the references in order to make the formation to react to the environment, e.g., to modify the trajectory to avoid obstacles (it should be included only when a centralized obstacle avoidance strategy is considered; in this work it is considered the decentralized obstacle avoidance).
- *The Formation Control layer* is responsible for generating the control signals to be sent to the mobile manipulators, working as a team, in order to reach the desired values established by the planning layers.
- *The Kinematic Control layer* is responsible for generating the control signals to the end-effector of the mobile manipulators considering different control objectives.
- *The Adaptive Dynamic Compensation layer* compensates the dynamics of each robot to reduce the velocity tracking error.
- *The Robot layer* represents the mobile manipulators (mobile manipulators with unicycle-like mobile platforms, car-like and/or omnidirectional type), and finally;
- *The Environment layer* represents all objects surrounding the mobile manipulators, including the mobile manipulators themselves, with their external sensing systems, necessary for implementing obstacle avoidance.

One of the main advantages of the proposed scheme is the independence of each layer, *i.e.*, changes within a layer do not cause structural changes in the other layers. As an example, several kinematic controllers or dynamic compensation approaches can be tested using the same formation control strategy and vice-versa. It is worth mentioning that a simple structure can be obtained from the presented scheme, that is, some layers can be eliminated whenever the basic structure is maintained and the absence of the eliminated layers do not affect the remaining layers. For example, the On-line Planning layer could be discarded in the case of trajectory tracking or path following by a multi-robot formation in a known environment free of obstacles, because the entire task accomplishment is controlled by the Formation Control layer. Also the Adaptive Dynamic Compensation layer can be suppressed, for applications demanding low velocities and light load transportation.

On the other hand, it is important to stress that some additional blocks are necessary to complete the multi-layer scheme, such as $\mathbf{J_F^{-1}}(\mathbf{r})$ and $f(\mathbf{x})$, which represents the inverse formation Jacobian matrix, and the forward kinematic transformation function for the formation, respectively.

Remark 2: The mobile manipulators can be different, *i.e.*, each mobile manipulator can be built by different types mobile platforms or/and different types robotic arms. Thus each mobile manipulator has its own configuration.

Remark 3: A mobile manipulator is defined as a redundant system because it has more degrees of freedom than required to achieve the desired end-effector motion. Hence, the redundancy of such systems can be effectively used for the achievement of additional performances.

4. Kinematic transformation

The proposed coordinated cooperative control method considers three or more mobile manipulators. In the first step, only three mobile manipulators are considered. In this case the control method is based on creating a regular or irregular prism defined by the position of the end-effector of each mobile manipulator. The location of the upper side of the prism in the plane X-Y of the global framework is defined by $\mathbf{P_F} = [x_F \quad y_F \quad \psi_F]$, where (x_F, y_F) represents the position of its centroid, and ψ_F represents its orientation with respect to the global Y-axis. The structure shape of the prism (regular or irregular) is defined by $\mathbf{S_F} = [p_F \quad q_F \quad \beta_F \quad z_{1F} \quad z_{2F} \quad z_{3F}]$, where, p_F represents the distance between $\mathbf{h}_1$ and $\mathbf{h}_2$, q_F the distance between $\mathbf{h}_1$ and $\mathbf{h}_3$, β_F the angle formed by $\mathbf{h}_2\hat{\mathbf{h}}_1\mathbf{h}_3$ and (z_{1F}, z_{2F}, z_{3F}) represents the height of the upper side of the prism. This situation is illustrated in Figure 2.

Remark 4: $\mathbf{h}_i$ represents the position the end-effector of the *i*-th mobile manipulator.

The relationship between the prism pose-orientation-shape and the end-effector positions of the mobile manipulators is given by the forward and inverse kinematics transformation, *i.e.*,

$$\mathbf{r} = f(\mathbf{x}) \text{ and } \mathbf{x} = f^{-1}(\mathbf{r}) \text{ , where } \mathbf{r} = [\mathbf{P_F} \quad \mathbf{S_F}]^T \text{ and } \mathbf{x} = [\mathbf{h}_1^T \quad \mathbf{h}_2^T \quad \mathbf{h}_3^T]^T.$$

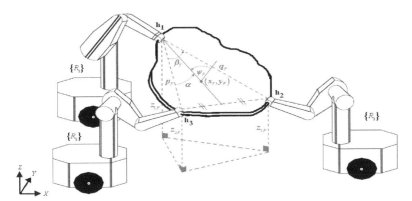

Fig. 2. Structure variables

The forward kinematic transformation $f(.)$, as shown in figure 2, is given by

$$\mathbf{P_F} = \begin{bmatrix} \dfrac{x_1 + x_2 + x_3}{3} \\[2mm] \dfrac{y_1 + y_2 + y_3}{3} \\[2mm] \arctan\dfrac{\frac{2}{3}x_1 - \frac{1}{3}(x_2 + x_3)}{\frac{2}{3}y_1 - \frac{1}{3}(y_2 + y_3)} \end{bmatrix}^T \quad ; \quad \mathbf{S_F} = \begin{bmatrix} \sqrt{(x_1 - x_2)^2 + (y_1 - y_2)^2} \\[2mm] \sqrt{(x_1 - x_3)^2 + (y_1 - y_3)^2} \\[2mm] \arccos\dfrac{p_F^2 + q_F^2 - r_F^2}{2 p_F q_F} \\[2mm] z_{1F} \\ z_{2F} \\ z_{3F} \end{bmatrix}^T$$

where, $r_F = \sqrt{(x_2 - x_3)^2 + (y_2 - y_3)^2}$. In turn, for the inverse kinematic transformation $f^{-1}(.)$, two representations are possible, depending on the disposition of the mobile manipulators in the prism shape (clockwise or counter-clockwise). Such disposition can be referred to as $R_1 R_2 R_3$ or $R_1 R_3 R_2$ sequence (R_i represents the i-th mobile manipulator robot). Considering the first possibility, $\mathbf{x} = f_{R_1 R_2 R_3}^{-1}(\mathbf{r})$ is given by,

$$\mathbf{x} = \begin{bmatrix} \mathbf{h}_1 \\ \mathbf{h}_2 \\ \mathbf{h}_3 \end{bmatrix} = \begin{bmatrix} x_F + \frac{2}{3}h_F \sin\psi_F \\ y_F + \frac{2}{3}h_F \cos\psi_F \\ z_{1F} \\ x_F + \frac{2}{3}h_F \sin\psi_F - p_F \sin(\alpha + \psi_F) \\ y_F + \frac{2}{3}h_F \cos\psi_F - p_F \cos(\alpha + \psi_F) \\ z_{2F} \\ x_F + \frac{2}{3}h_F \sin\psi_F + q_F \sin(\beta_F - \alpha - \psi_F) \\ y_F + \frac{2}{3}h_F \cos\psi_F - q_F \cos(\beta_F - \alpha - \psi_F) \\ z_{3F} \end{bmatrix}$$

where, $h_F = \sqrt{\frac{1}{2}\left(p_F^2 + q_F^2 - \frac{1}{2}r_F^2\right)}$ represents the distance between the end-effector $\mathbf{h}_1$ and the

point in the middle of the segment $\overline{\mathbf{h}_2\mathbf{h}_3}$, passing through (x_F, y_F), and

$\alpha = \arccos \dfrac{p_F^2 + h_F^2 - \frac{1}{4}r_F^2}{2p_F h_F}$. On the other hand, $\mathbf{x} = f^{-1}_{R_1R_3R_2}(\mathbf{r})$ is given by

$$\mathbf{x} = \begin{bmatrix} \mathbf{h}_1 \\ \mathbf{h}_2 \\ \mathbf{h}_3 \end{bmatrix} = \begin{bmatrix} x_F + \frac{2}{3}h_F \sin\psi_F \\ y_F + \frac{2}{3}h_F \cos\psi_F \\ z_{1F} \\ x_F + \frac{2}{3}h_F \sin\psi_F + p_F \sin(\alpha - \psi_F) \\ y_F + \frac{2}{3}h_F \cos\psi_F - p_F \cos(\alpha - \psi_F) \\ z_{2F} \\ x_F + \frac{2}{3}h_F \sin\psi_F - q_F \sin(\beta_F - \alpha + \psi_F) \\ y_F + \frac{2}{3}h_F \cos\psi_F - q_F \cos(\beta_F - \alpha + \psi_F) \\ z_{3F} \end{bmatrix}$$

Figure 3 shows the control structure proposed in this Chapter for the coordinated cooperative control of mobile manipulators. Taking the time derivative of the forward and the inverse kinematics transformations we can obtain the relationship between the time variations of $\mathbf{x}(t)$ and $\mathbf{r}(t)$, represented by the Jacobian matrix $\mathbf{J_F}$, which is given by

$$\dot{\mathbf{r}} = \mathbf{J_F}(\mathbf{x})\dot{\mathbf{x}} \tag{4}$$

and in the inverse way is given by

$$\dot{\mathbf{x}} = \mathbf{J}_F^{-1}(\mathbf{r})\dot{\mathbf{r}} \tag{5}$$

where,

$$\mathbf{J_F}(\mathbf{x}) = \frac{\partial \mathbf{r}_{fx1}}{\partial \mathbf{x}_{ex1}} \quad \text{and} \quad \mathbf{J}_F^{-1}(\mathbf{r}) = \frac{\partial \mathbf{x}_{ex1}}{\partial \mathbf{r}_{fx1}} \quad \text{with } e,f = 1,2..,9 .$$

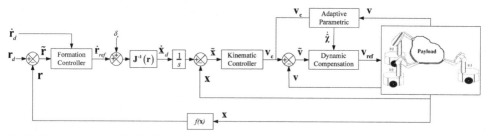

Fig. 3. Control system block diagram

5. Scalability for the cooperative control of multi-mobile manipulator

This Subsection proposes a way to generalize the control system associated to the coordinated cooperation of three mobile manipulators (virtual structure prism) to a coordinated cooperation of $n > 3$ mobile manipulators. Such proposition is based on the decomposition of a virtual 3-dimensional structure of n vertices into simpler components, i.e., $n - 2$ prisms. The idea is to take advantage of the control scheme proposed for a virtual prism to implement a coordinated cooperative control of $n > 3$ mobile manipulators using the same kinematics transformations presented in previous Section 3, thus not demanding to change the Jacobian (Figure 4).

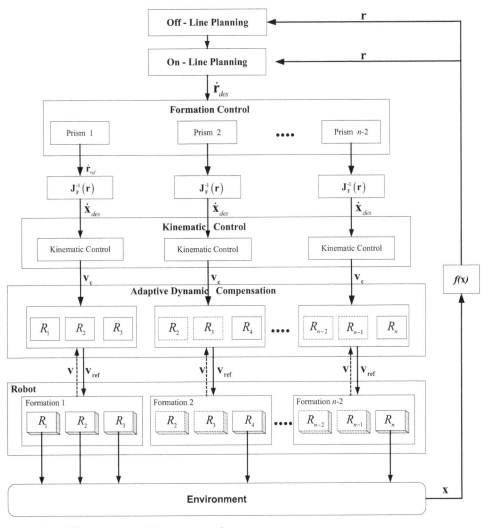

Fig. 4. Scalability in the multi-layer control

To do that, one should first label the mobile manipulators R_i, $i = 1,2,3...,n$ and determine the leader prism of the whole formation ($R_2R_1R_3$ or $R_3R_1R_2$, paying attention to the sequence **ABC** or **ACB**). After that, new prisms are formed with the remaining mobile manipulators, based on a simple algorithm: a new prism is formed with the last two mobile manipulators robots of the last prism already formed and the next mobile manipulator in the list of labelled mobile manipulators (in other words, $R_{j+1}R_jR_{j+2}$ or $R_{j+2}R_jR_{j+1}$ where $j = 1,2,...,n-2$ represents the current virtual structure prism). Additionally, from previous Section 6.3, a set of desired virtual structure variables $S_{F_j} = \begin{bmatrix} p_{F_j} & q_{F_j} & \beta_{F_j} & z_{1F_j} & z_{2F_j} & z_{3F_j} \end{bmatrix}$ is assigned to each virtual structure prism. Actually, the number of virtual structure variables is the same, but three of the variables has its value defined by the previous formation, i.e., $2(n-2)+1$, instead of $3(n-2)$, because it is assumed that $S_{F_j} = \begin{bmatrix} p_{F_{j-1}} & q_{F_j} & \beta_{F_j} & z_{1F_{j-1}} & z_{2F_{j-1}} & z_{3F_j} \end{bmatrix}$.

One point that deserves to be mentioned here is the control signals generated: there will always be a redundancy in the virtual structures with more than three mobile manipulators. For example, the mobile manipulators R_2 and R_3, in a virtual structures of four mobile manipulators, will receive control signals associated to the errors of the two virtual prisms ($R_2R_1R_3$ and $R_3R_2R_4$, for example). In this work, however, the implementation chosen is one in which the mobile manipulators R_{j+2} will receive control signals only from the controllers associated to the $j \geq 2$ virtual prisms, while the mobile manipulators R_1, R_2 and R_3 will receive the signals generated by the controller associated to the leader prism $(j = 1)$.

Remark 5: the proposed structure is also modular in the horizontal sense, i.e., it grows horizontally whenever a new robot is added to the formation.

Remark 6: the proposed structure is not centralized, since a controller is associated to each robot, except for the three first robots, which are governed by a single controller.

6. Controllers design

In this section it is presented the design of the controllers for the following control layers: Formation Control, Kinematic Control and Adaptive Dynamic Compensation. It is worth remark that both the kinematic control and adaptive dynamic compensation are performed separately for each mobile manipulator robot.

6.1 Formation controller

The Control Layer receives from the upper layer the desired formation pose and shape $\mathbf{r}_d = \begin{bmatrix} \mathbf{P}_{F_d} & \mathbf{S}_{F_d} \end{bmatrix}^T$ and its desired variations $\dot{\mathbf{r}}_d = \begin{bmatrix} \dot{\mathbf{P}}_{F_d} & \dot{\mathbf{S}}_{F_d} \end{bmatrix}^T$. It generates the pose and shape variation references $\dot{\mathbf{r}}_{ref} = \begin{bmatrix} \dot{\mathbf{P}}_{F_{ref}} & \dot{\mathbf{S}}_{F_{ref}} \end{bmatrix}^T$, where the subscripts d and *ref* represent the desired and reference signals, respectively. Defining the formation error as $\tilde{\mathbf{r}}(t) = \mathbf{r}_d(t) - \mathbf{r}(t)$ and taking its first time derivative, the following expression is obtained,

$$\dot{\tilde{\mathbf{r}}} = \dot{\mathbf{r}}_d - \dot{\mathbf{r}}. \tag{6}$$

Defining $\tilde{\mathbf{r}}(t) = 0$ as the control objective (an equilibrium point of the system), in order to prove its stability, it is proposed a controller in the sense of Lyapunov as follows. Defining the positive definite candidate function

$$V(\tilde{\mathbf{r}}) = \tfrac{1}{2}\tilde{\mathbf{r}}^{\mathrm{T}}\tilde{\mathbf{r}} > 0,$$

taking its first time derivative and replacing (6) and $\dot{\mathbf{r}}_{ref} = \mathbf{J_F}\dot{\mathbf{x}}_d$, assuming -by now- perfect velocity tracking, i.e., $\dot{\mathbf{r}} \equiv \dot{\mathbf{r}}_{ref}$, one gets

$$\dot{V}(\tilde{\mathbf{r}}) = \tilde{\mathbf{r}}^{\mathrm{T}}\dot{\tilde{\mathbf{r}}} = \tilde{\mathbf{r}}^{\mathrm{T}}\left(\dot{\mathbf{r}}_d - \mathbf{J_F}\dot{\mathbf{x}}_d\right).$$

Now, the proposed formation control law is defined as

$$\dot{\mathbf{x}}_d = \mathbf{J_F}^{-1}\left(\dot{\mathbf{r}}_d + K_1 \tanh\left(K_2\tilde{\mathbf{r}}\right)\right) = \mathbf{J_F}^{-1}\dot{\mathbf{r}}_{ref} \tag{7}$$

where K_1 and K_2 are diagonal positive gain matrix. Introducing (7) into the time derivative of $\dot{V}(\tilde{\mathbf{r}})$, it is obtained

$$\dot{V}(\tilde{\mathbf{r}}) = -\tilde{\mathbf{r}}^{\mathrm{T}}K_1 \tanh\left(K_2\tilde{\mathbf{r}}\right) < 0. \tag{8}$$

Thus, the equilibrium point is asymptotically stable, i.e. $\tilde{\mathbf{r}}(t) \to 0$ asymptotically.

Remark 7: Equation (7) represents the desired reference velocity vector for each mobile manipulator's end-effector.

Now, relaxing the assumption of perfect velocity tracking, it is considered a difference $\delta_{\dot{\mathbf{r}}}(t)$ between the desired and the real formation variations, such as $\delta_{\dot{\mathbf{r}}} = \dot{\mathbf{r}}_{ref} - \dot{\mathbf{r}}$. Then, (8) should be written as

$$\dot{V}(\tilde{\mathbf{r}}) = \tilde{\mathbf{r}}^{\mathrm{T}}\delta_{\dot{\mathbf{r}}} - \tilde{\mathbf{r}}^{\mathrm{T}}K_1 \tanh\left(K_2\tilde{\mathbf{r}}\right). \tag{9}$$

A sufficient condition for $\dot{V}(\tilde{\mathbf{r}})$ to be negative definite is,

$$\left|\tilde{\mathbf{r}}^{\mathrm{T}}K_1 \tanh\left(K_2\tilde{\mathbf{r}}\right)\right| > \left|\tilde{\mathbf{r}}^{\mathrm{T}}\delta_{\dot{\mathbf{r}}}\right|. \tag{10}$$

For large values of $\tilde{\mathbf{r}}$, it can be considered that: $K_1 \tanh\left(K_2\tilde{\mathbf{r}}\right) \approx K_1$. $\dot{V}(\tilde{\mathbf{r}})$ will be negative definite only if: $\|K_1\| > \|\delta_{\dot{\mathbf{r}}}\|$, thus making the errors $\tilde{\mathbf{r}}$ to decrease. Now, for small values of $\tilde{\mathbf{r}}$, it can be expressed: $K_1 \tanh\left(K_2\tilde{\mathbf{r}}\right) \approx K_1 K_2\tilde{\mathbf{r}}$, and (10) can be written as,

$$\|\tilde{\mathbf{r}}\| > \frac{\|\delta_{\dot{\mathbf{r}}}\|}{\lambda_{\min}(K_1)\lambda_{\min}(K_2)}$$

thus implying that the error $\tilde{\mathbf{r}}$ is bounded by,

$$\|\tilde{\mathbf{r}}\| \leq \frac{\|\boldsymbol{\delta}_{\tilde{\mathbf{r}}}\|}{\varsigma \lambda_{\min}(K_1)\lambda_{\min}(K_2)} \quad ; \qquad \text{with } 0<\varsigma<1 \tag{11}$$

Hence, with $\boldsymbol{\delta}_{\tilde{\mathbf{r}}}(t) \neq \mathbf{0}$, the formation error $\tilde{\mathbf{r}}(t)$ is ultimately bounded by (11).

6.2 Kinematic controller

This layer receives the desired position and velocities for each mobile manipulator $\mathbf{x}_d = \begin{bmatrix} \mathbf{h}_{d1} & \mathbf{h}_{d2} & \cdots & \mathbf{h}_{d\,i} & \cdots & \mathbf{h}_{d\,n} \end{bmatrix}^T$ and $\dot{\mathbf{x}}_d = \begin{bmatrix} \dot{\mathbf{h}}_{d1} & \dot{\mathbf{h}}_{d2} & \cdots & \dot{\mathbf{h}}_{d\,i} & \cdots & \dot{\mathbf{h}}_{d\,n} \end{bmatrix}^T$, respectively, and it generates the desired kinematic velocities $\mathbf{v}_c = \begin{bmatrix} \mathbf{v}_{c1} & \mathbf{v}_{c2} & \cdots & \mathbf{v}_{ci} & \cdots & \mathbf{v}_{cn} \end{bmatrix}^T$ for all robots. In other words, the desired operational motion of the n mobile manipulators is an application $\big(\mathbf{x}_d(t)\,|\,t \in [t_0,t_f]\big)$. Thus, the problem of control is to find the control vector of maneuverability $\big(\mathbf{v}_c(t)\,|\,t \in [t_0,t_f]\big)$ to achieve the desired operational motion (7). The corresponding evolution of the whole system is given by the actual generalized motion $\big(\mathbf{q}(t)\,|\,t \in [t_0,t_f]\big)$.

The design of the kinematic controller is based on the kinematic model of each mobile manipulator robot that belongs to the work team. The kinematic model (1) of the whole mobile manipulators can be represented by,

$$\dot{\mathbf{x}}(t) = \mathbf{J}(\mathbf{q})\mathbf{v}(t)$$

with

$$\dot{\mathbf{x}}(t) = \begin{bmatrix} \dot{\mathbf{h}}_1(t) & \dot{\mathbf{h}}_2(t) & \cdots & \dot{\mathbf{h}}_i(t) & \cdots & \dot{\mathbf{h}}_n(t) \end{bmatrix}^T \in \Re^{3.n},$$

$$\mathbf{v}(t) = \begin{bmatrix} \mathbf{v}_1(t) & \mathbf{v}_2(t) & \cdots & \mathbf{v}_i(t) & \cdots & \mathbf{v}_n(t) \end{bmatrix}^T \in \Re^{n.\delta_{n'}},$$

$$\mathbf{q}(t) = \begin{bmatrix} \mathbf{q}_1(t) & \mathbf{q}_2(t) & \cdots & \mathbf{q}_i(t) & \cdots & \mathbf{q}_n(t) \end{bmatrix}^T \in \Re^{n.n'}, \text{ and finally}$$

$$\mathbf{J}(\mathbf{q}) = \begin{bmatrix} \mathbf{J}_1(\mathbf{q}_1) & \mathbf{J}_2(\mathbf{q}_3) & \cdots & \mathbf{J}_i(\mathbf{q}_i) & \cdots & \mathbf{J}_n(\mathbf{q}_n) \end{bmatrix}^T \in \Re^{(3.n \times n.n')}.$$

where n' represents the dimensions of the generalized spaces associated to the robotic arms and to the mobile platforms of all mobile manipulators; i.e., $n' = n_1' + n_2' \ldots + n_i' \ldots + n_n'$ (see Remark 2).

It is worth noting, that the kinematic controller is performed separately for each robot. Hence, to obtain the vector of maneuverability $\mathbf{v}_i(t)$ that correspond to the i-th mobile manipulator, the right pseudo-inverse Jacobian matrix $\mathbf{J}_i(\mathbf{q}_i)$ is used

$$\mathbf{v}_i = \mathbf{J}_i^{\#} \dot{\mathbf{h}}_i \tag{12}$$

where, $\mathbf{J}_i^{\#} = \mathbf{W}_i^{-i}\mathbf{J}_i^T\left(\mathbf{J}_i\mathbf{W}_i^{-1}\mathbf{J}_i^T\right)^{-1}$, being $\mathbf{W}_i$ a definite positive matrix that weighs the control actions of the system,

$$\mathbf{v}_i = \mathbf{W}_i^{-i}\mathbf{J}_i^T\left(\mathbf{J}_i\mathbf{W}_i^{-1}\mathbf{J}_i^T\right)^{-1}\dot{\mathbf{h}}_i.$$

The following control law is proposed for the i-th mobile manipulator. It is based on a minimal norm solution, which means that, at any time, the mobile manipulator will attain its navigation target with the smallest number of possible movements,

$$\mathbf{v}_{ci} = \mathbf{J}_i^{\#}\left(\dot{\mathbf{h}}_{d_i} + \mathbf{L_{K}}_i\tanh\left(\mathbf{L_{K}}_i^{-1}\mathbf{K}_i\,\tilde{\mathbf{h}}_i\right)\right) + \left(\mathbf{I}_i - \mathbf{J}_i^{\#}\mathbf{J}_i\right)\mathbf{L_{D}}_i\tanh\left(\mathbf{L_{D}}_i^{-1}\mathbf{D}_i\,\Lambda_i\right) \tag{13}$$

where, $\dot{\mathbf{h}}_{d_i} = \begin{bmatrix} \dot{h}_{xd} & \dot{h}_{yd} & \dot{h}_{zd} \end{bmatrix}^T$ is the desired velocities vector of the end-effector, $\tilde{\mathbf{h}}_i$ is the control errors vector defined by $\tilde{\mathbf{h}}_i = [\mathbf{h}_{di} - \mathbf{h}_i]$, $\mathbf{K}_i$, $\mathbf{D}_i$, $\mathbf{L_{K}}_i$ and $\mathbf{L_{D}}_i$ are positive definite diagonal gain matrices that weigh the error vector $\tilde{\mathbf{h}}_i$ and vector Λ_i. The first term of the right hand side in (13) describes the primary task of the end effector which minimizes $\left\|\mathbf{v}_i - \mathbf{J}_i^{\#}\dot{\mathbf{h}}_i\right\|$. The second term defines self motion of the mobile manipulator in which the matrix $\left(\mathbf{I}_i - \mathbf{J}_i^{\#}\mathbf{J}_i\right)$ projects an arbitrary vector Λ_i onto the null space of the manipulator Jacobian $\mathcal{N}\left(\mathbf{J}_i\right)$ such that the secondary control objectives not affect the primary task of the end-effector. Therefore, any value given to Λ_i will have effects on the internal structure of the manipulator only, and will not affect the final control of the end-effector at all. By using this term, different secondary control objectives can be achieved effectively, as described in the next subsection.

In order to include an analytical saturation of velocities in the i-th mobile manipulator, the $\tanh(.)$ function, which limits the error in $\tilde{\mathbf{h}}_i$ and the magnitude of the vector Λ_i, is proposed. The expressions $\tanh\left(\mathbf{L_{K}}_i^{-1}\mathbf{K}_i\,\tilde{\mathbf{h}}_i\right)$ and $\tanh\left(\mathbf{L_{D}}_i^{-1}\mathbf{D}_i\,\Lambda_i\right)$ denote a component by component operation.

On the other hand, the behaviour of the control error of the i-th end-effector $\tilde{\mathbf{h}}_i$ is now analysed assuming -by now- perfect velocity tracking $\mathbf{v}_i \equiv \mathbf{v}_{ci}$. By substituting (13) in (12) it is obtained

$$\dot{\tilde{\mathbf{h}}}_i + \mathbf{L_{K}}_i\tanh\left(\mathbf{L_{K}}_i^{-1}\mathbf{K}_i\,\tilde{\mathbf{h}}_i\right) = 0. \tag{14}$$

For the stability analysis the following Lyapunov candidate function is considered $V\left(\tilde{\mathbf{h}}_i\right) = \frac{1}{2}\tilde{\mathbf{h}}_i^T\tilde{\mathbf{h}}_i > 0$. Its time derivative on the trajectories of the system is

$$\dot{V}\left(\tilde{\mathbf{h}}_i\right) = -\tilde{\mathbf{h}}_i^T\mathbf{L_{K}}_i\tanh\left(\mathbf{L_{K}}_i^{-1}\mathbf{K}_i\,\tilde{\mathbf{h}}_i\right) < 0,$$

which implies that the equilibrium point of the closed-loop (14) is asymptotically stable, thus the position error of the i-th end-effector verifies $\tilde{\mathbf{h}}_i(t) \to \mathbf{0}$ asymptotically with $t \to \infty$.

6.2.1 Secondary control objectives

A mobile manipulator is defined as redundant because it has more degrees of freedom than required to achieve the desired end-effector motion. The redundancy of such mobile manipulators can be effectively used for the achievement of additional performances such as: avoiding obstacles in the workspace and singular configuration, or to optimize various performance criteria. In this Chapter two different secondary objectives are considered: the avoidance of obstacles by the mobile platform and the singular configuration prevention through the system's manipulability control.

Manipulability

One of the main requirements for an accurate task execution by the robot is a good manipulability, defined as the robot configuration that maximizes its ability to manipulate a target object. Therefore, one of the secondary objectives of the control is to maintain maximum manipulability of the mobile manipulator during task execution. Manipulability is a concept introduced by Yoshikawa (1985) to measure the ability of a fixed manipulator to move in certain directions. Bayle and Fourquet (2001) present a similar analysis for the manipulability of mobile manipulators and extend the concept of manipulability ellipsoid as the set of all end-effector velocities reachable by robot velocities $\mathbf{v}_i$ satisfying $\|\mathbf{v}_i\| \leq 1$ in the Euclidean space. A global representative measure of manipulation ability can be obtained by considering the volume of this ellipsoid which is proportional to the quantity w called the *manipulability measure,*

$$w = \sqrt{\det\left(\mathbf{J}_i(\mathbf{q}_i)\mathbf{J}_i^T(\mathbf{q}_i)\right)} \tag{15}$$

Therefore, the mobile manipulator will have maximum manipulability if its internal configuration is such that maximizes the manipulability measure w.

Obstacle Avoidance

The main idea is to avoid obstacles which maximum height does not interfere with the robotic arm. Therefore the arm can follow the desired path while the mobile platform avoids the obstacle by resourcing to the null space configuration. The angular velocity and the longitudinal velocity of the mobile platform will be affected by a fictitious repulsion force. This force depends on the incidence angle on the obstacle α, and the distance d to the obstacle. This way, the following control velocities are proposed:

$$u_{obs} = Z^{-1}\left(k_{uobs}\left(d_o - d\right)\left[\pi/2 - |\alpha|\right]\right) \tag{16}$$

$$\omega_{obs} = Z^{-1}\left(k_{\omega obs}\left(d_o - d\right)sign(\alpha)\left[\pi/2 - |\alpha|\right]\right) \tag{17}$$

where, d_o is the radius which determines the distance at which the obstacle starts to be avoided, k_{uobs} and $k_{\omega obs}$ are positive adjustment gains, the sign function allows defining to which side the obstacle is to be avoided being $sign(0)=1$. Z represents the mechanical impedance characterizing the robot-environment interaction, which is calculated as $Z = Is^2 + Bs + K$ with I, B and K being positive constants representing, respectively, the effect of the inertia, the damping and the elasticity. The closer the platform is to the obstacle, the bigger the values of ω_{obs} and u_{obs}.

Taking into account the maximum manipulability (15) and the obstacle avoidance (16) and (17), the vector Λ_i is now defined as,

$$\Lambda_i = \left[-u_{iobs} \quad \omega_{iobs} \quad k_{vi1}\left(\theta_{i1d} - \theta_{i1}\right) \quad k_{vi2}\left(\theta_{i2d} - \theta_{i2}\right) \quad \cdots \quad k_{vina}\left(\theta_{inad} - \theta_{ina}\right) \right]^T \tag{18}$$

where $k_{vi}\left(\theta_{id} - \theta_i\right)$ – being $i = 1, 2, .., n_a$ and $k_{vi} > 0$ – are joint velocities proportional to the configuration errors of the mobile robotic arm, in such a way that the manipulator joints will be pulled to the desired θ_{id} values that maximize manipulability.

6.3 Adaptive dynamic compensation controller

The objective of the Adaptive Dynamic Compensation layer is to compensate for the dynamics of each mobile manipulator, thus reducing the velocity tracking error. This layer receives the desired velocities $\mathbf{v}_c = \begin{bmatrix} \mathbf{v}_{c1} & \mathbf{v}_{c2} & \cdots & \mathbf{v}_{ci} & \cdots & \mathbf{v}_{cn} \end{bmatrix}^T$ for all robots, and generates velocity references $\mathbf{v}_{ref} = \begin{bmatrix} \mathbf{v}_{ref1} & \mathbf{v}_{ref2} & \cdots & \mathbf{v}_{ref\,i} & \cdots & \mathbf{v}_{ref\,n} \end{bmatrix}^T$ to be sent to the mobile manipulators.

The adaptive dynamic compensation is performed separately for each robot. Each one of the controllers receive the velocities references $\mathbf{v}_{ci}(t)$ from the Kinemtic Control layer, and generates another velocities commands $\mathbf{v}_{ref\,i}(t)$ for the servos of the i-th mobile manipulator.

Thus, if there is no perfect velocity tracking of the i-th mobile manipulator, as assumed in Subsection 6.2, the will be a velocity error $\tilde{\mathbf{v}}_i(t)$ defined as, $\tilde{\mathbf{v}}_i = \mathbf{v}_{ci} - \mathbf{v}_i$ this velocity error motivates to design an adaptive dynamic compensation controller with a robust parameter updating law. It is consider the exact model of the i-th mobile manipulator without including disturbances (3),

$$\mathbf{M}_i \dot{\mathbf{v}}_i + \mathbf{C}_i \mathbf{v}_i + \mathbf{G}_i = \mathbf{v}_{refi} \tag{19}$$

Hence, the following control law is proposed for the i-th mobile manipulator is,

$$\mathbf{v}_{ref\,i} = \Phi_i \hat{\chi}_i = \Phi_i \chi_i + \Phi_i \tilde{\chi}_i = \mathbf{M}_i \sigma_i + \mathbf{C}_i \mathbf{v}_{ci} + \mathbf{G}_i + \Phi_i \tilde{\chi}_i \tag{20}$$

where $\Phi(\mathbf{q}, \mathbf{v}, \sigma) \in \mathfrak{R}^{\delta_{ni} \times l}$, $\chi_i = \begin{bmatrix} \chi_1 & \chi_2 & \cdots & \chi_l \end{bmatrix}$ and $\hat{\chi}_i = \begin{bmatrix} \hat{\chi}_1 & \hat{\chi}_2 & \cdots & \hat{\chi}_l \end{bmatrix}$ are respectively the unknown vector, real parameters vector and estimated parameters vector of the i-th mobile manipulator, whereas $\tilde{\chi}_i = \hat{\chi}_i - \chi_i$ represents the vector of parameter errors and $\sigma_i = \dot{v}_{ci} + L_{vi} \tanh\left(L_{vi}^{-1}K_{vi}\tilde{v}_i\right)$.

In order to obtain the closed-loop equation for the inverse dynamics with uncertain model (19) is equated to (20),

$$M_i \dot{v}_i + C_i v_i + G_i = M_i \sigma_i + C_i v_{ci} + G_i + \Phi_i \tilde{\chi}_i$$

$$M_i\left(\sigma_i - \dot{v}_i\right) = -C_i \tilde{v}_i - \Phi_i \tilde{\chi}_i \tag{21}$$

and next, σ_i is introduced in (21)

$$\dot{\tilde{v}}_i = -M_i^{-1}\Phi_i\tilde{\chi}_i - M_i^{-1}C_i\tilde{v}_i - L_{vi}\tanh\left(L_{vi}^{-1}K_{vi}\tilde{v}_i\right) \tag{22}$$

A Lyapunov candidate function is proposed as

$$V\left(\tilde{v}_i, \tilde{\chi}_i\right) = \tfrac{1}{2}\tilde{v}_i^T H_i M_i \tilde{v}_i + \tfrac{1}{2}\tilde{\chi}_i^T \gamma_i \tilde{\chi}_i \tag{23}$$

where $\gamma_i \in \Re^{l \times l}$ is a positive definite diagonal matrix and $H_i M_i$ is a symmetric and positive definite matrix. The time derivative of the Lyapunov candidate function on the system's trajectories is,

$$\dot{V}\left(\tilde{v}_i, \tilde{\chi}_i\right) = -\tilde{v}_i^T H_i M_i L_{vi} \tanh\left(L_{vi}^{-1}K_{vi}\tilde{v}_i\right) - \tilde{v}_i^T H_i C_i \tilde{v}_i - \tilde{v}_i^T H_i \Phi_i \tilde{\chi}_i + \tilde{\chi}_i^T \gamma_i \dot{\tilde{\chi}}_i + \tfrac{1}{2}\tilde{v}_i^T H_i \dot{M}_i \tilde{v}_i$$

Now, recalling that $M_i\left(q_i\right) = H_i^{-1}\left(\bar{M}_i + D_i\right)$ and $C_i\left(q_i, v_i\right) = H_i^{-1}\left(\bar{C}_i + P_i\right)$,

$$\dot{V}\left(\tilde{v}_i, \tilde{\chi}_i\right) = -\tilde{v}_i^T H_i M_i L_{vi} \tanh\left(L_{vi}^{-1}K_{vi}\tilde{v}_i\right) - \tilde{v}_i^T\left(\bar{C}_i + P_i\right)\tilde{v}_i - \tilde{v}_i^T H_i \Phi_i \tilde{\chi}_i + \tilde{\chi}_i^T \gamma_i \dot{\tilde{\chi}}_i + \tfrac{1}{2}\tilde{v}_i^T \dot{\bar{M}}_i \tilde{v}_i$$

Due to the well known skew-symmetric property of $\left(\dot{\bar{M}}_i - 2\bar{C}_i\right)$, $\dot{V}\left(\tilde{v}_i, \tilde{\chi}_i\right)$ reduces to,

$$\dot{V}\left(\tilde{v}_i, \tilde{\chi}_i\right) = -\tilde{v}_i^T H_i M_i L_{vi} \tanh\left(L_{vi}^{-1}K_{vi}\tilde{v}_i\right) - \tilde{v}_i^T P_i \tilde{v}_i - \tilde{v}_i^T H_i \Phi_i \tilde{\chi}_i + \tilde{\chi}_i^T \gamma_i \dot{\tilde{\chi}}_i \tag{24}$$

The following parameter-updating law is proposed for the adaptive dynamic compensation controller. It is based on a leakage term, or σ - modification (Kaufman et al., 1998; Sastry and Bodson, 1989). Reference (Nasisi and Carelli, 2003) presented an adaptive visual servo controller with σ -modification applied to a robot manipulator. By including such term, the robust updating law is obtained

$$\dot{\hat{\chi}}_i = \gamma_i^{-1}\Phi_i^T H_i \tilde{v}_i - \gamma_i^{-1}\Gamma_i \hat{\chi}_i \tag{25}$$

where $\Gamma_i \in \Re^{l \times l}$ is a diagonal positive gain matrix. Equation (25) is rewritten as

$$\dot{\hat{\chi}}_i = \gamma_i^{-1}\Phi_i^T H_i \tilde{v}_i - \gamma_i^{-1}\Gamma_i \tilde{\chi}_i - \gamma_i^{-1}\Gamma_i \chi_i \tag{26}$$

Let us consider that the dynamic parameters can vary, i.e., $\chi_i = \chi_i(t)$ and $\dot{\tilde{\chi}}_i = \dot{\hat{\chi}}_i - \dot{\chi}_i$. Substituting (23) in (24),

$$\dot{V}\left(\tilde{\mathbf{v}}_i, \tilde{\chi}_i\right) = -\tilde{\mathbf{v}}_i^{\mathsf{T}} \mathbf{H}_i \mathbf{M}_i \mathbf{L}_{\mathbf{v}i} \tanh\left(\mathbf{L}_{\mathbf{v}i}^{-1} \mathbf{K}_{\mathbf{v}i} \tilde{\mathbf{v}}_i\right) - \tilde{\mathbf{v}}_i^{\mathsf{T}} \mathbf{P}_i \tilde{\mathbf{v}}_i - \tilde{\chi}_i^{\mathsf{T}} \mathbf{\Gamma}_i \tilde{\chi}_i - \tilde{\chi}_i^{\mathsf{T}} \mathbf{\Gamma}_i \chi_i - \tilde{\chi}_i^{\mathsf{T}} \mathbf{Y}_i \dot{\chi}_i . \tag{27}$$

Considering small values of $\tilde{\mathbf{v}}_i$, then $\mathbf{L}_{\mathbf{v}i} \tanh\left(\mathbf{L}_{\mathbf{v}i}^{-1} \mathbf{K}_{\mathbf{v}i} \tilde{\mathbf{v}}_i\right) \approx \mathbf{K}_{\mathbf{v}i} \tilde{\mathbf{v}}_i$. The following constants are defined: $\upsilon_\Gamma = k_{max}(\mathbf{\Gamma}_i)$, $\upsilon_\gamma = k_{max}(\mathbf{Y}_i)$, $\mu_\Gamma = \chi(\mathbf{\Gamma}_i)$, $\mu_{MK_vP} = \chi_i(\mathbf{H}_i \mathbf{M}_i \mathbf{K}_{\mathbf{v}i}) + \chi_i(\mathbf{P}_i)$, where $\chi(\mathbf{Z}) = \sqrt{\lambda_{min}\left(\mathbf{Z}^{\mathsf{T}}\mathbf{Z}\right)}$ is the minimum singular value to $\mathbf{Z}$, $k_{max}(\mathbf{Z}) = \sqrt{\lambda_{max}\left(\mathbf{Z}^{\mathsf{T}}\mathbf{Z}\right)}$ denotes the maximum singular value of $\mathbf{Z}$, and $\lambda_{min}(.)$ and $\lambda_{max}(.)$ represent the smallest and the biggest eigenvalues of a matrix, respectively. Then, $\dot{V}$ can be rewritten as,

$$\dot{V}\left(\tilde{\mathbf{v}}_i, \tilde{\chi}_i\right) = -\mu_{MK_vP} \left\|\tilde{\mathbf{v}}_i\right\|^2 - \mu_\Gamma \left\|\tilde{\chi}_i\right\|^2 + \upsilon_\Gamma \left\|\tilde{\chi}_i\right\|\left\|\chi_i\right\| + \upsilon_\gamma \left\|\tilde{\chi}_i\right\|\left\|\dot{\chi}_i\right\| . \tag{28}$$

Considering $\zeta \in \Re^+$ in the difference square,

$$\left(\frac{1}{\zeta}\left\|\tilde{\chi}_i\right\| - \zeta\left\|\chi_i\right\|\right)^2 = \frac{1}{\zeta^2}\left\|\tilde{\chi}_i\right\|^2 - 2\left\|\tilde{\chi}_i\right\|\left\|\chi_i\right\| + \zeta^2\left\|\chi_i\right\|^2$$

can be written as,

$$\left\|\tilde{\chi}_i\right\|\left\|\chi_i\right\| \le \frac{1}{2\zeta^2}\left\|\tilde{\chi}_i\right\|^2 + \frac{\zeta^2}{2}\left\|\chi_i\right\|^2 . \tag{29}$$

By applying a similar reasoning with $\eta \in \Re^+$, it can be obtained

$$\left\|\tilde{\chi}_i\right\|\left\|\dot{\chi}_i\right\| \le \frac{1}{2\eta^2}\left\|\tilde{\chi}_i\right\|^2 + \frac{\eta^2}{2}\left\|\dot{\chi}_i\right\|^2 . \tag{30}$$

Substituting (30) and (29) in (28)

$$\dot{V}\left(\tilde{\mathbf{v}}_i, \tilde{\chi}_i\right) \le -\mu_{MK_vP} \left\|\tilde{\mathbf{v}}_i\right\|^2 - \mu_\Gamma \left\|\tilde{\chi}_i\right\|^2 + \upsilon_\Gamma\left(\frac{1}{2\zeta^2}\left\|\tilde{\chi}_i\right\|^2 + \frac{\zeta^2}{2}\left\|\chi_i\right\|^2\right) + \upsilon_\gamma\left(\frac{1}{2\eta^2}\left\|\tilde{\chi}_i\right\|^2 + \frac{\eta^2}{2}\left\|\dot{\chi}_i\right\|^2\right) \tag{31}$$

Equation (31) can be written in compact form as

$$\dot{V}\left(\tilde{\mathbf{v}}_i, \tilde{\chi}_i\right) \le -\varepsilon_1 \left\|\tilde{\mathbf{v}}_i\right\|^2 - \varepsilon_2 \left\|\tilde{\chi}_i\right\|^2 + \delta \tag{32}$$

where, $\varepsilon_1 = \mu_{MK_vP} > 0$, $\varepsilon_2 = \mu_\Gamma - \frac{\upsilon_\Gamma}{2\zeta^2} - \frac{\upsilon_\gamma}{2\eta^2} > 0$ and $\delta = \upsilon_\Gamma \frac{\zeta^2}{2}\left\|\chi_i\right\|^2 + \upsilon_\gamma \frac{\eta^2}{2}\left\|\dot{\chi}_i\right\|^2$, with ζ and η conveniently selected. Now, from the Lyapunov candidate function $V\left(\tilde{\mathbf{v}}_i, \tilde{\chi}_i\right) = \frac{1}{2}\tilde{\mathbf{v}}_i^{\mathsf{T}} \mathbf{H}_i \mathbf{M}_i \tilde{\mathbf{v}}_i + \frac{1}{2}\tilde{\chi}_i^{\mathsf{T}} \mathbf{Y}_i \tilde{\chi}_i$ it can be stated that

$$V \le \beta_1 \|\tilde{\mathbf{v}}_i\|^2 + \beta_2 \|\tilde{\chi}_i\|^2 \tag{33}$$

where $\beta_1 = \frac{1}{2}\vartheta_{\bar{M}}$, $\beta_2 = \frac{1}{2}\vartheta_\gamma$, $\vartheta_{\bar{M}} = k_{max}(\mathbf{H}_i\mathbf{M}_i)$, $\vartheta_\gamma = k_{max}(\mathbf{\gamma}_i)$. Then,

$$\dot{V} \le -\Lambda V + \delta \tag{34}$$

with $\Lambda = \left\{ \frac{\varepsilon_1}{\beta_1}, \frac{\varepsilon_2}{\beta_2} \right\}$. Since δ is bounded, (34) implies that $\tilde{\mathbf{v}}_i(t)$ and $\tilde{\chi}_i(t)$ are finally bounded. Therefore, the σ-modification term makes the adaptation law more robust at the expense of increasing the error bound. As δ is a function of the minimum singular value of the gain matrix $\mathbf{\Gamma}_i$ of the σ-modification term, and its values are arbitrary, then the error bound can be made small. Note that the proposed adaptive dynamic controller does not guarantee that $\tilde{\chi}_i(t) \to 0$ as $t \to \infty$. In other words, estimated parameters might converge to values that are different from their true values. Actually, it is not required that $\tilde{\chi}_i(t) \to 0$ in order to make $\tilde{\mathbf{v}}_i(t)$ converge to a bounded value.

Remark 8: Note that the updating law (25) needs the $\mathbf{H}_i$ matrix. This matrix includes parameters of the actuators, which can be easily known and remain constant. Therefore, this is not a relevant constraint within the adaptive control design.

6.4 Stability analysis considering $\tilde{\mathbf{h}}_i(t)$ and $\tilde{\mathbf{v}}_i(t)$

The behaviour of the tracking error of the end-effector $\mathbf{h}_i(t)$ of the i-th mobile manipulator is now analysed relaxing the assumption of perfect velocity tracking. Therefore, the (14) is now written as,

$$\dot{\tilde{\mathbf{h}}}_i + \mathbf{L}_{\mathbf{K}i} \tanh\left(\mathbf{L}_{\mathbf{K}i}^{-1}\mathbf{K}_i \, \tilde{\mathbf{h}}_i\right) = \mathbf{J}_i \tilde{\mathbf{v}}_i \tag{35}$$

where, $\tilde{\mathbf{v}}_i$ is the velocity error of the i-th mobile manipulator and $\mathbf{J}_i$ the Jacobian matrix. It is considered a Lyapunov candidate function $V(\tilde{\mathbf{h}}_i) = \frac{1}{2}\tilde{\mathbf{h}}_i^T\tilde{\mathbf{h}}_i$ and its time derivative,

$$\dot{V}(\tilde{\mathbf{h}}_i) = \tilde{\mathbf{h}}_i^T\mathbf{J}_i\tilde{\mathbf{v}}_i - \tilde{\mathbf{h}}_i^T\mathbf{L}_{\mathbf{K}i}\tanh\left(\mathbf{L}_{\mathbf{K}i}^{-1}\mathbf{K}_i \, \tilde{\mathbf{h}}_i\right).$$

A sufficient condition for $\dot{V}(\tilde{\mathbf{h}}_i)$ to be negative definite is

$$\left|\tilde{\mathbf{h}}_i^T\mathbf{L}_{\mathbf{K}i}\tanh\left(\mathbf{L}_{\mathbf{K}i}^{-1}\mathbf{K}_i \, \tilde{\mathbf{h}}_i\right)\right| > \left|\tilde{\mathbf{h}}_i^T\mathbf{J}_i\tilde{\mathbf{v}}_i\right|. \tag{36}$$

Following a similar analysis to the one in Section 6.1, it can be concluded that, if $\|\mathbf{L}_{\mathbf{K}i}\| > \|\mathbf{J}_i\|\|\tilde{\mathbf{v}}_i\|$ thus implying that the error $\tilde{\mathbf{h}}_i$ is bounded by,

$$\left\| \tilde{\mathbf{h}}_i \right\| \leq \frac{\left\| \mathbf{J}_i \tilde{\mathbf{v}}_i \right\|}{\varsigma \lambda_{\min}\left(\mathbf{K}_i\right)} \; ; \qquad \text{with } 0 < \varsigma < 1 \tag{37}$$

Hence, if $\mathbf{v}_i \neq \mathbf{v}_{ci}$ it is concluded that $\tilde{\mathbf{h}}_i(t)$ is ultimately bounded by (37).

Now finally, generalizing (35) for whole multi-layer control scheme, it is obtained

$$\dot{\tilde{\mathbf{x}}} + \mathbf{L_K}\tanh\left(\mathbf{L_K^{-1}}\mathbf{K}\,\tilde{\mathbf{x}}\right) = \mathbf{J}\,\tilde{\mathbf{v}} \tag{38}$$

where $\dot{\tilde{\mathbf{x}}} = \begin{bmatrix} \dot{\tilde{\mathbf{h}}}_1 & \dot{\tilde{\mathbf{h}}}_2 & \cdots & \dot{\tilde{\mathbf{h}}}_i & \cdots & \dot{\tilde{\mathbf{h}}}_n \end{bmatrix}^T$ represents the position error vector of all mobile manipulators, $\mathbf{K}$ and $\mathbf{L_K}$ are positive definite diagonal matrices defined as $\mathbf{K} = diag\left(\mathbf{K}_1, \mathbf{K}_2.., \mathbf{K}_i.., \mathbf{K}_n\right)$ and $\mathbf{L_K} = diag\left(\mathbf{L_{K1}}, \mathbf{L_{K2}}.., \mathbf{L_{Ki}}.., \mathbf{L_{Kn}}\right)$, respectively, while the function tanh(.) denote a component by component operation. By applying a similar reasoning as in (35), it is can concluded that $\tilde{\mathbf{x}}(t)$ is bounded by,

$$\left\| \tilde{\mathbf{x}} \right\| \leq \frac{\left\| \mathbf{J}\tilde{\mathbf{v}} \right\|}{\varsigma \lambda_{\min}\left(\mathbf{K}\right)} \; ; \qquad \text{with } 0 < \varsigma < 1 \tag{39}$$

For the case of perfect velocity tracking $\mathbf{v} \equiv \mathbf{v}_c$, i.e., $\tilde{\mathbf{v}} \equiv \mathbf{0}$, it is concluded that $\tilde{\mathbf{x}}(t) \to \mathbf{0}$, which implies that $\tilde{\mathbf{r}}(t) \to \mathbf{0}$ asymptotically with $t \to \infty$, thus accomplishing the control objective. Nevertheless, this is not always possible in real contexts, therefore if it is considered the velocity error $\tilde{\mathbf{v}}(t) \neq \mathbf{0}$, consequently it has that $\tilde{\mathbf{x}}(t) \neq \mathbf{0}$. This implies that the formation error is nonzero $\delta_{\tilde{\mathbf{r}}}(t) \neq \mathbf{0}$, i.e., the formation error $\delta_{\tilde{\mathbf{r}}}(t)$ is related with the errors $\tilde{\mathbf{x}}(t)$ and $\tilde{\mathbf{v}}(t)$. Therefore from (4), (5), (7) and (38) it is obtained the following error expression,

$$\delta_{\tilde{\mathbf{r}}}(t) = \mathbf{J_F}(\mathbf{x})\left[\mathbf{J}(\mathbf{q})\tilde{\mathbf{v}} - \mathbf{L_K}\tanh(\tilde{\mathbf{x}})\right] \tag{40}$$

Thus, it is concluded that the adaptive dynamic compensation reduces the velocity error $\tilde{\mathbf{v}}(t)$ and consequently the error $\tilde{\mathbf{x}}(t)$, hence formation error $\delta_{\tilde{\mathbf{r}}}(t)$ is also reduced. Finally, with this results and the conclusion previously obtained from (11), the adaptive dynamic compensation controller reduces the height limit of the formation error $\tilde{\mathbf{r}}(t)$.

7. Simulation result and discussion

In order to assess and discuss the performance of the proposed coordinated cooperative controller, it was developed a simulation platform for multi-mobile manipulators with Matlab interface, see the Figure 5. It is important mention that the developed simulator has incorporated the dynamic model of the robot. This is an online simulator, which allows users to view three-dimensional environment navigation of mobile manipulators. Ours simulation platform is based on the MRSiM platform presented by Brandao et al., 2008.

The i-th 6 DOF mobile manipulator used in the simulation is shown in Figure 5, which is composed by a non holonomic mobile platform, a laser rangefinder mounted on it, and a 3 DOF robotic arm. In order to illustrate the performance of the proposed multi-layer control scheme, several simulations were carried out for coordinated cooperative control of mobile manipulators. Most representative results are presented in this section.

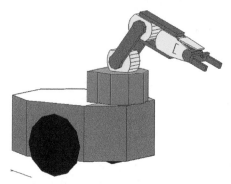

Fig. 5. Mobile manipulator robot used by the simulation platform

The simulation experiments consist of a team of three or more mobile mamipulator tracking a desired trajectory while carrying a payload cooperatively. Also, the obstacle avoidance and the singular configuration prevention through the system's manipulability control are considered in the simulations. It is assumed the existence of several obstacles, which have a maximum height that does not interfere with the robotic arms. That is, the obstacles only affect the platform navigation. Hence, the task for each mobile manipulator is divided into two subtasks. The first subtask: *to carry a payload cooperatively*; and the second subtask: *obstacle avoidance and the singular configuration prevention*.

It is important to remark that for all experiments in this section it was considered that there is an error of 30% in dynamic model parameters of each one mobile manipulator robot.

In the first one it is considered both the position and orientacion of the virtual structure. For this case, the desired positions of the arm joints are, Robot_1: $\theta_{1d} = 0[\text{rad}]$, $\theta_{2d} = -0.6065[\text{rad}]$, $\theta_{3d} = 1.2346[\text{rad}]$. Robot_2 and Robot_3: $\theta_{1d} = 0[\text{rad}]$, $\theta_{2d} = 0.6065[\text{rad}]$, and $\theta_{3d} = -1.2346[\text{rad}]$. Also, the desired virtual structure is selected as, $p_{F_d} = 1.75\,[\text{m}]$, $q_{F_d} = 1.2\,[\text{m}]$ and $\beta_{F_d} = 1.4\,[\text{rad}]$, $z_{1F_d} = 0.4\,[\text{m}]$ and $z_{2F_d} = z_{3F_d} = 0.3\,[\text{m}]$; while the desired trajectory for the prism centroid is described by:

$$x_{F_d} = 0.2\,t + 3.56 \,,$$

$$y_{F_d} = 3\cos(0.1\,t) + 3 \quad \text{and}$$

$$\psi_{F_d} = \tfrac{\pi}{2} + \gamma \,,$$

where $\gamma = \arctan\left(\frac{dy_{Fd}}{dt} \Big/ \frac{dx_{Fd}}{dt}\right)$. It is worth noting that this trajectory was chosen in order to excite the dynamics of the robots by changing their acceleration. The values of the gains matrices were adjusted considering the system performance with the dynamic compensation deactivated. After this gain setting, the values obtained were used in all simulations.

Figures 6 - 9 show the results of the simulation experiment. Figure 6 shows the stroboscopic movement on the X-Y-Z space. It can be seen that the proposed controller works correctly, where three mobile manipulators work in coordinated and cooperative form while transporting a common object. It can be noticed in Figure 6 that there are three triangles of different colours representing the upper side of a virtual structure. The yellow triangle represents the shape-position of the virtual structure that describes the end-effector of the mobile manipulator robots, while the pink triangle represents the location and shape of the upper side of the desired virtual structure, and the orange triangle indicates that both previously mentioned position-shapes are equal. While, figures 7 - 9 show that the control errors $\tilde{r}(t)$ achieve values close to zero. Figure 7 shows the errors of position and orientation of the virtual structure and in Figures 8 and 9 illustrate the errors of the virtual structure shape.

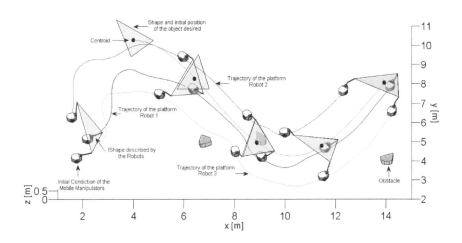

Fig. 6. Cooperative Coordinated Control of Mobile Manipulators

In order to show the scalability of the control structure for $n > 3$ mobile manipulators, the following simulations were carried out for coordinated cooperative control of multi-mobile manipulators.

In this context, the second simulation experiment shows a coordinated and cooperative control between four mobile manipulators. In this simulation the robots should navigate while carrying a payload, following a desired previously defined trajectory. It is considered a partially structured environment containing several obstacles in the trajectory.

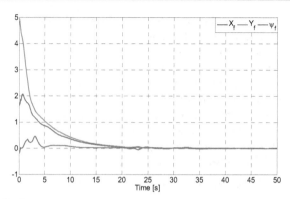

Fig. 7. Position and orientation errors

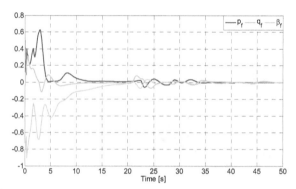

Fig. 8. Structure shape errors

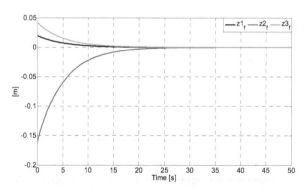

Fig. 9. Structure shape errors (heights)

Figures 10 show the stroboscopic movement on the X-Y-Z space, where it can be seen not only the good performance of the proposed cooperative control, but also the scalability of the proposal for multi-mobile manipulators. The pose and shape errors of the simulation experiments are shown in Figures 11 – 14.

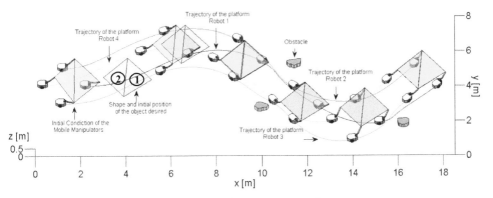

Fig. 10. Cooperative Coordinated Control of Mobile Manipulators

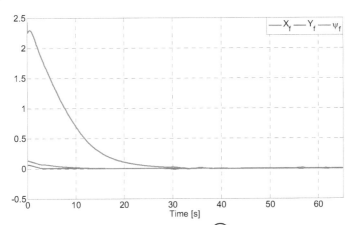

Fig. 11. Position and orientation errors of the triangle ①

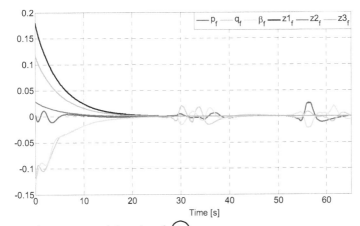

Fig. 12. Structure shape errors of the triangle ①

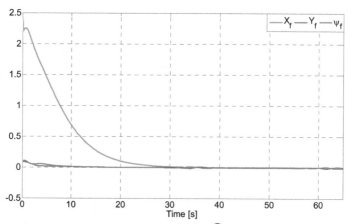

Fig. 13. Position and orientation errors of the triangle ②

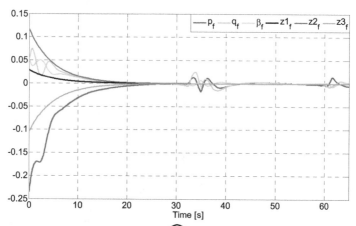

Fig. 14. Structure shape errors of the triangle ②

8. Conclusion

A multi-layer control scheme for adaptive cooperative coordinated control of $n \geq 3$ mobile manipulators, for transporting a common object was presented in this Chapter. Each control layer works as an independent module, dealing with a specific part of the problem of the adaptive coordinated cooperative control. On the other hand, the i-th mobile manipulator redundancy is used for the achievement of secondary objectives such as: the singular configuration prevention through the control of the system's manipulability and the avoidance of obstacles by the mobile platforms. Stability of the system has been analytically proved, concluding that the formation errors are ultimately bounded. The results, which were obtained by simulation, show a good performance of the proposed multi-layer control scheme.

9. References

Andaluz V. Roberti Flavio, Carelli Ricardo. (2010). Robust Control with Redundancy Resolution and Dynamic Compensation for Mobile Manipulators. *IEEE-ICIT Internacional Conference on Industrial Technology*, pp. 1449-1454, 2010.

Antonelli G., Arrichiello F., and Chiaverini S. (2008). The Entrapment/Escorting Mission. *IEEE Robotics & Automation Magazine*, Vol. 15, no. 1, pp. 22–29.

Bayle B. and J.-Y. Fourquet, "Manipulability Analysis for Mobile Manipulators", IEEE International Conference on Robots & Automation, pp. 1251-1256, 2001.

Bicchi A., Danesi A., Dini G., Porta S., Pallottino L., Savino I., and Schiavi R. (2008). "Heterogeneous Wireless Multirobot System. *IEEE Robotics & Automation Magazine*, Vol. 15, no. 1, pp. 62–70.

Brandao A. S., Carelli R., Sarcinelli-Filho M., and BastosFilho T. F. (2008). MRSiM: An environment for simulating mobile robots navigation. *Jornadas Argentinas de Robótica*, (written in Spanish).

Cao Y. U., Fukunaga A. S., and Kahng A. B. (1997). Cooperative mobile robotics: Antecedents and directions. *Autonomous Robots*, no. 4, pp. 1-23.

Consolini L., Morbidi F., Prattichizzo D., and Tosques M. (2007). A geometric characterization of leader-follower formation control. *IEEE International Conference on Robotics and Automation (ICRA07)*, pp. 2397-2402.

Fujii M., Inamura W., Murakami H. and Tanaka K. (2007). Cooperative Control of Multiple Mobile Robots Transporting a Single Object with Loose Handling. *IEEE International Conference on Robotics and Biomimetics*. pp. 816-822.

Hao Su and Venkat Krovi. (2008). Decentralized Dynamic Control of a Nonholonomic Mobile Manipulator Collective: a Simulation Study. *ASME Dynamic Systems and Control Conference*. pp.1-8.

Hirata Y., Kume Y., Sawada T., Wang Z., and Kosuge K. (2004). Handling of an object by multiple mobile manipulators in coordination based on casterlike dynamics. *IEEE Int. Conf. Robot. Autom.* Vol. 26, pp. 807–812.

Hougen D., Benjaafar S., B. J.C., Budenske J., Dvoraktt M., Gini M., French H., Krantz D., Li P., Malver F., Nelson B., Papanikolopoulos N., Rybski P., Stoeter S., Voyles R., and Yesin K. (2000). A miniature robotic system for reconnaissance and surveillance. *IEEE International Conference on Robotics and Automation (ICRA'00)*, pp. 501-507.

Jennings J. S., Whelan G., and Evans W. F. (1997). Cooperative search and rescue with a team of mobile robots. *8th IEEE-ICAR* , pp. 193-200.

Kaufman H., Barkana I. and Sobel K. (1998). Direct adaptive control algorithms, theory and applications. New York, USA: Springer-Verlag.

Khatib O., Yokoi K., Chang K., Ruspini D., Holmberg R., and Casal A. (1996). Vehicle/Arm Coordination and Multiple Mobile Manipulator. *Intelligent Robots and Systems*, Vol.2 pp. 546 – 553.

Martins F. N., Celeste W., Carelli R., Sarcinelli-Filho M., and Bastos-Filho T. (2008). An Adaptive Dynamic Controller for Autonomous Mobile Robot Trajectory Tracking. *Control Engineering Practice*. Vol. 16, pp. 1354-1363.

Mas I., Petrovic O., and Kitts C. (2008). Cluster space specification and control of a 3-robot mobile system. *IEEE International Conference on Robotics and Automation – ICRA*. pp. 3763–3768.

Nasisi O. and Carelli R. (2003). Adaptive servo visual robot control. *Robotics and Autonomous Systems*, 43 (1), pp. 51-78.

Sastry S., and Bodson M. (1989). *Adaptive control-stability, convergence and robustness.* Englewood Cliffs, NJ, USA: Prentice-Hall.

Sciavicco, L. and Siciliano, B. (2000). *Modeling and Control of Robot Manipulators.* Great Britain: Springer-Verlag London Limited.

Spong M. and Vidyasagar M., (1989). Robot Dynamics and Control.

Stouten B. and De Graaf A. (2004). Cooperative transportation of a large object-development of an industrial application. *IEEE - ICRA*, vol. 3, pp. 2450-2455.

Tanner H. G., Loizou S., and Kyriakopoulos K. J. (2003). Nonholonomic navigation and control of cooperating mobile manipulators. *IEEE Trans. Robot. Autom.* Vol.19, no. 1, pp. 53–64.

Thomas G. Sugar and Vijay Kumar. (2002). Control of Cooperating Mobile Manipulators. *IEEE Transactions on Robotics and Automation.* Vol. 18, no. 1, pp. 94-103.

Xin Chen and Yangmin Li. (2006). Cooperative Transportation by Multiple Mobile Manipulators Using Adaptive NN Control. *International Joint Conference on Neural Networks.* pp. 4193-4200.

Yamamoto Y., Hiyama Y., and Fujita A. (2004). Semi-autonomous reconfiguration of wheeled mobile robots in coordination. *IEEE Int. Conf. Robot. Autom.*, pp. 3456–3461.

Yasuhisa Hirata, Youhei Kume, Zhi-Dong Wang, Kszuhiro Kosiige. (2003). Coordinated Motion Control of Multiple Mobile Manipulators based on Virtual 3-D Caster. *International Conference on Robotics,Intelligent Systcms and Signal Processing.* pp. 19-24.

Yoshikawa T. (1985). Dynamic manipulability of robot manipulators. *International Journal of Robotic Systems*, pp. 113-124.

Yoshikawa T., (1990). *Foundations of Robotics: Analysis and Control.* The MIT Press, Cambridge, MA.

Zhijun Lia, b, Shuzhi Sam Gea, , and Zhuping Wanga. (2008). Robust adaptive control of coordinated multiple mobile manipulators. *Mechatronics.* Vol. 18, pp. 239-250.

Zhijun Li, Pey Y. Tao, Shuzhi S. Ge, Martin Adams, and Wijerupage S. W. (2009). Robust Adaptive Control of Cooperating Mobile Manipulators With Relative Motion. *IEEE Transactions on Systems, Man, and Cybernetics.*   Part B: Cybernetics, Vol. 39, no. 1, pp. 103-116.

8

PID Controller Tuning Based on the Classification of Stable, Integrating and Unstable Processes in a Parameter Plane

Tomislav B. Šekara and Miroslav R. Mataušek
University of Belgrade/Faculty of Electrical Engineering,
Serbia

1. Introduction

Classification of processes and tuning of the PID controllers is initiated by Ziegler and Nichols (1942). This methodology, proposed seventy years ago, is still actual and inspirational. Process dynamics characterization is defined in both the time and frequency domains by the two parameters. In the time domain, these parameters are the velocity gain K_v and dead-time L of an Integrator Plus Dead-Time (IPDT) model $G_{ZN}(s)=K_v\exp(-Ls)/s$, defined by the reaction curve obtained from the open-loop step response of a process. In the frequency domain these parameters are the ultimate gain k_u and ultimate frequency ω_u, obtained from oscillations of the process in the loop with the proportional controller $k=k_u$. The relationship between parameters in the time and frequency domains is determined by Ziegler and Nichols as

$$L=\frac{\pi}{2\omega_u},\ K_v=\varepsilon\frac{\omega_u}{k_u},\ \varepsilon=\varepsilon_{ZN}=\frac{4}{\pi}\ . \tag{1}$$

However, for the process $G_p(s)=G_{ZN}(s)$ in the loop with the proportional controller k, one obtains from the Nyquist stability criterion the same relationship (1) with $\varepsilon=1$. As a consequence, from (1) and the Ziegler-Nichols frequency response PID controller tuning, where the proportional gain is $k=0.6k_u$, one obtains the step response tuning $k=0.3\varepsilon\pi/(K_vL)$. Thus, for $\varepsilon=\varepsilon_{ZN}$ one obtains $k=1.2/(K_vL)$, as in (Ziegler & Nichols, 1942), while for $\varepsilon=1$ one obtains $k=0.9425/(K_vL)$, as stated in (Aström & Hägglund, 1995a). According to (1), the same values of the integral time $T_i=\pi/\omega_u$ and derivative time $T_d=0.25\pi/\omega_u$ are obtained in both frequency and time domains, in (Ziegler & Nichols, 1942) and from the Nyquist analysis. This will be discussed in more detail in Section 2.

Tuning formulae proposed by Ziegler and Nichols, were improved in (Hang et al., 1991; Aström & Hägglund, 1995a; 1995b; 2004). Besides the ultimate gain k_u and ultimate frequency ω_u of process $G_p(s)$, the static gain $K_p=G(0)$, for stable processes, and velocity gain $K_v = \lim_{s\to 0} sG_p(s)$, for integrating processes, are used to obtain better process dynamics characterization and broader classification (Aström et al.,1992). Stable processes are approximated by the First-Order Plus Dead-Time (FOPDT) model $G_{FO}(s)=K_p\exp(-Ls)/(Ts+1)$ and classified into four categories, by the normalized gain $\kappa_1=K_pk_u$ and normalized dead-time $\theta_1 = L/T$. Integrating processes are approximated by the Integrating First-Order Plus Dead-Time (IFOPDT) model $G_{IF}(s) = K_v \exp(-Ls)/(s(T_v s + 1))$ and classified into two categories, by the normalized gain $\kappa_2 = K_v k_u / \omega_u$ and normalized dead-time $\theta_2 = L/T_v$. The idea of classification proposed in (Aström et al., 1992) was to predict the achievable closed-loop performance and to make possible performance evaluation of feedback loops under closed-loop operating conditions.

In the present chapter a more ambitious idea is presented: define in advance the PID controller parameters in a classification plane for the purpose of obtaining a PID controller guaranteeing the desired performance/robustness tradeoff for the process classified into the desired region of the classification plane. It is based on the recent investigations related to: I) the process modeling of a large class of stable processes, processes having oscillatory dynamics, integrating and unstable processes, with the ultimate gain k_u (Šekara & Mataušek, 2010a; Mataušek & Šekara, 2011), and optimizations of the PID controller under constraints on the sensitivity to measurement noise, robustness, and closed-loop system damping ratio (Šekara & Mataušek, 2009,2010a; Mataušek & Šekara, 2011), II) the closed-loop estimation of model parameters (Mataušek & Šekara, 2011; Šekara & Mataušek, 2011b, 2011c), and III) the process classification and design of a new Gain Scheduling Control (GSC) in the parameter plane (Šekara & Mataušek, 2011a).

The motive for this research was the fact that the thermodynamic, hydrodynamic, chemical, nuclear, mechanical and electrical processes, in a large number of plants with a large number of operating regimes, constitutes practically an infinite batch of transfer functions $G_p(s)$, applicable for the process dynamics characterization and PID controller tuning. Since all these processes are nonlinear, some GSC must be applied in order to obtain a high closed-loop performance/robustness tradeoff in a large domain of operating regimes. A direct solution, mostly applied in industry, is to perform experiments on the plant in order to define GSC as the look-up tables relating the controller parameters to the chosen operating regimes. The other solution, more elegant and extremely time-consuming, is to define nonlinear models used for predicting accurately dynamic characteristics of the process in a large domain of operating regimes and to design a continuous GSC (Mataušek et al., 1996). However, both solutions are dedicated to some plant and to some region of operating regimes in the plant. The same applies for the solution defined by a nonlinear controller, for example the one based on the neural networks (Mataušek et al., 1999).

A real PID controller is defined by Fig. 1, with $C(s)$ and $C_{ff}(s)$ given by

$$C(s) = \frac{k_d s^2 + ks + k_i}{s(T_f s + 1)} F_C(s), \quad C_{ff}(s) = \frac{k_{ff} s + k_i}{s} F_C(s), \quad k_{ff} = bk, \quad F_C(s) \equiv 1, \quad 0 \le b. \tag{2}$$

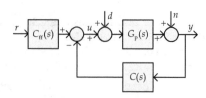

Fig. 1. Process $G_p(s)$ with a two-degree-of-freedom controller.

An effective implementation of the control system (2) is defined by relations

$$U(s) = \left(k\big(bR(s) - Y_f(s)\big) + \frac{k_i}{s}\big(R(s) - Y_f(s)\big) - k_d s Y_f(s) \right) F_C(s), \quad Y_f(s) = \frac{Y(s)}{T_f s + 1}, \tag{3}$$

for $F_C(s) \equiv 1$ as in (Panagopoulos et al., 2002; Mataušek & Šekara, 2011). When the proportional, integral, and derivative gains (k, k_i, k_d) and derivative (noise) filter time constant T_f are determined, parameter b can be tuned as proposed in (Panagopoulos et al., 2002). The PID controller (2), $F_C(s) \equiv 1$, can be implemented in the traditional form, when noise filtering affects the derivative term only if some conditions are fulfilled (Šekara & Mataušek, 2009). The derivative filter time constant T_f must be an integral part of the PID optimization and tuning procedures (Isaksson & Graebe, 2002; Šekara & Mataušek, 2009).

For $F_C(s)$ given by a second-order filter, one obtains a new implementation of the Modified Smith Predictor (Mataušek & Micić, 1996, 1999). The MSP-PID controller (3) guarantees better performance/robustness tradeoff than the one obtained by the recently proposed Dead-Time Compensators (DTC's), optimized under the same constraints on the sensitivity to measurement noise and robustness (Mataušek & Ribić, 2012).

Robustness is defined here by the maximum sensitivity M_s and maximum complementary sensitivity M_p. The sensitivity to measurement noise M_n, M_s, and M_p are given by

$$M_s = \max_\omega \left| \frac{1}{1 + L(i\omega)} \right|, \quad M_p = \max_\omega \left| \frac{L(i\omega)}{1 + L(i\omega)} \right|, \quad M_n = \max_\omega \left| C_{nu}(i\omega) \right|, \tag{4}$$

where $L(s)$ is the loop transfer function and $C_{nu}(s)$ is the transfer function from the measurement noise to the control signal. In the present chapter, the sensitivity to the high frequency measurement noise is used $M_n = M_{n\infty}$, where $M_{n\infty} = |\, C_{nu}(s)\, |_{s \to \infty}$.

2. Modeling and classification of stable, integrating, and unstable plants

A generalization of the Ziegler-Nichols process dynamics characterization, proposed by Šekara and Mataušek (2010a), is defined by the model

$$G_m(s) = \frac{A\omega_u \exp(-\tau s)}{s^2 + \omega_u^2 - A\omega_u \exp(-\tau s)} \frac{1}{k_u}, \quad \tau = \frac{\varphi}{\omega_u}, \quad A = \frac{\omega_u k_u G_p(0)}{1 + k_u G_p(0)}, \tag{5}$$

where φ is the angle of the tangent to the Nyquist curve $G_p(i\omega)$ at ω_u and $G_p(0)$ is the gain at the frequency equal to zero. Thus, for integrating processes $G_p(0) = \pm\infty$ and $A = \omega_u$. Adequate approximation of $G_p(s)$ by the model $G_m(s)$ is obtained for $\omega_u = \omega_\pi$, where $\arg\{G_p(i\omega_\pi)\} = -\pi$. It is demonstrated in (Šekara & Mataušek, 2010a; Mataušek & Šekara, 2011, Šekara & Mataušek, 2011a) that this extension of the Ziegler-Nichols process dynamics characterization, for a large class of stable processes, processes with oscillatory dynamics, integrating and unstable processes, guarantees the desired performance/robustness tradeoff if optimization of the PID controller, for the given maximum sensitivity M_s and given sensitivity to measurement noise M_n, is performed by applying the frequency response of the model (5) instead of the exact frequency response $G_p(i\omega)$.

Ziegler and Nichols used oscillations, defined by the impulse response of the system

$$G_p^*(s) = \frac{k_u G_p(s)}{1 + k_u G_p(s)}, \tag{6}$$

to determine k_u and ω_u, and to define tuning formulae for adjusting parameters of the P, PI and PID controllers, based on the relationship between the quarter amplitude damping ratio and the proportional gain k. Oscillations defined by the impulse response of the system (6) are used in (Šekara & Mataušek, 2010a) to define model (5), obtained from $G_m(s) \approx G_p(s)$ and the relation

$$\frac{k_u G_m(s)}{1 + k_u G_m(s)} = \frac{A \omega_u \exp(-\tau s)}{s^2 + \omega_u^2}. \tag{7}$$

Then, by analyzing these oscillations, it is obtained in (Šekara & Mataušek, 2010a) that amplitude $A = \omega_u \kappa / (1 + \kappa)$, $\kappa = k_u G_p(0)$, and dead-time τ is defined by ω_u and a parameter φ, given by

$$\varphi = \arg\left(\frac{\partial G_p(i\omega)}{\partial \omega} \right)\bigg|_{\omega = \omega_u}. \tag{8}$$

Other interpretation of amplitude $A = A_0$, obtained in (Mataušek & Šekara, 2011), is defined by

$$A_0 = \frac{2}{k_u} \left| \frac{\partial G_p(i\omega)}{\partial \omega} \right|^{-1}\bigg|_{\omega = \omega_u}. \tag{9}$$

Amplitudes A and A_0 are not equal, but they are closely related for stable and unstable processes, as demonstrated in (Mataušek & Šekara, 2011) and Appendix. Parameter A_0 is not used for integrating processes, since for these processes $A = \omega_u$.

The quadruplet $\{k_u, \omega_u, \varphi, A\}$ is used for classification of stable processes, processes with oscillatory dynamics, integrating and unstable processes in the ρ-φ parameter plane, defined by the normalized model (5), given by

$$G_n(s_n, \rho, \varphi) = \frac{\rho \exp(-\varphi s_n)}{s_n^2 + 1 - \rho \exp(-\varphi s_n)}, \quad s_n = \frac{s}{\omega_u}, \tag{10}$$

where $\rho = A / \omega_u$. From the Nyquist criterion it is obtained that the region of stable processes is defined by $0 < \varphi < \pi / \sqrt{\rho + 1}, 0 < \rho < 1$ (Šekara & Mataušek, 2011a). Integrating processes, since $A = \omega_u$, are classified as $\rho = 1, 0 < \varphi < \pi / \sqrt{2}$ processes, while unstable processes are outside these regions. It is demonstrated that a large test batch of stable and integrating processes used in (Aström & Hägglund, 2004) covers a small region in the ρ-φ plane.

To demonstrate that besides k_u and ω_u, parameters φ and $G_p(0)$ must by used for the classification of processes, Nyquist curves are presented in Fig. 2 for stable, integrating and unstable processes having the same values $k_u=1$ and $\omega_u=1$. For processes having also the same values of φ, the distinction of the Nyquist curves in the broader region around the critical point requires the information about gain $G_p(0)$, as demonstrated in Fig. 2-a. On the other hand, the results presented in Fig. 2-b to Fig. 2-d demonstrate that for the same values of k_u, ω_u, and $G_p(0)$ the distinction of the Nyquist curves in the region around the critical point is obtained by applying parameter φ. This fact confirms importance of parameter φ in process modeling for controller tuning, taking into account that optimization of the PID controller under constraints on the robustness is performed in the region around ω_u.

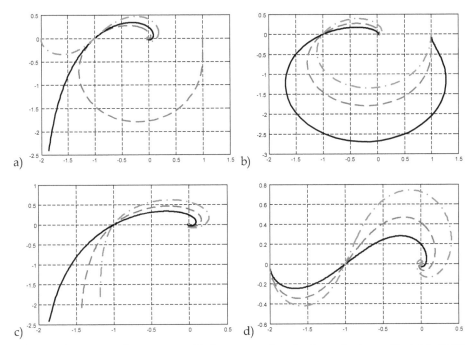

Fig. 2. Nyquist curves of processes with the same values $k_u=1$, $\omega_u=1$: a) $\varphi=\pi/4$, stable $G_p(0)=1$ (dashed), integrating $G_p(0)=\infty$ (solid), unstable $G_p(0)=-2$ (dashed-dotted); b) stable processes with $G_p(0)=1$, for $\varphi=\pi/4$ (dashed), $\varphi=\pi/6$ (solid), $\varphi=\pi/3$ (dashed-dotted); c) integrating processes with $\varphi=1$ (dashed), $\varphi=\pi/4$ (solid), $\varphi=1.2$ (dashed-dotted); d) unstable processes with $G_p(0)=-2$, for $\varphi=\pi/4$ (dashed), $\varphi=\pi/6$ (solid), $\varphi=\pi/3$ (dashed-dotted).

For the lag dominated process

$$G_{p1}(s) = 1 / \cosh \sqrt{2s} , \qquad (11)$$

and the corresponding models, the step and impulse responses, with the Nyquist curves around ω_u, are presented in Fig. 3. The models are Ziegler-Nichols IPDT model $G_{ZN}(s)=K_v\exp(-Ls)/s$ and model (5), with $A=\omega_u k_u G_p(0)/(1+k_u G_p(0))$ and $A=A_0$. The set-point and load disturbance step responses of this process, in the loop with the optimal PID controller (Mataušek & Šekara, 2011) and PID controller tuned as proposed by Ziegler and Nichols (1942), are compared in Fig. 4-a. In this case $k_u=11.5919$, $\omega_u=9.8696$ and $K_v=0.9251$,

L=0.1534. The PID controller tuned as proposed by Ziegler and Nichols is implemented in the form

$$U(s) = k\big(bR(s) - Y(s)\big) + \frac{k_i}{s}\big(R(s) - Y(s)\big) - \frac{k_d s}{T_f s + 1}Y(s), \; b = 0, k_i = \frac{k}{T_i}, k_d = kT_d, T_f = \frac{T_d}{N_d}, \quad (12)$$

where k=0.6k_u, T_i=π/ω_u, T_d= $\pi/(4\omega_u)$, for the frequency domain ZN tuning (ZN PID1). For the time domain ZN tuning (ZN PID2) the parameters are k=1.2$/(K_v L)$, T_i=2L, T_d=L/2, or, as suggested by the earlier mentioned Nyquist analysis, proportional gain k is adjusted to k=0.943$/(K_v L)$, denoted as the modified time domain ZN tuning (ZN ModifPID2). In $M_n = (N_d + 1)|k|$ parameter N_d is adjusted to obtain the same value of M_n=76.37 used in the constrained optimization of the PID in (3), $F_C(s)\equiv 1$, where M_n= $|k_d| / T_f$.

Parameters of the PID controllers and performance/robustness tradeoff are compared in Table 1. It is impressive that Ziegler and Nichols succeeded in defining seventy years ago an excellent experimental tuning for the process $G_{p1}(s)$, which is an infinite-order system that can be represented in simulation by the following high-order system $G_{p1}(s) \approx \exp(-Ls) / \Pi_{k=1}^{20}(T_k s + 1)$, L=0.01013 (Mataušek & Ribić, 2009). Also, it should be noted here, that Ziegler and Nichols succeeded seventy years ago in obtaining an excellent tuning with the IPDT model defined by K_v=0.9251, L=0.1534, which is an extremely crude approximation of the real impulse response of the process $G_{p1}(s)$, as in Fig. 3-b.

Tuning method	k	k_i	k_d	T_f	N_d	IAE	M_n	M_s	M_p
optPID	6.5483	18.4321	0.6345	0.0094	-	0.0609	76.37	2.00	1.45
ZN PID1	6.9551	21.8502	0.5535	0.0080	9.980	0.0538	76.37	2.20	1.72
ZN PID2	8.4560	27.5621	0.6486	0.0096	8.031	0.0429	76.37	2.82	2.23
ZN ModifPID2	6.6450	21.6592	0.5097	0.0073	10.49	0.0587	76.37	2.16	1.78

Table 1. Process $G_{p1}(s)$: comparison of the optimization (optPID) and the Ziegler-Nichols tuning in the frequency domain (ZN PID1) and time domain (ZN PID2, ZN ModifPID2).

The Nyquist curves of $G_{p1}(s)$, $G_{m1}(s)$, and $G_{m2}(s)$ are almost the same around ω_u. This is important since the PID controller optimization, based on the experimentally determined frequency response of the process, under constraints on M_s or on M_s and M_p, is performed around the ultimate frequency ω_u. Amplitudes A and A_0 are closely related for the stable and unstable processes, as demonstrated in (Mataušek & Šekara, 2011) and Appendix. For integrating processes A=ω_u. This means, that the Ziegler–Nichols parameters k_u and ω_u, and the Šekara-Mataušek parameters φ and A=A_0, for the stable and unstable processes, and A=ω_u, for integrating processes, constitute the minimal set of parameters, measurable in the frequency domain, necessary for obtaining PID controller tuning for the desired performance/robustness tradeoff. This will be demonstrated in the subsequent sections.

3. Optimization of PI/PID controllers under constraints on the sensitivity to measurement noise, robustness, and closed-loop system damping ratio

PID controllers are still mostly used control systems in the majority of industrial applications (Desborough & Miller, 2002) and "it is reasonable to predict that PID control

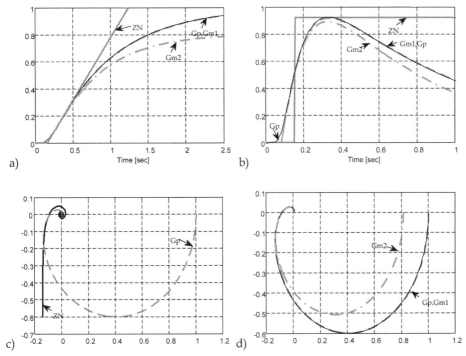

Fig. 3. Process $G_{p1}(s)$, denoted as (G_p), and models $G_{mj}(s)$, $j=1,2$, $k_u=11.5919$, $\omega_u=9.8696$, $\tau=0.0796$ for $A=9.0858$ (G_{m1}) and $A=A_0=8.9190$ (G_{m2}), and $G_{ZN}(s)=K_v\exp(-Ls)/s$, $K_v=0.9251$, $L=0.1534$ (ZN): a) step responses, b) impulse responses, c) Nyquist curves of $G_{p1}(s)$ and $G_{ZN}(s)$, d) Nyquist curves of $G_{p1}(s)$, $G_{m1}(s)$ and $G_{m2}(s)$ are almost the same around ω_u.

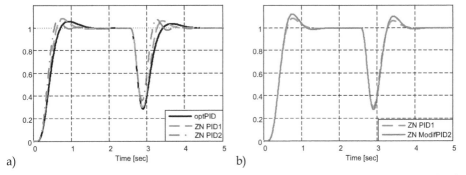

Fig. 4. Comparison of the optimization and the Ziegler-Nichols (ZN) tuning. Process $G_{p1}(s)$ in the loop with the optPID or ZN PID, tuned by using the rules: frequency domain (ZN PID1), time domain (ZN PID2), and time domain with the modified proportional gain $k=0.943/(K_vL)$ (ZN ModifPID2). In all controllers $b=0$ and $D(s)=-5\exp(-2.5s)/s$.

will continue to be used in the future" (Aström & Hägglund, 2001). They operate mostly as regulators (Aström & Hägglund, 2001) and rejection of the load step disturbance is of primary importance to evaluate PID controller performance under constraints on the robustness (Shinskey, 1990), measured by the Integrated Absolute Error (IAE). Inadequate tuning and sensitivity to measurement noise are the reasons why derivative action is often excluded in the industrial process control. This is the main reason why PI controllers predominate (Yamamoto & Hashimoto, 1991). However, for lag-dominated processes, processes with oscillatory dynamics and integrating/unstable processes PID controller guarantees considerably better performance than PI controller, if adequate tuning of the PID controller is performed (Mataušek & Šekara, 2011). Moreover, PID controller is a "prerequisite for successful advanced controller implementation" (Seki & Shigemasa, 2010).

Besides PI/PID controllers, in single or multiple loops (Jevtović & Mataušek, 2010), only Dead-Time Compensators (DTC) are used in the process industry with an acceptable percentage (Yamamoto & Hashimoto, 1991). They are based on the Smith predictor (Smith, 1957; Mataušek & Kvaščev, 2003) or its modifications. However, the area of application of PID controllers overlaps deeply with the application of DTC's, as confirmed by the Modified Smith Predictor, which is a PID controller in series with a second-order filter, applicable to a large class of stable, integrating and unstable processes (Mataušek & Ribić, 2012).

Optimization of the performance may by carried out under constraints on the maximum sensitivity to measurement noise M_n, the maximum sensitivity M_s and maximum complementary sensitivity M_p, as done in (Mataušek & Ribić, 2012). In this case it is recommended to use some algorithm for global optimization, such as Particle Swarm Optimization algorithm (Rapaić, 2008), requiring good estimates of the range of unknown parameters. Other alternatives, presented here, are recently developed in (Šekara & Mataušek, 2009, 2010a; Mataušek & Šekara, 2011). For the PID controller (3), for $F_C(s) \equiv 1$ defined by four parameters k, k_i, k_d and T_f, optimization under constraints on M_n and M_s is reduced in (Šekara & Mataušek, 2009) to the solution of a system of three algebraic equations with adequate initial values of the unknown parameters. The adopted values of M_n and M_s are satisfied exactly for different values of ζ_z. Thus, by repeating calculations for a few values of the damping ratio of the controller zeros ζ_z in the range $0.5 \leq \zeta_z$, the value of ζ_z corresponding to the minimum of IAE is obtained. Optimization methods from (Šekara & Mataušek, 2009) are denoted as max(k) and max(ki) methods.

The improvement of the max(k) method is proposed in (Šekara & Mataušek, 2010a). It consists of avoiding repetition of calculations for different values of ζ_z in order to obtain the minimal value of the IAE for a desired value of M_s. In this method, denoted here as method optPID, the constrained optimization is based on the frequency response of model (5).

For the PI optimization, an improvement of the performance/robustness tradeoff is obtained by applying the combined performance criterion $J_c = \beta k_i + (1-\beta)\omega$ (Šekara & Mataušek, 2008). Thus, one obtains

$$\max_{k_i, \omega} J_c \, , \tag{13}$$

$$F(\omega, k, k_i) = 0 \, , \ \partial F(\omega, k, k_i) / \partial \omega = 0 \, , \tag{14}$$

where $0 \leq \omega < \infty$ and β is a free parameter in the range $0 < \beta \leq 1$. The calculations are repeated for a few values of β, in order to find β corresponding to the minimum of IAE. The optimization in this method, denoted here as opt2, is performed for the desired value of M_s. For $\beta=1$ one obtains the same values of parameters k and k_i as obtained by the method proposed in (Aström et al., 1998), denoted here as opt1.

The most general is the new tuning and optimization procedure proposed in (Mataušek & Šekara, 2011). Besides the tuning formulae, the optimization procedure is derived. For the PID and PI controllers it requires only obtaining the solution of two nonlinear algebraic equations with adequate initial values of the unknown parameters. PID optimization is performed for the desired closed-loop system of damping ratio ζ and under constraints on M_n and M_s. Thus, for $\zeta=1$ the critically damped closed-loop system response is obtained. PI optimization is performed under constraint on M_s for the desired value of ζ. The procedure proposed in (Mataušek & Šekara, 2011) will be discussed here in more details, since it is entirely based on the concept of using oscillators (6)-(7) for dynamics characterization of the stable processes, processes having oscillatory dynamics, integrating and unstable processes. The method is derived by defining a complex controller $C(s)=k_u(1+C^*(s))$, where the controller $C^*(s)$, given by

$$C^*(s) = \frac{s^2 + \omega_u^2}{A\omega_u \Lambda(s)} \frac{E(s)/\Lambda(s)}{1 - E(s)\exp(-\tau s)/\Lambda(s)^2}, \quad E(s) = \eta_2 s^2 + \eta_1 s + 1, \quad \Lambda(s) = \lambda^2 s^2 + 2\zeta\lambda s + 1, \quad (15)$$

is obtained by supposing that in Fig. 1 process $G_p(s)$ is defined by oscillator $G_p^*(s)$ in (6), approximated by (7). Complex controller $C(s)=k_u(1+C^*(s))$ is defined by the parameters k_u, ω_u, τ, A and by the two tuning parameters λ and ζ, with the clear physical interpretation. Parameter λ is proportional to the desired closed-loop system time constant. Parameter ζ is the desired closed-loop system damping ratio. Then, by applying Maclaurin series expansion, the possible internal instability of the complex controller $C(s)$ is avoided and parameters of PID controller $C(s)$ in Fig. 1 are obtained, defined by:

$$T_f = \frac{\eta_2 - \beta_2(\eta_1 - \beta_2) - \beta_3 + 1/\omega_u^2}{\beta_2 - \eta_1 - (1 - M_n/|k_u|)/\beta_1}, \quad (16)$$

$$k = k_u\left(\beta_1(T_f + \eta_1 - \beta_2) + 1\right), k_i = k_u\beta_1, \quad (17)$$

$$k_d = k_u\beta_1\left(\eta_2 + (T_f - \beta_2)(\eta_1 - \beta_2) - \beta_3 + 1/\omega_u^2\right) + k_u T_f. \quad (18)$$

Parameters η_1, η_2, β_1, β_2 and β_3, from (Mataušek & Šekara, 2011), depends on λ, ζ and k_u, ω_u, τ, A. They are given in Appendix. Generalization of this approach is presented in (Šekara & Trifunović, 2010; Šekara et al., 2011).

For the desired closed-loop damping ratio $\zeta=1$, $\lambda=1/\omega_u$, and for

$$T_f = 1/(N\omega_u), \quad (19)$$

one obtains (Mataušek & Šekara, 2011) the PID tuning that guarantees set-point and load disturbance step responses with negligible overshoot for a large class of stable processes, processes with oscillatory dynamics, integrating and unstable processes. Tuning formulae defined by (17)-(19) are denoted here as method tunλ_u. Absolute value of the Integrated Error (IE), approximating almost exactly the obtained IAE, is given by $|IE|=1/(|k_u|\beta_1)$. Here the value $T_f=1/(10\omega_u)$ is used, as in (Mataušek & Šekara, 2011).

To demonstrate the relationship between PID controller, tuned by using the method tunλ_u, and complex controller $C(s)=k_u(1+C^*(s))$, obtained for $\lambda=1/\omega_u$ and $\zeta=1$, the frequency responses of these controllers, tuned for the process

$$G_{p2}(s) = 1/(s+1)^4,\qquad(20)$$

are presented in Fig. 5-a. For this process, parameters k_u, ω_u, τ, A, ρ and φ are given in Appendix. The load disturbance unite step responses, obtained for $G_{p2}(s)$ in the loop with the PID controller and complex controller $C(s)$, are presented in Fig. 5-b. Further details about the relationship between these controllers are presented in (Mataušek & Šekara, 2011; Trifunović & Šekara, 2011; Šekara et al., 2011).

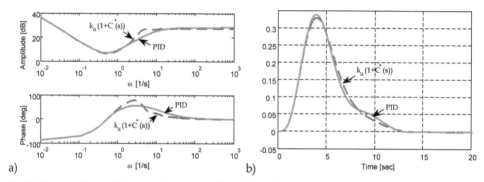

a)   b)

Fig. 5. Comparison of the complex controller $C(s)=k_u(1+C^*(s))$ with PID controller, both tuned for $G_{p2}(s)$: a) Bode plots of the controllers and b) the load unite step disturbance responses of $G_{p2}(s)$ in the loop with these controllers.

By applying tuning formulae (17)-(19), the desired closed-loop damping ratio $\zeta=1$ is obtained with the acceptable values of maximum sensitivity M_s and maximum sensitivity to measurement noise M_n. However, when a smaller value of M_n is required for a desired value of M_s and the desired closed-loop damping ratio ζ, the other possibility is to determine the closed-loop time constant λ and the corresponding ω_0, by using (16)-(18) and by solving two algebraic equations:

$$\left|1+C(i\omega)G_m(i\omega)\right|^2 -1/M_s^2 = 0,\qquad(21)$$

$$\partial\left(\left|1+C(i\omega)G_m(i\omega)\right|^2\right)/\partial\omega = 0.\qquad(22)$$

In this case, the PID controller in (3), $F_C(s)\equiv1$, is obtained for the desired critical damping ratio $\zeta=1$ of the closed-loop system and the desired values of M_n and M_s. This is the unique possibility of the procedure (16)-(18) and (21)-(22) proposed in (Mataušek & Šekara, 2011). Moreover, by repeating the calculations for a few values of ζ, the value of ζ is obtained guaranteeing, for desired M_n and M_s, almost the same value of the IAE as obtained by the constrained PID optimization based on the exact frequency response $G_p(i\omega)$. This PID optimization method is denoted here as the method opt2A, when the quadruplet $\{k_u, \omega_u, \varphi, A\}$ is used, or opt2A$_0$, when the quadruplet $\{k_u, \omega_u, \varphi, A_0\}$ is used. It should be noted here, that for $k_d=0$ and $T_f=0$, by relations (17) and (21)-(22) a new effective constrained PI controller optimization is obtained, denoted here as opt3. It is successfully compared (Mataušek & Šekara, 2011) with the procedure proposed in (Aström et al., 1998), opt1.

Now, tuning defined by (17)-(19) with $N=10$, $\lambda=1/\omega_u$ and $\zeta=1$, method tunλ_u, will be compared with the optimization defined by (16)-(18), (21)-(22), method opt2A. Both procedures guarantee desired critical damping $\zeta=1$, however only the second one guarantees the desired values of M_n and M_s. Thus, for $\zeta=1$ and for the maximum sensitivity M_s obtained by applying method tunλ_u, the smaller value of sensitivity to measurement noise M_n will be used by applying PID optimization method opt2A. The results of this analysis are presented in Table 2 and Fig. 6. As in Table 1, controller is tuned by using the model $G_m(s)$ in (5) and then applied to processes $G_{p3}(s)$ to obtain IAE, M_s and M_p, where

$$G_{p3}(s) = \frac{1.507(3.42s+1)(1-0.816s)}{(577s+1)(18.1s+1)(0.273s+1)(104.6s^2+15s+1)}. \tag{23}$$

Lower value of IAE is obtained, for almost the same robustness, by using higher value of the sensitivity to measurement noise. However, for the lower value of M_n the controller and, as a result, the actuator activity is considerably reduced. Thus, the comparison of the IAE, obtained by the PID controllers with the same robustness, is meaningless if the sensitivity to measurement noise M_n is not specified, as demonstrated in Fig. 6. This fact is frequently ignored.

method	λ	k	k_i	k_d	T_f	IAE	M_n	M_s	M_p
tunλ_u	17.3310	22.3809	0.2778	377.2723	1.7331	3.62	217.7	2.14	1.58
opt2A	20.4849	18.2791	0.1944	345.5996	5.0893	5.17	67.91	2.12	1.51

Table 2. Process $G_{p3}(s)$ in the loop with the PID controllers. Tuning method (17)-(19), tunλ_u and optimization (16)-(18), (21)-(22), opt2A for $\zeta=1$.

Concluding this section, the constrained PI/PID controller optimization methods proposed in (Mataušek & Šekara, 2011) is compared with the constrained PID controller optimization method proposed in (Šekara & Mataušek, 2010a), optPID1, and the constrained PI controller optimization method proposed in (Šekara & Mataušek, 2008), opt2. The test batch of stable processes, processes having oscillatory dynamics, integrating and unstable processes used in this analysis is defined by transfer functions $G_{p1}(s)$, $G_{p2}(s)$, $G_{p3}(s)$ and

$$G_{p4}(s) = \frac{e^{-5s}}{(s+1)^3}, \quad G_{p5}(s) = \frac{e^{-s}}{9s^2+0.24s+1}, \tag{24}$$

$$G_{p6}(s) = \frac{e^{-5s}}{s(s+1)(0.5s+1)(0.25s+1)(0.125s+1)}, \quad G_{p7}(s) = \frac{2e^{-5s}}{(10s-1)(2s+1)}, \tag{25}$$

with parameters k_u, ω_u, τ, A, A_0, ρ, φ presented in Appendix. Comparison of the methods for PID controller tuning is presented in Table 3. Comparison of the methods for PI controller tuning is presented in Table 4 and Fig. 7.

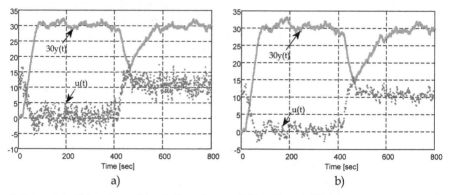

a)   b)

Fig. 6. Set-point, $R(s)=1/s$, and load disturbance, $D(s)=-10\exp(-400s)/s$, step responses. $G_{p3}(s)$ and PID controllers tuned by: a) tunλ_u, $b=0.5$; b) opt2A, $b=0.6$. Measurement noise is obtained by passing uniform random noise ±1 through a low-pass filter $F(s)=0.5/(10s+1)$.

Process/ method	k	k_i	k_d	T_f	IAE	M_n	M_s	M_p	ζ_z	ζ
G_{p3}/max(k)	17.0778	0.2372	320.06	4.7131	4.83	67.91	2.00	1.56	0.98	-
G_{p3}/optPID	17.1037	0.2303	315.14	4.6407	4.84	67.91	2.00	1.54	-	-
G_{p3}/opt2A	17.1994	0.1788	316.59	4.6621	5.62	67.91	2.00	1.41	-	1
G_{p3}/opt2A$_0$	16.9411	0.2670	312.65	4.6040	4.87	67.91	2.00	1.69	-	0.75
G_{p3}/opt2A	16.8802	0.2083	268.32	3.9513	4.92	67.91	2.00	1.59	-	0.80
G_{p5}/max(ki)	-0.3090	0.0654	0.8640	1.7597	21.17	0.49	2.00	1.03	0.65	-
G_{p5}/optPID	-0.3032	0.0651	0.8280	1.6864	21.87	0.49	2.00	1.07	-	-
G_{p5}/opt2A	-0.4139	0.0336	0.9398	1.9140	30.04	0.49	2.00	1.04	-	1
G_{p5}/opt2A$_0$	-0.3369	0.0583	0.8948	1.8223	20.29	0.49	2.00	1.02	-	0.65
G_{p5}/opt2A	-0.3542	0.0528	0.8860	1.8044	20.30	0.49	2.00	1.02	-	0.70
G_{p6}/max(k)	0.1177	0.0063	0.3961	0.8353	207.22	0.47	2.00	1.76	1.18	-
G_{p6}/optPID	0.1181	0.0054	0.3736	0.7878	208.65	0.47	2.00	1.63	-	-
G_{p6}/opt2A	0.1133	0.0043	0.2373	0.5003	234.73	0.47	2.01	1.60	-	1
G_{p6}/opt2A	0.1160	0.0043	0.2709	0.5712	233.50	0.47	2.01	1.55	-	1.05
G_{p7}/max(k)	0.8608	0.0158	3.3101	0.1418	73.50	23.35	3.61	3.39	1.88	-
G_{p7}/optPID	0.8609	0.0150	3.2946	0.1411	75.00	23.35	3.61	3.33	-	-
G_{p7}/opt2A	0.8543	0.0106	2.9385	0.1258	93.96	23.35	3.54	3.18	-	1.3
G_{p7}/opt2A$_0$	0.8060	0.0093	2.3759	0.1017	107.41	23.35	3.61	3.77	-	1.1

Table 3. PID controllers, obtained by applying model $G_m(s)$ and tuning methods: max(k), max(ki); (31)-(35) optPID; (16)-(18), (21)-(22) opt2A and opt2A$_0$.

In Table 3 optimization (16)-(18), (21)-(22) is performed for stable $G_{p3}(s)$, $G_{p5}(s)$ and unstable $G_{p7}(s)$ processes by using $G_m(s)$ with two quadruplets: $\{k_u, \omega_u, \varphi, A\}$, denoted as opt2A, and

$\{k_u, \omega_u, \varphi, A_0\}$, denoted opt2A$_0$. As mentioned previously, for integrating processes $A=\omega_u$. Almost the same performance/robustness tradeoff is obtained for A and A_0, as supposed in Section 2. This result is important since it confirms that an adequate approximation of the frequency response of the stable and unstable processes around ω_u can be used in the optimization (16)-(18) and (21)-(22), instead of the model $G_m(i\omega)$ in (5). Obviously, the same applies for integrating processes. The advantage of the constrained PID controller optimization (16)-(18) and (21)-(22) is that only two nonlinear algebraic equations have to be solved, with very good initial conditions for the unknown parameters λ and ω_0. Moreover, the optimization is performed for the desired values of M_s, M_n and for the desired closed-loop system damping ratio ζ.

Finally, the results of the PI controller optimization are demonstrated in Table 4 and in Fig. 7. By repeating calculations for a few values of ζ, for the same values of M_s and M_p, the same (minimal) value of the IAE is obtained by applying method opt3, defined by (17) and (21)-(22), and the method opt2, defined by (13)-(14). As mentioned previously, method opt2 is an improvement of the method proposed in (Aström et al., 1998), denoted here as method opt1.

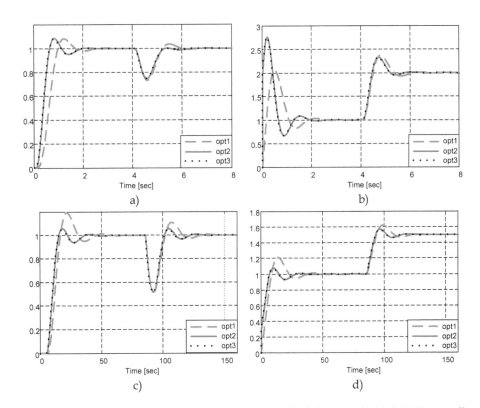

Fig. 7. Set-point and load disturbance step responses: $y(t)$ (left) and $u(t)$ (right). PI controllers from Table 4: opt1 $b=0$, opt2 $b=0.6$, opt3 $b=0.6$. In a) and b) $G_{p1}(s)$, $D(s)=-\exp(-4s)/s$; in c) and d) $G_{p4}(s)$, $D(s)=-0.5\exp(-80s)/s$.

Process/method	k	k_i	IAE	M_s	M_p	β	ζ
G_{p1}/opt1	2.6707	6.4739	0.19	1.98	1.58	1	-
G_{p1}/opt2	3.1874	6.1391	0.16	1.99	1.48	0.52	-
G_{p1}/opt3	3.2119	6.1083	0.16	1.99	1.48	-	0.95
G_{p3}/opt1	7.4060	0.0692	15.61	1.92	1.65	1	-
G_{p3}/opt2	8.1456	0.0680	14.72	1.94	1.60	0.48	-
G_{p3}/opt3	8.1355	0.0679	14.73	1.94	1.60	-	0.90
G_{p4}/opt1	0.3248	0.1259	12.04	2.16	1.35	1	-
G_{p4}/opt2	0.4608	0.1137	10.23	2.11	1.18	0.69	-
G_{p4}/opt3	0.4651	0.1128	10.19	2.10	1.18	-	0.90

Table 4. PI controllers, obtained for M_s=2 by applying model (5) and methods: (Aström et al., 1998) opt1, (13)-(14) opt2, and (17), (21)-(22) opt3.

4. Closed-loop estimation of model parameters

Approximation of process dynamics, around the operating regime, can be defined by some transfer function $G_p(s)$ obtained from the open-loop or closed-loop process identification. One two step approach (Hjalmarsson, 2005) is based on the application of the high-order ARX model identification in the first step. In the second step, to reduce the variance of the obtained estimate of frequency response of the process, caused by the measurement noise, this ARX model is reduced to a low-order model $G_p(s)$. By applying this procedure an adequate approximation $G_p(i\omega)$ of the unknown Nyquist curve can be obtained in the region around the ultimate frequency ω_u. As demonstrated for the Ziegler-Nichols tuning, in Fig. 3-c and Fig. 4-b, such approximation of the unknown Nyquist curve is of essential importance for designing an adequate PID controller. The same applies for the successful PID optimization under constraints on the desired values of M_n and M_s, as demonstrated in Table 5 for the value of A defined as in (5) and for $A=A_0$.

The Closed-Loop (CL) system identification can be performed by using indirect or direct identification methods. In indirect CL system identification methods it is assumed that the controller in operation is linear and a priory known. Direct CL system identification methods are based only on the plant input and output data (Agüero et al., 2011). Finally, the identification can be based on the simple tests, as initiated by Ziegler and Nichols (1942), to obtain an IPDT model (1). Later on, this approach is extended to obtain FOPDT model and the Second-Order Plus Dead-Time (SOPDT) model, for integrating processes characterized by the IFOPDT model. The SOPDT model can be obtained from k_u, ω_u, φ, A. In this case it is defined by

$$G_{SO}(s) = \frac{e^{-Ls}}{as^2 + bs + c},$$ (26)

where parameters a, b, c and L are functions of k_u, ω_u, φ and A, obtained from the tangent rule (Šekara & Mataušek, 2010a). This model (26) is an adequate SOPDT approximation of the Nyquist curve $G_p(i\omega)$ in the region around the ultimate frequency ω_u, for a large class of stable processes, processes with oscillatory dynamics, integrating and unstable processes.

The recently proposed new Phase-Locked Loop (PLL) estimator (Mataušek & Šekara, 2011), its improvement (Šekara & Mataušek, 2011c), and new relay SheMa estimator (Šekara & Mataušek 2011b) make possible determination of parameters k_u, ω_u, φ and A_0 of the model $G_m(s)$ in the closed-loop experiments, without breaking the control loop in operation. This property of the proposed PLL and SheMa estimators is important for practice, since breaking of control loops in operation is mainly ignored by plant operators, especially in the case of controlling processes with oscillatory dynamics, integrating or unstable processes. The PLL estimator can be applied in the case when the controller in operation is an unknown linear controller, while the SheMa estimator can be applied when the controller in operation is unknown and nonlinear. In that sense, the SheMa estimator belongs to the direct CL system identification methods, based only on the plant input and output data, as in (Agüero et al., 2011). Both procedures, SheMa and PLL, are based on the parameterization presented in (Šekara & Mataušek, 2010a; Mataušek & Šekara, 2011). Estimates of parameters k_u^-, k_u, k_u^+ and ω_u^-, ω_u, ω_u^+, obtained for $\arg G_p(i\omega) = -\pi + \varphi$, $\varphi = 0$ and $\varphi = \varphi^\pm = \pm\pi/36$, are used for determining φ and A_0, as defined in (Mataušek & Šekara, 2011).

In this section, an improvement of the new PLL estimator from (Mataušek & Šekara, 2011) is presented in Fig. 8. The improvement, proposed by Šekara and Mataušek (2011c), consists of adding two integrators at the input to the PLL estimator from (Mataušek & Šekara, 2011). Inputs to these integrators are defined by outputs of the band-pass filters AF_1, used to eliminate the load disturbance. Outputs of these integrators are passed through a cascade of the band-pass filters AF_m, $m=2,3,4$. All filters AF_m, $m=1,2,3,4$, are tuned to the ultimate frequency. Such implementation of the PLL estimator eliminates the effects of the high measurement noise and load disturbance. Blocks AF_m, $j=1,2,3,4$, are implemented as presented in (Mataušek & Šekara, 2011), while implementation of blocks for determining $\arg\{G_p(i\omega)\}$ and $|G_p(i\omega)|$ are presented in (Šekara & Mataušek, 2011c).

PLL estimator from Fig. 8 is applied to processes $G_{p8}(s)=\exp(-s)/(2s+1)$ and $G_{p9}(s)=4\exp(-2s)/(4s-1)$ in the loop with the known PID controller. Estimation of parameters k_u^-, k_u, k_u^+ and ω_u^-, ω_u, ω_u^+ is presented in Fig. 9. Highly accurate estimates of k_u^-, k_u, k_u^+ and ω_u^-, ω_u, ω_u^+ are obtained in the presence of the high measurement noise and load disturbance. Since these parameters are used to determine φ and A_0, this experiment demonstrates that highly accurate estimate of the quadruplet $\{k_u, \omega_u, \varphi, A_0\}$ can be obtained, in the presence of the high measurement noise and load disturbance, by the PLL estimator from (Šekara & Mataušek 2011c). In Fig. 10, estimation of the unknown Nyquist curve of the unstable process in the loop with the PID controller is demonstrated.

The PLL estimator from (Mataušek & Šekara, 2011; Šekara & Mataušek 2011c) is a further development of the idea firstly proposed in (Crowe & Johnson, 2000) and used in (Clarke & Park, 2003). The SheMa estimator is a further development of the estimator proposed by Aström and Hägglund (1984) as an improvement of the Ziegler-Nichols experiment.

The Ziegler and Nichols (1942) experiment, used to determine k_u and ω_u of a process is performed by setting the integral and derivative gains to zero in the PID controller $C(s)$ in operation. However, in this approach the amplitude of oscillations is not under control. This drawback is eliminated by Aström and Hägglund (1984). The factors influencing the critical point estimation accuracy in this conventional relay setup are: the use of describing function method is faced with the fact that higher harmonics are not efficiently filtered out by the process, presence of the load disturbance d, and presence of the measurement noise n. The first drawback of the conventional relay experiment is eliminated by the modified relay setup (Lee et al., 1995).

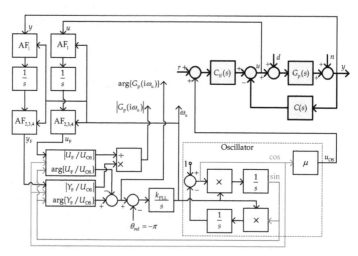

Fig. 8. Improved PLL estimator. $AF_{2,3,4}$ is the cascade of band-pass filters AF_m, $m=1,2,3,4$.

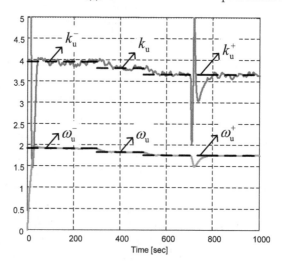

Fig. 9. PLL estimates of k_u^-, k_u, k_u^+ and ω_u^-, ω_u, ω_u^+, in the presence of the high measurement noise and step load disturbance at $t=700$ s. Process $G_{p8}(s)=\exp(-s)/(2s+1)$, for: $\phi^- = -\pi/36$ for $0 \leq t \leq 300$ s, $\phi = 0$ for $300 < t \leq 500$ s and $\phi^+ = \pi/36$ for $500 < t \leq 1000$ s.

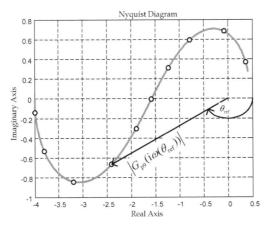

Fig. 10. Estimates (circles) of the Nyquist curve (solid) obtained by the PLL estimator for the desired values $\theta_{ref}= \arg\{G_{p9}(i\omega)\}$. Process $G_{p9}(s)=4\exp(-2s)/(4s-1)$, the noise-free case.

Due to its simplicity, the relay-based setup proposed by Aström and Hägglund (1984) is still a basic part of different methods developed in the area of process dynamics characterization. For example, it is used to generate signals to be applied for determining FOPDT and SOPDT models, using a biased relay (Hang et al., 2002). However, from the viewpoint of the process control system in operation, the estimation based on this setup, and its modifications, is performed in an open-loop configuration: the loop with the controller $C(s)$ in operation is opened and the process output is connected in feedback with a relay.

In the paper (Šekara & Mataušek, 2011b) a new relay-based setup is developed, with the controller $C(s)$ in operation. It consists of a cascade of variable band-pass filters AF_m, from (Clarke & Park, 2003), a new variable band-pass filter F_{mod} proposed by Šekara and Mataušek (2011b) and a notch filter $F_{NF}=1-F_{mod}$. Center frequencies of variable band-pass filters AF_m and F_{mod} are at ω_u.

Highly accurate estimates of ω_u and k_u are obtained in the presence of the measurement noise and load disturbance. Also, highly accurate estimates of the Nyquist curve $G_p(i\omega)$ at the desired values of $\arg\{G_p(i\omega)\}$ are obtained by including into the SheMa the modified relay instead of the ordinary relay. The amplitude μ of both relays is equal to $\mu=\pi k_{u,0}y_{ref}\varepsilon_0/4$, where $k_{u,0}$ is the ultimate gain obtained in the previous activation of the SheMa, y_{ref} is the amplitude of the set-point r and ε_0 is a small percent of y_{ref}, for example $\varepsilon_0=0.1\%$ in the examples presented in (Šekara & Mataušek, 2011b). The proposed closed-loop procedure can be activated or deactivated with small impact on the controlled process output. Further details of the SheMa estimator, including the stability and robustness analyses, and implementation details, are presented in (Šekara & Mataušek 2011b).

5. Gain scheduling control of stable, integrating, and unstable processes, based on the controller optimization in the classification parameter plane

For a chosen region in the ρ-φ classification plane, presented in Fig. 11, the normalized parameters $k_n(\rho,\varphi)$, $k_{in}(\rho,\varphi)$, $k_{dn}(\rho,\varphi)$ and $T_{fn}=|k_{dn}(\rho,\varphi)|/m_n$ of a virtual PID_n controller are calculated in advance by using the process-independent model $G_n(i\omega_n, \rho, \varphi)$ in (10).

Then, parameters k, k_i, k_d and T_f of the PID controller (3), $F_C(s)\equiv1$, are obtained, for the process classified in the chosen region of the ρ-φ plane, by using the estimated k_u, ω_u, φ, A and the following relations

$$k = k_u k_n, \quad k_i = k_u \omega_u k_{in}, \quad k_d = k_u k_{dn} / \omega_u, \quad T_f = T_{fn} / \omega_u. \tag{27}$$

Depending on the method applied to obtain parameters k_n, k_{in}, k_{dn} and $T_{fn}=|k_{dn}|/m_n$ of a PID_n controller, parameters k, k_i, k_d and T_f of the PID controller (3), $F_C(s)\equiv1$, guarantee the desired M_s and the sensitivity to measurement noise equal to $M_n=|k_u|m_n$, or guarantee the

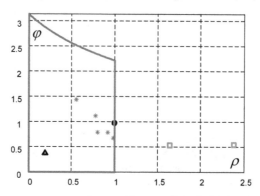

Fig. 11. Classification ρ-φ parameter plane, with processes $G_{pj}(s)$, $j=1,2,...,9$. Stable processes are classified in the region $0 < \rho < 1, 0 < \varphi < \pi / \sqrt{\rho+1}$, integrating processes are classified as $\rho = 1, 0 < \varphi < \pi / \sqrt{2}$ processes. Unstable processes are classified outside this region.

desired M_s, ζ and $M_n=|k_u|m_n$. Since parameters k_n, k_{in}, k_{dn} and $T_{fn}=|k_{dn}|/m_n$ are determined in advance, they can be memorized as look-up tables in the ρ-φ plane. Besides, this can be done for different values of M_s, m_n and ζ. These look-up tables define a new Gain Scheduling Control (GSC) concept. Important feature of this GSC is that these look-up tables, obtained for some values of M_s, m_n and ζ from the model $G_n(i\omega_n, \rho, \varphi)$, are process-independent. Enormous resources are avoided, required for performing experiments on the plant in order to define the standard GSC as the look-up tables of PID controller parameters for this plant and the desired region of operating regimes. Thus, the important and exclusive feature of the new GSC is that a desired performance/robustness tradeoff can be obtained for a large region of dynamic characteristics of processes in different plants and different operating regimes, covered by the look-up tables of parameters k_n, k_{in}, k_{dn} in the ρ-φ classification plane.

Now, this GSC PID controller tuning, performed by using (27), will be demonstrated by the two different procedures applied for obtaining parameters k_n, k_{in}, k_{dn} and $T_{fn}=|k_{dn}|/m_n$ of the PID_n controller for integrating and stable processes. Stable processes having a weakly damped impulse response are denoted as processes having oscillatory dynamics, while processes with damped impulse response are denoted as stable processes.

For integrating processes, parameters k_n, k_{in}, k_{dn} and $T_{fn}=|k_{dn}|/m_n$ of the PID_n controller depend only on angle φ, since $\rho=1$. In this case, for desired values of M_s and m_n, PID controller parameters (27) are obtained from tuning formulae for $k_n(\varphi)$, $k_{in}(\varphi)$ and $k_{dn}(\varphi)$

(Šekara & Mataušek, 2011a). Thus, for integrating process $G_{p6}(s)$ parameters of the PID_n controller are obtained by applying angle $\varphi=0.9716$ in the tuning formulae defined in (Šekara & Mataušek, 2011a) for $M_s=2$ and $m_n=2$, given in Appendix as tun1. The results are presented in Table 5, G_{p6}-tun1.

For processes having the oscillatory dynamics look-up tables and tuning formulae are derived in (Šekara & Mataušek, 2011a) for $M_s=2$ and $m_n=40$, in the region $0.1\leq\rho\leq0.2$, $0.1745\leq\varphi\leq1.0472$ of the ρ-φ classification plane of Fig. 11. These tuning formulae, in Appendix denoted as tun2, are applied to determine parameters k, k_i, k_d and T_f for the process having the oscillatory dynamics $G_{p5}(s)$, classified as process $\rho=0.1971$, $\varphi=0.3679$ (Table 5, G_{p5}-tun2). To illustrate the direct application of the look-up tables from (Šekara & Mataušek, 2011a, Table A4) and interpolation procedure defined in Appendix, Fig. 17, since this process is classified as $\rho=0.1971$, $\varphi=21.0791°$ (0.3679), the following points are determined from (Šekara & Mataušek, 2011a, Table A4) and Appendix, Fig. 17: $\rho_{1,1}=0.15$, $\varphi_{1,1}=20°$, $\rho_{1,2}=0.2$, $\varphi_{1,2}=20°$ and $\rho_{2,2}=0.2$, $\varphi_{2,2}=30°$. Parameters (k_n, k_{in}, k_{dn}) are defined by: (-2.4122, 0.5988, 3.9353) for $\rho_{1,1}$, $\varphi_{1,1}$, (-1.7022, 0.4125, 2.8783) for $\rho_{1,2}$,$\varphi_{1,2}$ and (-1.6626, 0.4164, 2.3017) for $\rho_{2,2}$,$\varphi_{2,2}$. Then, by using three point interpolation from Appendix, upper triangle ($a_{ru}=0.0578$, $\beta_{ru}=0.1971$), one obtains parameters in Table 5, G_{p5}-GSC: $k=-0.4220$, $k_i=0.0384$, $k_d=1.9116$, $T_f=1.947$.

For stable processes, in a large region of the ρ-φ plane, look-up tables of parameters k_n, k_{in} and k_{dn} are defined for $M_s=2$ and $m_n=2$ (Šekara & Mataušek, 2011a, Tables A1-A3). These look-up tables are applied in the present paper to determine parameters k, k_i, k_d and T_f for the stable process $G_{p3}(s)$. This process is classified as process $\rho=0.9808$, $\varphi=0.6783$ (38.8637°). Thus, for $G_{p3}(s)$ parameters (k_n, k_{in}, k_{dn}) can be obtained from the three points in the ρ-φ classification plane (Appendix, Fig. 17): $\rho_{1,1}=0.95$, $\varphi_{1,1}=30°$; $\rho_{2,1}=0.95$, $\varphi_{2,1}=40°$ and $\rho_{2,2}=1$, $\varphi_{2,2}=40°$ (0.6981). Two points are used for stable processes (0.5086, 0.1349, 0.6569) for $\rho_{1,1}$,$\varphi_{1,1}$ and (0.5013, 0.1261, 0.5332) for $\rho_{2,1}$,$\varphi_{2,1}$ from the look-up tables (Šekara & Mataušek, 2011a, Tables A1-A3), while data (0.5036, 0.1109, 0.5332) for $\rho_{2,2}$, $\varphi_{2,2}$ are obtained from tuning formulae derived for integrating processes in (Šekara & Mataušek, 2011a), given in Appendix as tun1. Then, by using three point interpolation from Appendix, Fig. 17 lower triangle ($a_{ll}=0.6166$, $\beta_{ll}=0.1136$), one obtains parameters presented in Table 5, G_{p3}-GSC: $k=17.0973$, $k_i=0.2307$, $k_d=315.2928$ and $T_f=4.6430$.

Process-method	k	k_i	k_d	T_f	IAE	M_n	M_s	M_p
G_{p3}-GSC	17.0973	0.2307	315.29	4.6430	4.84	67.91	2.00	1.54
G_{p5}-tun2	-0.4220	0.0380	1.8758	0.1910	26.32	9.82	1.99	1.08
G_{p5}-GSC	-0.4269	0.0384	1.9116	0.1947	26.04	9.82	2.01	1.09
G_{p6}-tun1	0.1182	0.0054	0.3746	0.7970	209.10	0.47	2.00	1.62

Table 5. PID controllers: stable process $G_{p3}(s)$, method GSC-Appendix; stable process having oscillatory dynamics $G_{p5}(s)$, method tun2 and method GSC-Appendix; integrating process $G_{p6}(s)$, method tun1.

5.1 Experimental results

Experimental results, presented in Fig. 12, are obtained by using the laboratory thermal plant. It consists of a thin plate made of aluminum, $L_a=0.1m$ long and $h=0.03m$ wide (Mataušek & Ribić, 2012). Temperature $T(x,t)$ is distributed along the plate, from $x=0$ to

$x=L_a$, and measured by precision sensors LM35 (TO92), at $x=0$ and $x=L_a$. The plate is heated by a terminal adjustable regulator LM317 (TO 220) at position $x=0$. The manipulated variable is the dissipated power of the heater at $x=0$. The input to the heater is the control variable $u(t)$ (%), defined by the output of the PID controller. The controlled variable is $y(t)=T(L_a,t)$, measured by the sensor at position $x=L_a$. Temperature sensor at $x=0$ is used in the safety device, to prevent overheating when $70°C \leq T(0,t)$. The anti-windup implementation of the PID controller (3), $F_C(s)\equiv1$, is given by

$$u_C = T_{aw}\left(\frac{bks+k_i}{T_{aw}s+1}r - \frac{k_ds^2+ks+k_i}{(T_{aw}s+1)(T_fs+1)}y\right) + \frac{1}{T_{aw}s+1}u \; . \tag{28}$$

The saturation element is defined by the input $u_C(t)$ and output $u(t)$:

$$u = \begin{cases} l_{low}, & u_C \leq l_{low} \\ u_C, & l_{low} < u_C < l_{high} \\ l_{high}, & u_C \geq l_{high} \end{cases} \tag{29}$$

Obviously, in the linear region $l_{low}<u_C(t)< l_{high}$ of the saturation element, for $u_C(t)\equiv u(t)$ one obtains (3), $F_C(s)\equiv1$, from (28).

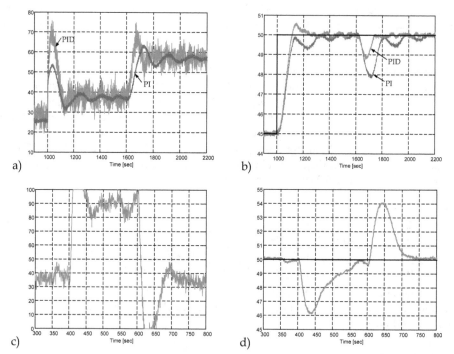

Fig. 12. Experimental results. Set-point and load step (-20% change of the controller output at $t=1600$ s) responses of the real plant, with the PI and PID controller: a) control variable $u(t)$ and b) controlled variable $y(t)$. The real plant, with the anti-windup PID controller under the disturbance induced by activating/deactivating the fan: c) $u(t)$ and d) $y(t)$.

Transfer function $G_{p3}(s)$, used for determining parameters of the PID controller applied in the real-time experiment, is obtained previously in (Mataušek & Ribić, 2012). By applying a Pseudo-Random-Binary-Sequence for $u(t)$, the open-loop response $y(t)$ of the laboratory thermal plant is obtained. From these $u(t)$ and $y(t)$ a 100-th order ARX model is determined and reduced then to the 5-th order transfer function $G_{p3}(s)$ in (Mataušek & Ribić, 2012). This model of the process is used here to determine the quadruplet $\{k_u, \omega_u, \varphi, A\}$ presented in the Appendix. Thus, the laboratory thermal plant is classified as the process $\rho=0.9808$, $\varphi=0.6783$. Then, PID controller applied to the real thermal plant is determined by using look-up tables of parameters $k_n(\rho,\varphi)$, $k_{in}(\rho,\varphi)$, $k_{dn}(\rho,\varphi)$, for stable processes, and parameters $k_n(\varphi)$, $k_{in}(\varphi)$, $k_{dn}(\varphi)$, for integrating processes, previously determined in (Šekara & Mataušek, 2011a). This procedure, used to obtain PID in Table 5, row G_{p3}-GSC, and results obtained by this PID controller, presented in Fig. 12, demonstrate that in advance determined look-up tables of parameters k_n, k_{in} and k_{dn} defines a process-independent GSC applicable for obtaining the desired performance/robustness tradeoff for a real plant classified in the ρ-φ parameter plane. For $T_i=k/k_i$ and $T_d=k_d/k$, parameter $T_{aw}=15s$ is obtained from $T_{aw}=pT_i+(1-p)T_d$, for $p=0.2$, and $l_{low}=0$, $l_{high}=100\%$, $b=0.25$.

Closed-loop experiment in Fig. 12-a and Fig. 12-b is used to demonstrate advantages of the designed PID controller, compared with the PI controller, from Table 4, row G_{p3}/opt3 defined by: $k=8.1355$, $k_i=0.0679$, and $b=0.5$. This experiment starts from temperature $T(L_a,t)\approx45^\circ C$, as presented in Fig. 12-b. Then at $t=1000$ s the set point is changed to $r=45^\circ C+r_0$, $r_0=5^\circ C$. At $t=1600$ s a load disturbance is inserted as a step change of the controller output equal to -20%. Improvement of the performance obtained by the PID controller is evident. As expected, this is obtained with the greater variation of the control signal $u_{PID}(t)$ than that obtained by $u_{PI}(t)$. This is the reason why PID controller from Table 2, row tunλ_u, having a greater value of $M_n=217.7$, is not applied to the real thermal plant.

The closed-loop experiment presented in Fig. 12-c and Fig. 12-d starts from the steady state temperature $T(L_a,t)\approx50^\circ C$ by activating a fan at $t=400$ s. Then, at $t=600$ s the fan is switched-off. Action of the fan induced a strong disturbance, as seen from the control signal $u(t)$ in Fig. 12-c. It should be observed that anti-windup action is activated two times, around 410 s and 625 s. Anti-windup action is effective and rejection of the disturbance is fast, as seen from Fig. 18-d.

6. Conclusion

The extension of the Ziegler-Nichols process dynamics characterization, developed in (Šekara & Mataušek, 2010a; Mataušek & Šekara, 2011), is defined by the model (5). Based on this model, a procedure is derived for classifying a large class of stable, integrating and unstable processes into a two-parameter ρ-φ classification plane (Šekara & Mataušek, 2011a). As a result of this classification, a new CSC concept is developed. In the ρ-φ classification plane, parameters $g_n(\rho,\varphi)=\{k_n(\rho,\varphi), k_{in}(\rho,\varphi), k_{dn}(\rho,\varphi)\}$ and $T_{fn}(\rho,\varphi)=|k_{dn}(\rho,\varphi)|/m_n$, of a virtual PID_n controller can be calculated in advance, to satisfy robustness defined by M_s and sensitivity to measurement noise defined by m_n. Also it is possible to satisfy M_s, m_n and the closed-loop system damping ratio ζ. Calculation of parameters $g_n(\rho,\varphi)$ and $T_{fn}(\rho,\varphi)$ is process-independent. The calculation is performed by using model $G_n(s_n,\rho,\varphi)$, defined by the values of ρ and φ for stable processes in the range $0<\rho<1, 0<\varphi<\pi/\sqrt{\rho+1}$, for integrating processes in the range $\rho=1, 0<\varphi<\pi/\sqrt{2}$, for unstable processes by the values of the ρ and φ outside these regions.

Parameters $g_n(\rho,\varphi)$, calculated for a given region in the ρ-φ classification plane, are memorized as process-independent look-up tables. Then, for the process $G_p(s)$ classified into this region of the ρ-φ classification plane, parameters of a real PID controller k, k_i, k_d, T_f are obtained directly from $g_n(\rho,\varphi)$, $T_{fn}(\rho,\varphi)$ and the estimated quadruplet $\{k_u, \omega_u, \varphi, A\}$ or $\{k_u, \omega_u, \varphi, A_0\}$ for stable/unstable processes, and the triplet $\{k_u, \omega_u, \varphi\}$ for integrating processes. It is demonstrated by simulations that for the real M_n equal to $M_n=|k_u|m_n$, the desired M_s and ζ are obtained when a real PID controller, obtained by the proposed GSC, is applied to the process $G_p(s)$. The desired performance/robustness tradeoff can be accurately predicted. Namely, performance index IAE and robustness index M_s, obtained on the model $G_m(s)$ in (5) are almost the same as those obtained for the process $G_p(s)$, as confirmed here and by a large test batch considered in (Šekara & Mataušek, 2010a; Mataušek & Šekara, 2011; Šekara & Mataušek, 2011a).

A set of new constrained PID optimization techniques is derived for determining the four parameters k, k_i, k_d, T_f of the PID controller. The one of them has a unique property. The unknown parameters are obtained as the solution of only two nonlinear algebraic equations, with the good initial values of the unknown two parameters, determined to satisfy the desired values M_s and M_n, given desired value of the closed-loop system damping ratio ζ. Thus, the critically damped closed-loop system response is obtained for $\zeta=1$. Two extensions of the PLL-based and relay-based procedures are derived in (Mataušek & Šekara, 2011; Šekara & Mataušek, 2010b; 2011c; 2011b) for determining the quadruplet $\{k_u, \omega_u, \varphi, A_0\}$. These procedures can be applied for the closed-loop PID controller tuning/retuning, in the presence of the measurement noise and load disturbance, without breaking the loop of the controller in operation.

Process-independent look-up tables of parameters $g_n(\rho,\varphi)$, defining the process-independent GSC, can be applied by using any process dynamics characterization defined by the estimated frequency response of the process around the ultimate frequency. This is demonstrated in the present chapter by applying a model obtained previously by a high-order ARX identification of a laboratory thermal plant, and reduced then to the fifth order $G_{p3}(s)$, used here to determine the quadruplet $\{k_u, \omega_u, \varphi, A\}$. This quadruplet is applied to determine parameters of the real PID, by using the look-up tables of parameters $g_n(\rho,\varphi)$ calculated previously in (Šekara & Mataušek, 2011a). As confirmed by the experimental results, the method of the proposed process-independent GSC is effective. Finally, it is believed that material presented in this chapter will initiate further development of the proposed process-independent GSC and its implementation in advanced controllers.

7. Appendix

Parameters η_1, η_2, β_1, β_2 and β_3

$$\eta_1 = \frac{\alpha_1 \sin(\omega_u \tau) + \alpha_2 \cos(\omega_u \tau)}{\omega_u^2}, \quad \eta_2 = \frac{\alpha_2 \sin(\omega_u \tau) - \alpha_1 \cos(\omega_u \tau) + 1}{\omega_u^2},$$

$$\alpha_1 = \lambda^4 \omega_u^4 - 2\lambda^2 \omega_u^2 (1 + 2\zeta^2) + 1, \quad \alpha_2 = 4\zeta\lambda\omega_u(1 - \lambda^2 \omega_u^2),$$

$$\beta_1 = \frac{\omega_u}{A(4\zeta\lambda + \tau - \eta_1)}, \quad \beta_2 = \frac{2\lambda^2(1 + 2\zeta^2) - \tau^2/2 + \eta_1\tau - \eta_2}{4\zeta\lambda + \tau - \eta_1}, \quad \beta_3 = \frac{4\zeta\lambda^3 + \tau^3/6 - \eta_1\tau^2/2 + \eta_2\tau}{4\zeta\lambda + \tau - \eta_1},$$

are presented here, to make possible to repeat the results obtained by the PID optimization from (Mataušek & Šekara, 2011).

Tuning formulae tun1, for integrating processes for $M_s=2$ and $m_n=2$, given by

$$
\begin{bmatrix} k_n \\ k_{in} \\ k_{dn} \end{bmatrix} = \begin{bmatrix} 0.5904 & -0.2707 & 0.3029 & -0.1554 & 0.0311 \\ 0.1534 & -0.0826 & 0.0409 & -0.0164 & 0.0033 \\ 1.2019 & -1.5227 & 1.0714 & -0.4944 & 0.0916 \end{bmatrix} \begin{bmatrix} 1 \\ \varphi \\ \varphi^2 \\ \varphi^3 \\ \varphi^4 \end{bmatrix},
$$

and tun2, for processes with the oscillatory dynamics for $M_s=40$ and $m_n=2$, given by

$$
\begin{bmatrix} k_n \\ k_{in} \\ k_{dn} \end{bmatrix} = \begin{bmatrix} -8.9189 & 63.0913 & 0.6494 & -135.2567 & 0.2806 & -3.5564 \\ 2.2218 & -16.5791 & 0.1361 & 37.5733 & 0.0136 & -0.6388 \\ 14.8966 & -82.7969 & -9.0810 & 145.2467 & 0.9056 & 25.1221 \end{bmatrix} \begin{bmatrix} 1 \\ \rho \\ \varphi \\ \rho^2 \\ \varphi^2 \\ \rho\varphi \end{bmatrix},
$$

are defined in (Šekara & Mataušek, 2011a). The angle φ is in radians.

Process	k_u	ω_u	τ	A	A_0	ρ	φ
Gp1	11.5919	9.8696	0.0796	9.0858	8.9190	0.9206	0.7854
Gp2	4	1	0.7854	0.8	0.7071	0.8	0.7854
Gp3	33.9538	0.0577	11.7521	0.0566	0.0519	0.9808	0.6783
Gp4	1.2494	0.4	3.5881	0.2222	0.2612	0.5555	1.4352
Gp5	0.2455	0.3695	0.9956	0.0728	0.0729	0.1971	0.3679
Gp6	0.2371	0.2291	4.2403	0.2291	-	1	0.9716
Gp7	0.8625	0.1333	4.1446	0.3173	0.3211	2.3793	0.5526
Gp8	3.8069	1.8366	0.6271	1.4545	1.6054	0.7920	1.1517
Gp9	0.6341	0.5828	0.9105	0.9621	1.0000	1.6509	0.5333

Table 6. Parameters of models $G_{mj}(s)$ of processes $G_{pj}(s)$, $j=1,2,...,9$.

Normalized parameters of the PID_n controller can be obtained by interpolation based on the three points in the ρ-φ look-up tables of the memorized parameters $k_n(\rho_i,\varphi_j)$, $k_{in}(\rho_i,\varphi_j)$ and $k_{dn}(\rho_i,\varphi_j)$, $i=1,2,...,I_m$, $j=1,2,...,J_m$, determined in advance. In the present paper the look-up tables from (Šekara & Mataušek, 2011a, Tables 1-4) are used. The four points mash in the ρ-φ look-up tables is presented in Fig. 13. The normalized parameters of the PID_n controller, for the lower triangle are given by:

$$
k_n = (1-\alpha-\beta)k_{n2,1} + \alpha k_{n2,2} + \beta k_{n1,1}, \qquad k_{in} = (1-\alpha-\beta)k_{in2,1} + \alpha k_{in2,2} + \beta k_{in1,1},
$$

$$
k_{dn} = (1-\alpha-\beta)k_{dn2,1} + \alpha k_{dn2,2} + \beta k_{dn1,1},
$$

where $a=a_{ll}$, $\beta=\beta_{ll}$. The normalized parameters of the PID_n controller, for the upper triangle are given by:

$$k_n = (1 - \alpha - \beta)k_{n1,2} + \alpha k_{n1,1} + \beta k_{n2,2}, \qquad k_{in} = (1 - \alpha - \beta)k_{in1,2} + \alpha k_{in1,1} + \beta k_{in2,2},$$

$$k_{dn} = (1 - \alpha - \beta)k_{dn1,2} + \alpha k_{dn1,1} + \beta k_{dn2,2},$$

where $a=a_{ru}$, $\beta =\beta_{ru}$. In both cases $T_{fn} = k_{dn} / m_n$. Then, parameters of the PID controller are obtained from (27).

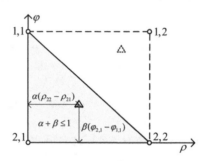

Fig. 13. The four point mash in the ρ-φ plane in (Šekara & Mataušek, 2011a, Tables 1-4). For lower triangle $\alpha_{ll}=(\rho_{est} - \rho_{2,1})/(\rho_{2,2} - \rho_{2,1})$ and $\beta_{ll}=(\varphi_{2,1} - \varphi_{est})/(\varphi_{2,1} - \varphi_{1,1})$. For the upper triangle $\alpha_{ru}=(\rho_{1,2} - \rho_{est})/(\rho_{1,2} - \rho_{1,1})$ $\beta_{ru}=(\varphi_{est} - \varphi_{1,2})/(\varphi_{2,2} - \varphi_{1,2})$. All angles are in degrees and $\varphi_{1,1} \leq \varphi_{2,1}$, $\varphi_{1,1} \leq \varphi_{est} \leq \varphi_{2,1}$, $\rho_{1,1} \leq \rho_{est} \leq \rho_{2,1}$.

10. Acknowledgement

Authors gratefully acknowledge discussion with Dr. Aleksandar Ribić and his help in implementing the anti-windup PID controller on the laboratory thermal plant. T.B. Šekara gratefully acknowledges the financial support from the Serbian Ministry of Science and Technology (Project TR33020). M.R. Mataušek gratefully acknowledges the financial support from TERI Engineering, Belgrade, Serbia.

11. References

Agüero, J.C.; Goodwin, G.C. & Van den Hof, P.M.J. (2011). A virtual closed loop method for closed loop identification. *Automatica*, Vol. 47, pp. 1626-1637.

Aström, K.J. & Hägglund, T. (1984). Automatic tuning of simple regulators with specifications on phase and amplitude margins, *Automatica*, Vol. 20, pp. 645-651.

Aström, K.J.; Hang, C.C., Persson, P. & Ho, W.K. (1992), Towards intelligent PID control, *Automatica*, Vol. 28, pp. 1-9.

Aström, K.J. & Hägglund, T. (1995a). *PID controllers: Theory, design and tuning*, 2nd edition, ISA, ISBN 1-55617-516-7, Research Triangle Park, NC 27709.

Aström, K.J. & Hägglund, T. (1995b), New tuning methods for PID controllers, *Proceedings European control conference*, Rome, Italy, pp. 2456-2462.

Aström, K.J.; Panagopoulos, H. & Hägglund, T. (1998). Design of PI controllers based on non-convex optimization. *Automatica*, Vol. 34, pp. 585-601.

Aström, K.J. & Hägglund, T. (2001). The future of PID control, *Control Engineering Practice*, Vol. 9, pp. 1163-1175.

Aström, K.J. & Hägglund, T. (2004). Revisiting the Ziegler-Nichols step response method for PID control, *Journal of Process Control*, Vol. 14, pp. 635-650.

Clarke, D.W. & Park, J.W. (2003). Phase-locked loops for plant tuning and monitoring. *IEE Proceedings on Control Theory and Applications*, Vol. 150, pp. 155-169.

Crowe, J. & Johnson, M.A. (2000). Process identifier and its application to industrial control, *IEE Proceedings Control Theory and Applications*, Vol. 147, pp. 196-204.

Desborough, L. & Miller, R. (2002). Increasing customer value of industrial control performance monitoring – Honeywell's experience, in *Sixth International Conference on Chemical Process Control*, AIChE Symposium Series Number 326, Vol. 98, pp. 172-192.

Hang, C.C.; Aström, K.J. & Ho, W.K. (1991). Refinements of the Ziegler-Nichols tuning formula, *IEE Proceedings of Control Theory and Applications,* Vol. 138, pp. 111-118.

Hang, C.C.; Aström, K.J. & Wang, Q.G. (2002). Relay feedback auto-tuning of process controllers – a tutorial review, *Journal of Process Control*, Vol. 12, pp. 143-162.

Hjalmarsson, H. (2005). From experiment design to closed-loop control, *Automatica*, Vol. 41, pp. 393-438.

Isaksson, A.J. & Graebe, S.F. (2002). Derivative filter is an integral part of PID design. *IEE Control Theory and Applications*, Vol. 179, pp. 41-45.

Jevtović, B.T. & Matausek, M.R. (2010). PID controller design of TITO system based on ideal decoupler, *Journal of Process Control*, Vol. 20, pp. 869-876.

Lee, H.; Wang, Q.G. & Tan, K.K. (1995). A modified relay-based technique for improved critical point estimation in process control, *IEEE Transactions on Control System Technology*, Vol. 3, pp. 330-337.

Matausek, M.R. & Micić, A.D. (1996). A modified Smith predictor for controlling a process with an integrator and long dead-time, *IEEE Transactions on Automatic Control*, Vol. 41, pp. 1199-1203.

Matausek, M.R.; Jeftenić, B.I.; Miljković, D.M. & Bebić, M.Z. (1996). Gain scheduling control of DC motor drive with field weakening, *IEEE Transactions of Industrial Electronics*, Vol. 43, pp. 153-162.

Matausek, M.R.; Miljković, D.M. & Jeftenić, B.I. (1998). Nonlinear multi-input multi-output neural network control of DC motor drive with field weakening, *IEEE Transactions of Industrial Electronics*, Vol. 45, pp. 185-187.

Matausek, M.R. & Micić, A.D. (1999). On the modified Smith predictor for controlling a process with an integrator and long dead-time, *IEEE Transactions on Automatic Control*, Vol. 44, pp. 1603-1606.

Matausek, M.R. & Kvaščev, G.S., 2003. A unified step response procedure for autotuning of PI controller and Smith predictor for stable processes, *Journal of Process Control*, 13, 787-800.

Matausek, M.R. & Ribić, A.I. (2009). Design and robust tuning of control scheme based on the PD controller plus Disturbance Observer and low-order integrating first-order plus dead-time model, *ISA Transactions*, Vol. 48, pp. 410-416.

Matausek, M.R. & Šekara, T.B. (2011). PID controller frequency-domain tuning for stable, integrating and unstable processes, including dead-time. *Journal of Process Control*, Vol. 21, pp. 17-27.

Mataušek, M.R. & Ribić, A.I. (2012). Control of stable, integrating and unstable processes by the Modified Smith Predictor, *Journal of Process Control*, Vol. 22, pp. 338-343.

Panagopoulos, H.; Aström, K.J. & Hägglund, T. (2002). Design of PID controllers based on constrained optimization. *IEE Control Theory and Applications*, Vol. 149, pp. 32-40.

Rapaić, M.R. (2008). Matlab implementation of the Particle Swarm Optimization (PSO) algorithm, *Matlab Central*, Available from http://www.mathworks.com/matlabcentral/fileexchange/22228-particleswarm-optimization-pso-algorithm

Seki, H. & Shigemasa, T. (2010). Retuning oscillatory PID control loops based on plant operation data. *Journal of Process Control*, Vol. 20, pp. 217-227.

Shinskey, F.G. (1990). How good are our controllers in absolute performance and robustness?. *Measurement and Control*, Vol. 23, pp. 114-121.

Smith, O.J. (1959). A controller to overcome dead time, *ISA Journal*, Vol. 6, pp. 28-33.

Šekara, T.B. & Mataušek, M.R. (2008). Optimal and robust tuning of the PI controller based on the maximization of the criterion J_C defined by the linear combination of the integral gain and the closed-loop system bandwidth. *ELECTRONICS*, Vol.12, pp. 41-45.

Šekara, T.B. & Mataušek, M.R. (2009). Optimization of PID controller based on maximization of the proportional gain under constraints on robustness and sensitivity to measurement noise. *IEEE Transactions on Automatic Control*, Vol. 54, pp. 184-189.

Šekara, T.B. & Mataušek, M.R. (2010a). Revisiting the Ziegler-Nichols process dynamics characterization. *Journal of Process Control*, Vol. 20, pp. 360-363.

Šekara, T.B. & Mataušek, M.R. (2010b). Comparative analysis of the relay and phase-locked loop experiment used to determine ultimate frequency and ultimate gain. *ELECTRONICS*, Vol. 14, pp. 77-81.

Šekara, T.B. & Trifunović, M.B. (2010). Optimization in the frequency domain of the PID controller in series with a lead-lag filter. (*in Serbian*), *Proceedings of Conference INDEL,* Bosnia and Herzegovina ,Vol. 8, pp. 258-261.

Šekara, T.B. & Mataušek, M.R. (2011a). Classification of dynamic processes and PID controller tuning in a parameter plane. *Journal of Process Control*, Vol. 21, pp. 620-626.

Šekara, T.B. & Mataušek, M.R. (2011b). Relay-based critical point estimation of a process with the PID controller in the loop. *Automatica*, Vol. 47, pp. 1084-1088.

Šekara, T.B. & Mataušek, M.R. (2011c). Robust process identification by using Phase-Locked-Loop (*in Serbian*), *Proceedings of Conference INFOTEH-JAHORINA*, Bosnia and Herzegovina, Vol. 10, pp. 18-21.

Šekara, T.B.; Trifunović, M.B. & Govedarica, V. (2011). Frequency domain design of a complex controller under constraints on robustness and sensitivity to measurement noise. *ELECTRONICS*, Vol. 15, pp. 40-44.

Trifunović, M.B. & Šekara, T.B. (2011). Tuning formulae for PID/PIDC controllers of processes with the ultimate gain and ultimate frequency (*in Serbian*), *Proceedings of Conference INFOTEH-JAHORINA*, Bosnia and Herzegovina, Vol. 10, pp. 12-17.

Yamamoto, S. & Hashimoto, I. (1991). Present status and future needs: the view form Japanese industry. In: *Chemical Process Control IV (CRC-IV)*, South Padre Island, US, pp.1-28

Ziegler, J.G. & Nichols, N.B. (1942). Optimum settings for automatic controllers, *Transaction of ASME*, Vol.64, pp. 759-768.

9

Iterative Learning - MPC: An Alternative Strategy

Eduardo J. Adam[1] and Alejandro H. González[2]
[1]Chemical Engineering Department - Universidad Nacional del Litoral
[2]INTEC (CONICET and Universidad Nacional del Litoral)
Argentina

1. Introduction

A repetitive system is one that continuously repeats a finite-duration procedure (operation) along the time. This kind of systems can be found in several industrial fields such as robot manipulation (Tan, Huang, Lee & Tay, 2003), injection molding (Yao, Gao & Allgöwer, 2008), batch processes (Bonvin et al., 2006; Lee & Lee, 1999; 2003) and semiconductor processes (Moyne, Castillo, & Hurwitz, 2003). Because of the repetitive characteristic, these systems have two count indexes or time scales: one for the time running within the interval each operation lasts, and the other for the number of operations or repetitions in the continuous sequence. Consequently, it can be said that a control strategy for repetitive systems requires accounting for two different objectives: a short-term disturbance rejection during a finite-duration single operation in the continuous sequence (this frequently means the tracking of a predetermined optimal trajectory) and the long-term disturbance rejection from operation to operation (i.e., considering each operation as a single point of a continuous process[1]). Since in essence, the continuous process basically repeats the operations (assuming that long-term disturbances are negligible), the key point to develop a control strategy that accounts for the second objective is to use the information from previous operations to improve the tracking performance of the future sequence.

Despite the finite-time nature of every individual operation, the within-operation control is usually handled by strategies typically used on continuous process systems, such as PID ((Adam, 2007)) or more sophisticated alternatives as Model Predictive Control (MPC) (González et al., 2009a;b). The main difficulty arising in these applications is associated to the stability analysis, since the distinctive finite-time characteristic requires an approach different from the traditional one; this was clearly established in (Srinivasan & Bonvin, 2007). The operations sequence control can be handled by strategies similar to the standard Iterative Learning Control (ILC), which uses information from previous operations. However, the ILC exhibits the limitation of running open-loop with respect to the current operation, since no feedback corrections are made during the time interval the operation lasts.

In order to handle batch processes (Lee et al., 2000) proposed the Q-ILC, which considers a model-based controller in the iterative learning control framework. As usual in the ILC literature, only the iteration-to-iteration convergence is analyzed, as the complete input and

---

[1] In this context, continuous process means one that has not an end time.

output profiles of a given operation are considered as fix vectors (open-loop control with respect to the current operation). Another example is an MPC with learning properties presented in (Tan, Huang, Lee & Tay, 2003), where a predictive controller that iteratively improves the disturbance estimation is proposed. Form the point of view of the learning procedure, any detected state or output disturbance is taken like parameters that are updated iteration to iteration. Then, in (Lee & Lee, 1997; 1999) and (Lee et al., 2000), a real-time feedback control is incorporated into the Q-ILC (BMPC). As the authors declare, some cares must be taken when combining ILC with MPC. In fact, as read in Lee and Lee 2003, a simple-minded combination of ILC updating the nominal input trajectory for MPC before each operation does not work.

The MPC proposed in this Chapter is formulated under a closed-loop paradigm (Rossiter, 2003). The basic idea of a closed-loop paradigm is to choose a stabilizing control law and assume that this law (underlying input sequence) is present throughout the predictions. More precisely, the MPC propose here is an Infinite Horizon MPC (IHMPC) that includes an underlying control sequence as a (deficient) reference candidate to be improved for the tracking control. Then, by solving on line a constrained optimization problem, the input sequence is corrected, and so the learning updating is performed.

1.1 ILC overview

Iterative Learning Control (ILC) associates three main concepts. The concept *Iterative* refers to a process that executes the same operation over and over again. The concept *Learning* refers to the idea that by repeating the same thing, the system should be able to improve the performance. Finally, the concept *control* emphasizes that the result of the learning procedure is used to control the plant.

The ILC scheme was initially developed as a feedforward action applied directly to the open-loop system ((Arimoto et al., 1984) ; (Kurek & Zaremba, 1993); among others). However, if the system is integrator or unstable to open loop, or well, it has wrong initial condition, the ILC scheme to open loop can be inappropriate. Thus, the feedback-based ILC has been suggested in the literature as a more adequate structure ((Roover, 1996); (Moon et al., 1998); (Doh et al., 1999); (Tayebi & Zaremba, 2003)). The basic idea is shown in Fig. 1.

This scheme, in its discrete version time, operates as follows. Consider a plant which is operated iteratively with the same set-point trajectory, $y^r(k)$, with k going from 0 to a final finite value T_f, over and over again, as a robot or an industrial batch process. During the i-th trail an input sequence $u^i(k)$, with k going from 0 to a final finite value T_f, is applied to the plant, producing the output sequence $y^i(k)$. Both sequences, that we will call $\mathbf{u}^i$ and $\mathbf{y}^i$, respectively, are stored in the memory devise. Thus, two vectors with length T_f are available for the next iteration. If the system of Fig. 1 operates in open loop, using $\mathbf{u}^i$ in the $(i+1)$-th trail, it is possible to obtain the same output again and again. But, if at the $i+1$ iteration information about both, $\mathbf{u}^i$ and $\mathbf{e}^i = \mathbf{y}^i - \mathbf{y}^r$, where $\mathbf{y}^r = \left[y^r(0), \cdots, y^r(T_f) \right]$, is considered, then new sequences $\mathbf{u}^{i+1}$ and $\mathbf{y}^{i+1}$, can be obtained. The key point of the input sequence modification is to reduce the tracking error as the iterations are progressively increased. The

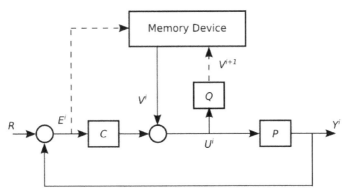

Fig. 1. Feedback-based ILC diagram. Here, continuous lines denote the sequence used during the i-th trail and dashed lines denote sequence that will be used in the next iteration.

purpose of an ILC algorithm is then to find a unique input sequence $\mathbf{u}^\infty$ which minimizes the tracking error.

The ILC formulation uses an iterative updating formula for the control sequence, given by

$$\mathbf{u}^{i+1} = f(\mathbf{y}^r, \mathbf{y}^i, ..., \mathbf{y}^{i-l}, \mathbf{u}^i, \mathbf{u}^{i-1}, ..., \mathbf{u}^{i-l}), \qquad i, l \geq 0. \tag{1}$$

This formula can be categorized according to how the information from previous iteration is used. Thus, (Norrlöf, 2000) among other authors define,

DEFINITION 0.1. *An ILC updating formula that only uses measurements from previous iteration is called* first order ILC. *On the other hand, when the ILC updating formula uses measurements from more than previous iteration, it is called a* high order ILC.

The most common algorithm suggested by several authors ((Arimoto et al., 1984); (Horowitz, 1993); (Bien & Xu, 1998); (Tayebi & Zaremba, 2003); among others), is that whose structure is given by

$$V^{i+1} = Q(z)(V^i + C(z)E^i), \tag{2}$$

where $V^1 = 0$, $C(z)$ denotes the controller transfer function and $Q(z)$ is a linear filter.

Six postulates were originally formulated by different authors ((Chen & Wen, 1999); (Norrlöf, 2000); (Scholten, 2000), among others).

1. Every iteration ends in a fixed discrete time of duration T_f.

2. The plant dynamics are invariant throughout the iterations.

3. The reference or set-point, $\mathbf{y}^r$, is given a priori.

4. For each trail or run the initial states are the same. That means that $x^i(0) = x^0(0), i \geq 0$.

5. The plant output $y(k)$ is measurable.

6. There exists a unique input sequence, $\mathbf{u}^\infty$, that yields the plant output sequence, $\mathbf{y}$, with a minimum tracking error with respect to the set-point, $\mathbf{e}^\infty$.

Regarding the last postulate, we present now the key concept of *perfect control*.

DEFINITION 0.2. *The perfect control input trajectory,*

$$u^{perf} = \left[u_0^{perf^T} \dots u_{T_f-1}^{perf^T} \right]^T,$$

is one that, if injected into the system, produces a null output error trajectory

$$e^i = \left[e_1^{i^T} \dots e_{T_f}^{i^T} \right]^T = [0 \dots 0]^T.$$

It is interesting to note that the impossibility of achieving discrete perfect control, at least for discrete nominal non-delayed linear models, is exclusively related to the input and/or states limits, which are always present in real systems and should be consistent with the control problem constraints. In this regard, a system with slow dynamic might require high input values and input increments to track an abrupt output reference change, producing in this way the constraint activation. If we assume a non-delayed linear model without model mismatch, the perfect control sequence can be found as the solution of the following (unconstrained) open-loop optimization problem

$$\mathbf{u}^{perf} = \arg\min_{\mathbf{u}^i} \sum_{k=1}^{T_f} \left\| e_k^i \right\|^2.$$

On the other hand, for the constrained case, the best possible input sequence, i.e., $\mathbf{u}^\infty$, is obtained from:

$$\mathbf{u}^\infty = \arg\{\min_{\mathbf{u}^i} \sum_{k=1}^{T_f} \left\| e_k^i \right\|^2, \ s.t. \ \mathbf{u} \in \mathbf{U}\},$$

where $\mathbf{U}$ represents the input sequence limits, and will be discussed later.

A no evident consequence of the theoretical concept of perfect control is that only a controller that takes into account the input constraints could be capable of actually approach the perfect control, i.e. to approximate the perfect control up to the point where some of the constraints become active. A controller which does not account for constraints can maintain the system apart from those limits by means of a conservative tuning only. This fact open the possibility to apply a constrained Model Predictive Control (MPC) strategy to account for this kind of problems.

1.2 MPC overview

As was already said, a promising strategy to be used to approach good performances in an iterative learning scheme is the constrained model predictive control, or receding horizon control. This strategy solves, at each time step, an optimization problem to obtain the control action to be applied to the system at the next time. The optimization attempt to minimizes the difference between the desired variable trajectories and a forecast of the system variables, which is made based on a model, subject to the variable constraints (Camacho & Bordons, 2009). So, the first stage to design an MPC is to choose a model. Here, the linear model will be given by:

$$x_{k+1} = Ax_k + Bu_k \tag{3}$$

$$d_{k+1} = d_k \qquad (4)$$

$$y_k = Cx_k + d_k \qquad (5)$$

where $x_k \in \mathbb{R}^n$ is the state at time k, $u_k \in \mathbb{R}^m$ is the manipulated input, $y_k \in \mathbb{R}^l$ is the controlled output, A, B and C are matrices of appropriate dimension, and $d_k \in \mathbb{R}^l$ is an integrating output disturbance (González, Adam & Marchetti, 2008).

Furthermore, and as a part of the system description, input (and possibly input increment) constraints are considered in the following inclusion:

$$u \in U, \qquad (6)$$

where U is given by:

$$U = \{u \in \mathbb{R}^m : u_{min} \le u \le u_{max}\}.$$

A simplified version of the optimization problem that solves on-line (at each time k) a typical stable MPC is as follows:

Problem P0

$$\min_{\{u_{k|k},\dots,u_{k+N-1|k}\}} V_k = \sum_{j=0}^{N-1} \ell \left(e_{k+j|k}, u_{k+j|k}\right) + F\left(e_{k+N|k}\right)$$

subjet to:

$$e_{k+j|k} = Cx_{k+j|k} + d_{k+j} - y_{k+j}^r, \qquad j = 0,\dots,N,$$

$$x_{k+j+1|k} = Ax_{k+j|k} + Bu_{k+j|k}, \qquad j = 0,\dots,N-1,$$

$$u_{k+j|k} \in U, \qquad j = 0,1,\dots,N-1,$$

where $\ell(e,u) := ||e||_Q^2 + ||u||_R^2$, $F(e) := ||e||_P^2$. Matrices Q and R are such that $Q > 0$ and $R \ge 0$. Furthermore, a terminal constraint of the form $x_{k+N|k} \in \Omega$, where Ω is a specific set, is usually included to assure stability. In this general context, some conditions should be fulfilled by the different "*components*" of the formulation (i.e., the terminal matrix penalization P, the terminal set, Ω, etc) to achieve the closed loop stability and the recursive feasibility [2] ((Rawlings and Mayne, 2009)). In the next sections, this basic formulation will be modified to account for learning properties in the context of repetitive systems.

2. Preliminaries

2.1 Problem index definition

As was previously stated, the control strategy proposed in this chapter consists of a basic MPC with learning properties. Then, to clarify the notation to be used along the chapter (that comes form the ILC and the MPC literature), we start by defining the following index variables:

- i: is the iteration or run index, where $i = 0$ is the first run. It goes from 0 to ∞.

---

[2] Recursive feasibility refers to the guarantee that once a feasible initial condition is provided, the controller will guide the system trough a feasible path

- k: is the discrete time into a single run. For a given run, it goes from 0 to T_{f-1} (that is, T_f time instants).
- j: is the discrete time for the MPC predictions. For a given run i, and a given time instant k, it goes from 0 to $H = T_f - k$. To clearly state that j represents the time of a prediction made at a given time instant k, the notation $k + j|k$, wich is usual in MPC literature, will be used.

The control objective for an individual run i is to find an input sequence defined by

$$\mathbf{u}^i := \left[u_0^{i^T} \ldots u_{T_f-1}^{i^T} \right]^T \tag{7}$$

which derives in an output sequence

$$\mathbf{y}^i = \left[y_0^{i^T} \ldots y_{T_f}^{i^T} \right]^T \tag{8}$$

as close as possible to a output reference trajectory

$$\mathbf{y}^r := \left[y_0^{r^T} \ldots y_{T_f}^{r^T} \right]^T. \tag{9}$$

Furthermore, assume that for a given run i there exists an input reference sequence (an input candidate) given by

$$\mathbf{u}^{i^r} := \left[u_0^{i^r T} \ldots u_{T_f-1}^{i^r T} \right]^T \tag{10}$$

and that the output disturbance profile,

$$\mathbf{d}^i = \left[d_0^{i^T} \ldots d_{T_f}^{i^T} \right]^T,$$

is known. During the learning process the disturbance profile is assumed to remain unchanged for several operations. Furthermore, the value $u_{T_f-1}^{i^r}$ represents a stationary input value, satisfying $u_{T_f-1}^r = G^{-1}(y_{T_f}^r - d_{T_f})^i$, for every i, with $G = [C(I - A)^{-1}B]$.

2.2 Convergence analysis

In the context of repetitive systems, we will consider two convergence analyses:

DEFINITION 0.3 (Intra-run convergence). *It concerns the decreasing of a Lyapunov function (associated to the output error) along the run time k, that is, $V\left(y_{k+1}^i - y_{k+1}^r\right) \leq V\left(y_{k+1}^i - y_k^r\right)$ for $k = 1, \ldots, T_{f-1}$, for every single run. If the execution of the control algorithm goes beyond T_f, with $k \to \infty$, and the output reference remains constant at the final reference value ($y_k^r = y_{T_f}^r$ for $T_f \leq k < \infty$) then the intra-run convergence concerns the convergence of the output to the final value of the output reference trajectory $\left(y_{k+1}^i \to y_k^r$ as $k \to \infty\right)$. This convergence was proved in (González et al., 2009a) and presented in this chapter.*

DEFINITION 0.4 (Inter-run convergence). *It concerns the convergence of the output trajectory to the complete reference trajectory from one run to the next one, that is, considering the output of a given run as a vector of T_f components ($\mathbf{y}^i \to \mathbf{y}^r$ as $i \to \infty$).*

3. Basic formulation

In this subsection, a first approach to a new MPC design, which includes learning properties, is presented. It will be assumed that an appropriate input reference sequence $\mathbf{u}^{i^r}$ is available (otherwise, it is possible to use a null constant value), and the disturbance d_k^i as well as the states $x_{k|k}^i$ are estimated. Given that the operation lasts T_f time instants, it is assumed here a *shrinking output horizon* defined as the distance between the current time k and the final time T_f, that is, $H := T_f - k$ (See Figure 2). Under these assumptions the optimization problem to be solved at time k, as part of the single run i, is described as follows:

Problem P1

$$\min_{\{\bar{u}_{k|k}^i,\ldots,\bar{u}_{k+N_s-1|k}^i\}} V_k^i = \sum_{j=0}^{H-1} \ell\left(e_{k+j|k}^i, \bar{u}_{k+j|k}^i\right) + F\left(e_{k+H|k}^i\right)$$

subject to:

$$e_{k+j|k}^i = Cx_{k+j|k}^i + d_{k+j}^i - y_{k+j}^r, \qquad j = 0,\ldots,H,$$

$$x_{k+j+1|k}^i = Ax_{k+j|k}^i + Bu_{k+j|k}^i, \qquad j = 0,\ldots,H-1,$$

$$u_{k+j|k}^i \in U, \qquad j = 0,1,\ldots,H-1, \tag{11}$$

$$u_{k+j|k}^i = u_{k+j}^{i^r} + \bar{u}_{k+j|k}^i, \qquad j = 0,1,\ldots,H-1, \tag{12}$$

$$\bar{u}_{k+j|k}^i = 0, \qquad j \geq N_s, \tag{13}$$

where the (also shrinking) control horizon N_s is given by $N_s = \min(H, N)$ and N is the fixed control horizon introduced before (it is in fact a controller parameter). Notice that predictions with indexes given by $k + H|k$, which are equivalent to $T_f|k$, are in fact prediction for a fixed future time (in the sense that the horizon does not recedes). Because this formulation contains some new concepts, a few remarks are needed to clarify the key points:

Remark 0.1. *In the ith-operation, T_f optimization problems P1 must be solved (from $k = 0$ to $k = T_f - 1$). Each problem gives an optimal input sequence $u_{k+j|k}^{i\,opt}$ for $j = 0, \cdots, H - 1$, and following the typical MPC policy, only the first input of the sequence, $u_{k|k}^{i\,opt}$, is applied to the system.*

Remark 0.2. *The decision variables $\bar{u}_{k+j|k}^i$, are a correction to the input reference sequence $u_{k+j}^{i^r}$ (see Equation (12)), attempting to improve the closed loop predicted performance. $u_{k+j}^{i^r}$ can be seen as the control action of an* underlying stabilizing controller *acting along the whole output horizon, which could be corrected, if necessary, by the control actions $\bar{u}_{k+j|k}^i$. Besides, because of constraints (13), $\bar{u}_{k+j|k}^i$ is different from zero only in the first N_s steps (or predictions) and so, the optimization problem P1 has N_s decision variables (See Figure 2). All along every single run, the input and output references, $u_{k+j}^{i^r}$ and y_{k+j}^r , as well as the disturbance d_{k+j}^i may be interpreted as a set of fixed parameters.*

Remark 0.3. *The convergence analysis for the operation sequence assumes that once the disturbance appears it remains unchanged for the operations that follow. In this way the cost remains bounded despite it represents an infinite summation; this happens because the model used to compute the predictions leads to a final input (and state) that matches $(y_{T_f}^r - d_{T_f}^i)$. Thus, the model output is guided to $(y_{T_f}^r - d_{T_f}^i)$, and the system output is guided to $y_{T_f}^r$.*

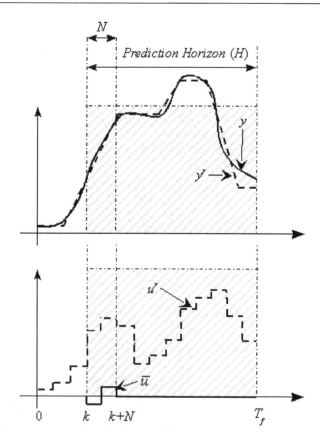

Fig. 2. Diagram representing the MPC optimization problem at a given time k.

3.1 Decreasing properties of the closed-loop cost for a single run

The concept of stability for a finite-duration process is different from the traditional one since, except for some special cases such finite-time escape, boundless of the disturbance effect is trivially guaranteed. In (Srinivasan & Bonvin, 2007), the authors define a quantitative concept of stability by defining a variability index as the induced norm of the variation around a reference (state) trajectory, caused by a variation in the initial condition. Here, we will show two controller properties (Theorem 0.1). 1) The optimal IHMPC cost monotonically decreases w.r.t time k, and 2) if the control algorithm execution goes beyond T_f with $k \to \infty$, and the output reference remains constant at the final reference value ($y_k^r = y_{T_f}^r$ for $k \geq T_f$) then, the IHMPC cost goes to zero as $k \to \infty$, which implies that $y_k^i \to y_{T_f}^r$ as $k \to \infty$.

Theorem 0.1 (intra-run convergence). *Let assume that the disturbance remains constant from one run to the next. Then, for the system (3-5), and the constraint (6), by using the control law derived from the on-line execution of problem P1 in a shrinking horizon manner, the cost is decreasing, that is,* $V_k^{i^{opt}} - V_{k-1}^{i^{opt}} + \ell(e_{k-1}^i, \bar{u}_{k-1}^i) \leq 0$, *for* $0 \leq k \leq T_f - 1$.

Furthermore, the last cost of a given operation "i" is given by:

$$V_{T_f-1}^{i^{opt}} = \ell(e_{T_f-1|T_f-1}^{i^{opt}}, \overline{u}_{T_f-1|T_f-1}^{i^{opt}}) + F(e_{T_f|T_f-1}),$$

and since current and one steps predictions are coincident with the actual values, it follows that:

$$V_{T_f-1}^{i^{opt}} = \ell(e_{T_f-1}^{i^{opt}}, \overline{u}_{T_f-1}^{i^{opt}}) + F(e_{T_f}). \tag{14}$$

Proof See the Appendix. □

Remark 0.4. *The cost $V_k^{i^{opt}}$ of Problem P1 is not a strict Lyapunov function, because the output horizon is not fixed and then, $V_k^{i^{opt}}(e_k^i)$ changes as k increases (in fact, as k increases the cost becomes less demanding because the output horizon is smaller). However, if a virtual infinite output horizon for predictions is defined, and stationary values of output and input references are assumed for $T_f < \infty$ (i.e. $u_{ss}^i = (C(I-A)^{-1}B)^{-1}(y_{ss}^r - d_{ss}^i)$, where d_{ss}^i is the output disturbance at T_f), then by selecting the terminal cost $F(e_{T_f|k}^i)$ to be the sum of the stage penalization $\ell(\cdot, \cdot)$ from T_f to ∞, it is possible to associate $V_k^{i^{opt}}(e_k^i)$ with a fixed (infinite) output horizon. In this way $V_k^{i^{opt}}(e_k^i)$ becomes a Lyapunov function since it is an implicit function of the actual output error e_k^i. To make the terminal cost the infinite tail of the output predictions, it must be defined as*

$$F(e_{T_f|k}^i) = \left\| Cx_{T_f|k}^i + d_{ss}^i - y_{ss}^r \right\|_P^2 = \left\| x_{T_f|k}^i + x_{ss}^{i^r} \right\|_{C^TPC}^2 = \sum_{j=T_f}^{\infty} \left\| x_{j|k}^i + x_{ss}^{i^r} \right\|_{C^TQC}^2$$

$$i = 0, 1, \ldots, T_f - 1,$$

where $x_{ss}^{i^r} = (I-A)^{-1}Bu_{ss}^{i^r}$ and C^TPC is the solution of the following Lyapunov equation: $A^TC^TPCA = C^TPC - C^TQC$. With this choice of the terminal matrix P, the stability results of Theorem 0.1 is stronger since the closed loop becomes Lyapunov stable.

3.2 Discussion about the stability of the closed-loop cost for a single run

Theorem 0.1, together with the assumptions of Remark 0.4, shows convergence characteristics of the Lyapunov function defined by the IHMPC strategy. These concepts can be extended to determine a variability index in order to establish a quantitative concept of stability (β-stability), as it was highlighted by (Srinivasan & Bonvin, 2007). To formulate this extension, the MPC stability conditions (rather than convergence conditions) must be defined, following the stability results presented in ((Scokaert et al., 1997)). An extension of this remark is shown below.

First, we will recall the following exponential stability results.

Theorem 0.2 ((Scokaert et al., 1997)). *Let assume for simplicity that state reference x_k^r is provided, such that $y_k^r = Cx_k^r$, for $k = 0, \ldots, T_f - 1$, and no disturbance is present. If there exist constants a_x, a_u, b_u, c_x, c_u and d_x such that the stage cost $\ell(x,u)$, the terminal cost $F(x)$, and the model matrices A, B and C, in Problem P1, fulfill the following conditions:*

$$\underline{\gamma}.\|x\|^\sigma \leq \ell(x,u) = \|x\|_Q^2 + \|u\|_R^2 \leq c_x.\|x\|^\sigma + c_u.\|u\|^\sigma \tag{15}$$

$$\|u^{iopt}_{k+j|k}\| \le b_u \|x_k\|, \ \textit{for } j = 0, \dots, H-1 \tag{16}$$

$$\|Ax + Bu\| \le a_x \|x\| + a_u \|u\| \tag{17}$$

$$F(x) \le d_x \|x\|^\sigma \tag{18}$$

then, the optimal cost $V^{iopt}_k(x_k)$ *satisfies*

$$\underline{\gamma} \cdot \|x_k\|^\sigma \le V^{iopt}_k(x_k) \le \overline{\gamma} \cdot \|x_k\|^\sigma \tag{19}$$

$$V^{iopt}_k(x_k) \le -\underline{\gamma} \cdot \|x_k\|^\sigma \tag{20}$$

with $\overline{\gamma} = \left(c_x \cdot \sum_{i=0}^{N-1} \alpha_i^\sigma + N \cdot c_u \cdot b_u^\sigma + d_x \cdot \alpha_N^\sigma \right)$, $\alpha_j = a_x \cdot \alpha_{j-1} + a_u \cdot b_u$ *and* $\alpha_0 = a_x + a_u \cdot b_u$.

Proof The proof of this theorem can be seen in (Scokaert et al., 1997). □

Condition (15) is easy to determine in terms of the eigenvalues of matrices Q and R. Condition (16), which are related to the Lipschitz continuity of the input, holds true under certain regularity conditions of the optimization problem.

Now, we define the following *variability index*, which is an induced norm, similar to the one presented in (Srinivasan & Bonvin, 2007):

$$\xi = \max_{V^{iopt}_0 = \delta} \left(\frac{\sum_{k=0}^{T_f - 1} V^{iopt}_k}{V^{iopt}_0} \right)$$

for a small value of $\delta > 0$. With the last definition, the concept of β-stability for finite-duration systems is as follows.

DEFINITION 0.5 ((Scokaert et al., 1997)). *The closed-loop system obtained with the proposed IHMPC controller is intra-run β-stable around the state trajectory x^r_k if there exists $\delta > 0$ such that $\xi \le \beta$.*

Theorem 0.3 (quantitative β-stability). *Let assume for simplicity that a state reference, x^r_k, is provided, such that $y^r_k = Cx^r_k, k = 0, \dots, T_f - 1$, and no disturbance is present. If there exist constants a_x, a_u, b_u, c_x, c_u and d_x as in Theorem 0.2, then, the closed-loop system obtained with system(3) -(5) and the proposed IHMPC controller law is intra-run β-stable around the state trajectory x^r_k, with*

$$\beta = \frac{\left[\overline{\gamma} + \sum_{n=1}^{T_f - 1} \left(\overline{\gamma} - \underline{\gamma} \right)^n \right]}{\underline{\gamma}}.$$

Proof See the Appendix. □

4. IHMPC with learning properties

In the last section we studied the single-operation control problem, where we have assumed that an input reference is available and the output disturbance is known. However, one alternative is defining the input reference and disturbance as the input and disturbance obtained during the last operation (i.e. the last implemented input and the last estimated disturbance, beginning with a constant sequence and a zero value, respectively). In this way, a dual MPC with learning properties accounting for the operations sequence control can be derived. The details of this development are presented next.

4.1 Additional MPC constraints to induce learning properties

For a given operation i, consider the problem P1 with the following additional constraints:

$$u^{i^r}_{k+j} = u^{i-1^{opt}}_{k+j|k+j}, \qquad k = 0, \ldots, T_f - 1, \qquad j = 0, \ldots, H - 1$$

$$d^i_{k+j} = \hat{d}^{i-1}_{k+j}, \qquad k = 1, \ldots, T_f, \qquad j = 0, \ldots, H \tag{21}$$

where $\hat{d}^{i-1}_{k+j}$ represents the disturbance estimation. The first constraint requires updating the input reference for operation i with the last optimal sequence executed in operation $i - 1$ (i.e. $\mathbf{u}^{i^r} = \mathbf{u}^{i-1}$, for $i = 1, 2, \cdots$, with an initial value given by $\mathbf{u}^0 := [G^{-1}y^r_{T_f} \cdots G^{-1}y^r_{T_f}]$). The second one updates the disturbance profile for operation i with the last estimated sequence in operation $i - 1$ (i.e. $\mathbf{d}^i = \hat{\mathbf{d}}^{i-1}$, for $i = 1, 2, \cdots$, with an initial value given by $\mathbf{d}^0 = [0 \cdots 0]$).

Besides, notice that the vector of differences between two consecutive control trajectories, $\boldsymbol{\delta}^i := \mathbf{u}^i - \mathbf{u}^{i-1}$, is given by $\boldsymbol{\delta}^i = \begin{bmatrix} \bar{u}^{i^{opt^T}}_{0|0} & \cdots & \bar{u}^{i^{opt^T}}_{T_f-1|T_f-1} \end{bmatrix}$, i.e., the elements of this vector are the collection of first control movements of the solutions of each optimization problem P1, for $k = 0, \cdots, T_f - 1$.

Remark 0.5. *The input reference update, together with the correction presented in Remark 0.2, has the following consequence: the learning procedure is not achieved by correcting the implemented input action with past information but, by correcting the predicted input sequence with the past input profile , which represents here the learning parameter. In this way better output forecast will be made because the optimization cost has predetermined input information. Figure 3 shows the difference between these two learning procedures.*

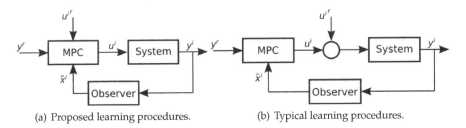

| (a) Proposed learning procedures. | (b) Typical learning procedures. |

Fig. 3. Learning procedures.

Remark 0.6. *The proposed disturbance update implies that the profile estimated by the observer at operation $i - 1$ is not used at operation $i - 1$, but at operation i. This disturbance update works properly when the disturbance remains unmodified for several operations, i.e., when permanent disturbances, or model mismatch, are considered. If the disturbance substantially changes from one operation to next (that is, the disturbance magnitude or the time instant in which the disturbance enter the system change), it is possible to use an additional "current" disturbance correction given by . This correction is then added to permanent disturbance profile at each time k of the operation i.*

4.2 MPC formulation and proposed run cost

Let us consider the following optimization problem:

Problem P2

$$\min_{\{\bar{u}_{k|k}^i,\dots,\bar{u}_{k+T_f-1|k}^i\}} V_k^i$$

subject to (3 - 13) and (21): Run to run convergence means that both, the output error trajectory e^i and the input difference between two consecutive implemented inputs, $\delta^i = u^i - u^{i-1}$, converges to zero as $i \to \infty$. Following an Iterative Learning Control nomenclature, this means that the implemented input, u^i, converges to the perfect control input u^{perf}.

To show this convergence, we will define a cost associated to each run, which penalizes the output error. As it was said, T_f MPC optimization problems are solved at each run i, that is, from $k = 0$ to $k = T_f - 1$. So, a candidate to describe the run cost is as follows:

$$J_i := \sum_{k=0}^{T_f-1} V_k^{i^{opt}}, \tag{22}$$

where $V_k^{i^{opt}}$ represents the optimal cost of the on-line MPC optimization problem at time k, corresponding to the run i.

Notice that, once the optimization problem P2 is solved and an optimal input sequence is obtained, this MPC cost is a function of only $e_{k|k}^{i^{opt}} = \left(y_{k|k}^{i^{opt}} - y_k^r\right) = e_k^i$. Therefore, it makes sense using (22) to define a run cost, since it represents a (finite) sum of positive penalizations of the current output error, i.e., a positive function of e^i. However, since the new run index is made of outputs predictions rather than of actual errors, some cares must be taken into consideration. Firstly, as occurs with usual indexes, we should demonstrate that null output error vectors produce null costs (which is not trivial because of predictions). Then, we should demonstrate that the perfect control input corresponds to a null cost. These two properties, together with an additional one, are presented in the next subsection.

4.3 Some properties of the formulation

One interesting point is to answer what happens if the MPC controller receives as input reference trajectory the *perfect control sequence* presented in the first section. The simplest answer is to associate this situation with a null MPC cost. However, since the proposed MPC controller does not add the input reference (given by the past control profile) to the implemented inputs but to the predicted ones, some care must be taken. Property 0.1, below, assures that for this input reference the MPC cost is null. Without loss of generality we consider in what follows that no disturbances enter the system.

Property 0.1. *If the MPC cost penalization matrices, Q and R, are definite positive ($Q \succ 0$ and $R \succ 0$) and the perfect control input trajectory is a feasible trajectory, then $u^{i^r} = u^{perf} \Leftrightarrow V_k^{i^{opt}} = 0$ for $k = 0, \dots, T_f - 1$; where*

$$V_k^{i^{opt}} = \sum_{j=0}^{H-1} l\left(e_{k+j|k}^{i^{opt}}, \bar{u}_{k+j|k}^{i^{opt}}\right) + F\left(x_{k+H|k}^{i^{opt}}\right).$$

Proof See the Appendix. □

This property allow as to formulate the following one:

Property 0.2. *If the MPC cost penalization matrices, Q and R, are definite positive ($Q \succ 0$ and $R \succ 0$) and perfect control input trajectory is a feasible trajectory, cost (12), which is an implicit function of e^i, is such that, $e^i = 0 \Leftrightarrow J_i = 0$.*

Proof See the Appendix. □

Finally, as trivial corollary of the last two properties, it follows that:

Property 0.3. *If the MPC cost penalization matrices, Q and R, are definite positive, then $u^{i^r} = u^{perf} \Leftrightarrow J^i = 0$. Otherwise, $u^{i^r} \neq u^{perf} \Rightarrow J^i \neq 0$.*

Proof It follows from Property 0.1 and Property 0.2. □

4.4 Main convergence result

Now, we are ready to establish the run to run convergence with the following theorem.

Theorem 0.4. *For the system (3)-(5), by using the control law derived from the on-line execution of problem P2 in a shrinking horizon manner, together with the learning updating (21), and assuming that a feasible perfect control input trajectory there exists, the output error trajectory e^i converges to zero as $i \to \infty$. In addition, δ^i converges to zero as $i \to \infty$ which means that the reference trajectory u^{i^r} converges to u^{perf}.*

Remark 0.7. *In most real systems a perfect control input trajectory is not possible to reach (which represents a system limitation rather than a controller limitation). In this case, the costs $V_k^{i^{opt}}$ will converge to a non-null finite value as $i \to \infty$,and then, since the operation cost J^i is decreasing (see previous proof), it will converge to the smallest possible value. Given that, as was already said, the impossibility to reach perfect control is exclusively related to the input and/or states limits (which should be consistent with the control problem constraints), the proposed strategy will find the best approximation to the perfect control, which constitutes an important advantage of the method.*

Remark 0.8. *In the same way that the intra-run convergence can be extended to determine a variability index in order to establish a quantitative concept of stability (β-stability), for finite-duration systems (Theroem 0.3); the inter-run convergence can be extended to establish stability conditions similar to the ones presented in (Srinivasan & Bonvin, 2007).*

5. Ilustrative examples

Example 1. In order to evaluate the proposed controller performance, we consider first a linear system (Lee & Lee, 1997) given by $G(s) = 1/15s^2 + 8s + 1$. The MPC parameters were tuned as $Q = 1500$, $R = 0.5$ and $T = 1$. Figure 4 shows the obtained performance in the controlled variable where the difference with the reference is undistinguished. Given that the problem assumes that no information about the input reference is available, the input sequence u and u are equals.

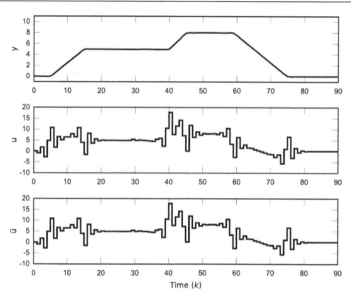

Fig. 4. Reference, output response according to the input variables u and $\bar{u}$

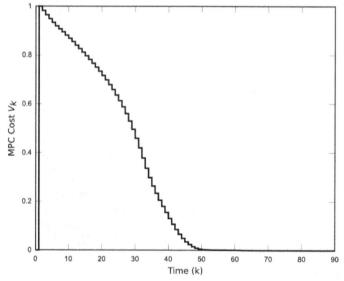

Fig. 5. Normalized MPC cost function. Here, the normalized cost function is obtained as $V_k / V_{k\,max}$.

The MPC cost function is showed in Fig. 5. According to the proof of Theorem 0.1 (nominal case), this cost function is monotonically decreasing.

Example 2. Consider now a nonlinear-batch reactor where an exothermic and irreversible chemical reaction takes place, (Lee & Lee, 1997). The idea is to control the reactor temperature by manipulating the inlet coolant temperature. Furthermore, the manipulated variable has minimum and maximum constrains given by: $Tc_{min} \leq Tc \leq Tc_{max}$, where $Tc_{min} = -25[°C]$, $Tc_{max} = 25[°C]$ and, Tc is written in deviation variable. In addition, to show how the MPC controller works, it is assumed that a previous information about the cooling jacked temperature ($u = Tc$) is available.

Here the proposed MPC was implemented and the MPC parameters were tuned as, $Q = 1000$, $R = 5$ and $T = 1[min]$. The nominal linear model used for predictions is the same proposed by (Adam, 2007).

Figure 6 shows both the reference and the temperature of the batch reactor are expressed in deviation variable. Furthermore, the manipulated variable and the correction made by the MPC, u are shown.

Notice that, 1) the cooling jacked temperature reaches the maximum value and as a consequence the input constraints becomes active in the time interval from 41 minutes to 46 minutes; 2) similarly, when the cooling jacked temperature reaches the minimum value, the other constraint becomes active in the time interval from 72 minutes to 73 minutes; 3) the performance is quite satisfactory in spite of the problem is considerably nonlinear and, 4) given that it is assumed that a previous information about the cooling jacked temperature is available, the correction u is somewhat smaller than u (Fig. 6).

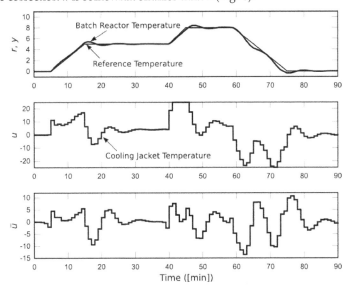

Fig. 6. Temperature reference and controlled temperature of the batch reactor. Also, the cooling jacked temperature (u) and the correction ($\bar{u}$) are showed.

Example 3. In order to evaluate the proposed controller performance we assume a true and nominal process given by (Lee et al., 2000; Lee & Lee, 1997) $G(s) = 1/15s^2 + 8s + 1$ and $G(s) = 0.8/12s^2 + 7s + 1$, respectively. The sampling time adopted to develop the discrete state space model is $T = 1$ and the final batch time is given by $T_f = 90T$. The proposed strategy achieves a good control performance in the first two or three iterations, with a rather

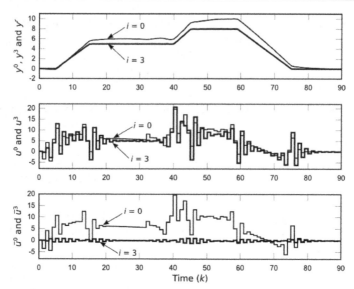

Fig. 7. Output and input responses.

reduced control horizon. The controller parameters are as follows: $Q = 1500, R = 0.05, N = 5$. Figure 7 shows the output response together with the output reference, and the inputs u^i and $\bar{u}^i$, for the first and third iteration. At the first iteration, since the input reference is a constant value ($u^{i_r}_{T_f-1} = 0$), u^i and $\bar{u}^i$ are the same, and the output performance is quite poor (mainly because of the model mismatch). At the third iteration, however, given that a disturbance state is estimated from the previous run, the output response and the output reference are undistinguishable. As expected, the batch error is reduced drastically from run 1 to run 3, while the MPC cost is decreasing (as was established in Theorem 0.1) for each run (Fig. 8a). Notice that the MPC cost is normalized taking into account the maximal value $\left(V_k^i / V_{\max}^i\right)$,

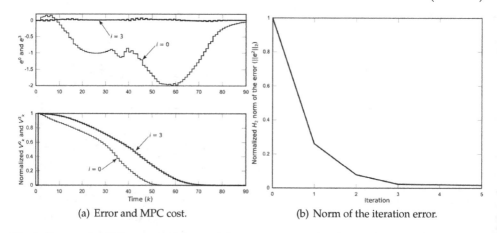

(a) Error and MPC cost.   (b) Norm of the iteration error.

Fig. 8. Error and MPC cost, and Norm of the iteration error for the example 3.

where $V_{max}^1 \approx 1.10^6$ and $V_{max}^1 \approx 286.5$. This shows that the MPC cost J^i decrease from one run to the next, as was stated in Theorem 0.4. Finally, Fig. 8b shows the normalized norm of the error corresponding to each run.

6. Conclusion

In this paper a different formulation of a stable IHMPC with learning properties applied to batch processes is presented. For the case in which the process parameters remain unmodified for several batch runs, the formulation allows a repetitive learning algorithm, which updates the control variable sequence to achieve nominal perfect control performance. Two extension of the present work can be considered. The easier one is the extension to linear-time- variant (LTV) models, which would allow representing the non-linear behavior of the batch processes better. A second extension is to consider the robust case (e.g. by incorporating multi model uncertainty into the MPC formulation). These two issues will be studied in future works.

7. Appendix

Proof of Theorem 0.1

Proof Let $\overline{u}_{k-1}^{i^{opt}} := \left\{ \overline{u}_{k-1|k-1}^{opt}, \cdots, \overline{u}_{k+N_s-2|k-1}^{i^{opt}}, 0, \ldots, 0 \right\}$ and $x_{k-1}^{i^{opt}} := \left\{ x_{k-1|k-1}^{i^{opt}}, \cdots, x_{k+T_f|k-1}^{i^{opt}} \right\}$ be the optimal input and state sequence that are the solution to problem P1 at time $k - 1$, with $k = 1, \cdots, T_f - N$ (that means that the last N optimization problem of a given run i are not considered). The cost corresponding to these variables are

$$
\begin{aligned}
V_{k-1}^{i^{opt}} &= \sum_{j=0}^{N_s-1} \ell \left(e_{k+j-1|k-1}^{i^{opt}}, \overline{u}_{k+j-1|k-1}^{i^{opt}} \right) + \sum_{j=N_s}^{H-1} \ell \left(e_{k+j-1|k-1}^{i^{opt}}, 0 \right) + F \left(x_{k+H-1|k-1}^{i^{opt}} \right) \\
&= \sum_{j=0}^{H-1} \ell \left(e_{k+j-1|k-1}^{i^{opt}}, \overline{u}_{k+j-1|k-1}^{i^{opt}} \right) + F \left(x_{T_f|k-1}^{i^{opt}} \right).
\end{aligned}
\tag{23}
$$

Notice that at time $k - 1$, $H = T_f - k + 1$, since H is a shrinking horizon. Now, let $\overline{u}_k^{i^{feas}} := \left\{ \overline{u}_{k|k-1}^{i^{opt}}, \cdots, \overline{u}_{k+N_s-2|k-1}^{opt}, 0, \ldots, 0 \right\}$ be a feasible solution to problem P1 at time k. Since no new input is injected to the system from time $k - 1$ to time k, and no unknown disturbance is considered, the predicted state at time k, using the feasible input sequence, will be given by $\overline{x}_k^{i^{feas}} := \left\{ x_{k|k-1}^{i^{opt}}, \cdots, x_{k+H-1|k-1}^{i^{opt}}, x_{k+H|k-1}^{i^{opt}} \right\} = \left\{ x_{k|k-1}^{i^{opt}}, \cdots, x_{k+T_f|k-1}^{i^{opt}}, x_{k+T_f+1|k-1}^{i^{opt}} \right\}$. Then, the cost at time k corresponding to the feasible solution $\overline{u}^{i^{feas}}$ is as follows:

$$
\begin{aligned}
V_k^{i^{feas}} &= \sum_{j=0}^{N_s-1} \ell \left(e_{k+j-1}^{i^{opt}}, \overline{u}_{k+j|k-1}^{i^{opt}} \right) + \sum_{j=N_s}^{H-1} \ell \left(e_{k+j-1}^{i^{opt}}, 0 \right) + F \left(e_{k+H|k-1}^{i^{opt}} \right) \\
&= \sum_{j=0}^{H-1} \ell \left(e_{k+j|k-1}^{i^{opt}}, \overline{u}_{k+j|k-1}^{i^{opt}} \right) + F \left(e_{T_f|k-1}^{i^{opt}} \right).
\end{aligned}
\tag{24}
$$

Notice that now $H = T_f - k$, because predictions are referred to time k. Now, subtracting (23) from (24) we have

$$
V_k^{i^{feas}} - V_{k-1}^{i^{opt}} = -\ell \left(e_{k-1|k-1}^{i^{opt}}, \overline{u}_{k-1|k-1}^{i^{opt}} \right).
$$

This means that the optimal cost at time k, which is not grater than the feasible one at the same time, satisfies

$$V_k^{i^{opt}} - V_{k-1}^{i^{opt}} + \ell\left(e_{k-1|k-1}^{i^{opt}}, \bar{u}_{k-1|k-1}^{i^{opt}}\right) \leq 0.$$

Finally, notice that $e_{k-1|k-1}^{i^{opt}}$ and $\bar{u}_{k-1|k-1}^{i^{opt}}$ represent actual (not only predicted) variables. Thus, we can write

$$V_k^{i^{opt}} - V_{k-1}^{i^{opt}} + \ell\left(e_{k-1}^{i^{opt}}, \bar{u}_{k-1}^{i^{opt}}\right) \leq 0. \tag{25}$$

This shows that, whatever the output error is different from zero, the cost decreases when time k increases.

Finally, the decreasing property for $k = T_f - N + 1, \cdots, T_f - 1$, and the last part of the theorem, can be proved following similar steps as before (i.e., finding a feasible solution). $\square$

Proof of Theorem 0.3

Proof From the recursive use of (25), together with (15), (19) and (20), we have

$$V_{k+1}^{i^{opt}} \leq V_k^{i^{opt}} - l\left(\bar{x}_k^i, \bar{u}_k^i\right) \leq \bar{\gamma}.\|\bar{x}_k^i\|^\sigma - \underline{\gamma}.\|\bar{x}_k^i\|^\sigma = (\bar{\gamma} - \underline{\gamma})\|\bar{x}_k^i\|^\sigma,$$

for $k = 0, \ldots, T_f - 2$. So we can write:

$$\sum_{k=0}^{T_f-1} V_k^{i^{opt}} \leq \left[r + \sum_{n=1}^{T_f-1}(\bar{\gamma} - \underline{\gamma})^n\right] \cdot \|\bar{x}_0^i\|^\sigma.$$

Therefore,

$$\frac{\sum_{k=0}^{T_f-1} V_k^{i^{opt}}}{V_0^{i^{opt}}} \leq \frac{\left[\bar{\gamma} + \sum_{n=1}^{T_f-1}(\bar{\gamma} - \underline{\gamma})^n\right]}{\underline{\gamma}},$$

since $\underline{\gamma}.\|x_0\|^\sigma$ is a lower bound of $V_0^{i^{opt}}$ (that is, $\underline{\gamma}.\|x_0\|^\sigma \leq V_0^{i^{opt}}$).

Finally,

$$\beta = \frac{\left[\bar{\gamma} + \sum_{n=1}^{T_f-1}\left(\bar{\gamma} - \underline{\gamma}\right)^n\right]}{\underline{\gamma}}.$$

$\square$

Proof of Property 0.1

Proof $\Leftarrow$) Let us assume that $V_k^{i^{opt}} = 0$, for $k = 0, ..., T_f - 1$. Then, the optimal predicted output error and input will be given by $e_{k+j|k}^{i^{opt}} = 0$, $j = 0, ..., T_f$ and $\bar{u}_{k+j|k}^{i^{opt}} = 0$, for $j = 0, ..., T_f - 1$, respectively. If $e_{k+j|k}^{i^{opt}}$ adn $\bar{u}_{k+j|k}^{i^{opt}} = 0$ simultaneously, it follows that $u_k^{i^r} = u_k^{perf}$ for $k = 0, \ldots, T_f - 1$, since it is the only input sequence that produces null predicted output error (otherwise, the optimization will necessarily find an equilibrium such that $\left\| e_{k+1|k}^{i^{opt}} \right\| > 0$ and $\left\| \overline{u}_{k|k}^{i^{opt}} \right\| > 0$, provided that $Q \succ 0$ and $R \succ 0$ by hypothesis). Consequently, $\mathbf{u}^{i^r} = \mathbf{u}^{perf}$.

$\Rightarrow$) Let us assume that $\mathbf{u}_k^{i^r} = \mathbf{u}_k^{perf}$. Because of the definition of the perfect control input, the optimization problem without any input correction will produce a sequence of null output error predictions given by

$$e_{k|k}^i = 0$$

$$e_{k+1|k}^i = Cx_{k+1|k}^i - y_{k+1}^r = C\left[Ax_{k|k}^i + Bu_k^{perf} \right] - y_{k+1}^r = 0$$

$$\vdots$$

$$e_{k+T_f|k}^i = Cx_{k+T_f|k}^i - y_{k+T_f}^r = C\left[A^{T_f} x_{k|k}^i + ABu_k^{perf} \cdots Bu_{k+T_f-1}^{perf} \right] - y_{k+T_f}^r = 0.$$

Consequently, the optimal sequence of decision variables (predicted inputs) will be $\overline{u}_{k+j|k}^{i^o pt} = 0$ for $k = 0, \ldots, T_f - 1$ and $j = 0, \ldots, T_f - 1$, since no correction is needed to achieve null predicted output error. This means that $V_k^{i^{opt}} = 0$ for $k = 0, \ldots, T_f - 1$. $\square$

Proof of Property 0.2

Proof $\Rightarrow$) Let us assume that $\mathbf{e}^i = 0$. This means that $e_{k|k}^{i^{opt}} = 0$, for $k = 0, \ldots, T_f$. Now, assume that the input reference vector is different from the perfect control input, $\mathbf{u}^{i^r} \neq \mathbf{u}^{perf}$, and consider the output error predictions necessary to compute the MPC cost V_k^i:

$$e_{k|k}^i = 0$$

$$e_{k+1|k}^i = Cx_{k+1|k}^i - y_{k+1}^r = C\left[Ax_{k|k}^i + Bu_k^{i^r} + B\overline{u}_{k|k}^i \right] - y_{k+1}^r$$

$$\vdots$$

Since u^{i^r} is not an element of the perfect control input, then $\left[Ax_{k|k}^i + Bu_k^{i^r} \right] \neq 0$. Consequently, (assuming that CB is invertible) the input $\overline{u}_{k|k}^{i^*}$ necessary to make $e_{k+1|k}^{i^{opt}} = 0$, will be given by:

$$\overline{u}_{k|k}^{i^*} = (CB)^{-1} \left(y_{k+1}^r - C\left[Ax_{k|k}^i + Bu_k^{i^r} \right] \right),$$

which is a non null value. However, the optimization will necessary find an equilibrium solution such that $\left\| e_{k+1|k}^{i^{opt}} \right\| > 0$ and $\left\| \overline{u}_{k|k}^{i^{opt}} \right\| < \left\| \overline{u}_{k|k}^{i^*} \right\|$, since $Q \succ 0$ and $R \succ 0$ by hypothesis. This implies that [3] $e_{k+1|k}^{i^{opt}} = e_{k+1|k+1}^{i^{opt}} \neq 0$, contradicting the initial assumption of null output error.

From this reasoning for subsequent output errors, it follows that the only possible input reference to achieve $e^i = 0$ will be the perfect control input ($\mathbf{u}^{i^r} = \mathbf{u}^{perf}$). If this is the case, it follows that $V_k^{i^{opt}} = 0$, for $k = 0, \ldots, T_f$ (Property 0.1), and so, $J^i = 0$

---

[3] Note that for the nominal case is $e_{k+1|k+1}^i = e_{k+1|k}^i$

$\Leftarrow$) Let us assume that $J^i = 0$. Then, $V_k^{i^{opt}} = 0$, which implies that $e_{k+j|k}^{i^{opt}} = 0$, for $k = 0, ..., T_f$ and for $j = 0, ..., T_f$. Particularly, $e_{k|k}^{i^{opt}} = 0$, for $k = 0, ..., T_f$, which implies $\mathbf{e}^i = 0$. $\square$

Proof of Theorem 0.4

Proof The idea here is to show that $V_k^{i^{opt}} \leq V_k^{i-1^{opt}}$ for $k = 0, ..., T_f - 1$ and so, $J_i \leq J_{i-1}$. First, let us consider the case in which the sequence of T_f optimization problems P2 *do nothing* at a given run i. That is, we will consider the case in which

$$\delta^i = \left[\bar{u}_{0|0}^{i^{opt T}} \; ... \; \bar{u}_{T_f-1|T_f-1}^{i^{opt T}} \right] = [0 ... 0]^T,$$

for a given run i. So, for the nominal case, the total actual input will be given by

$$\mathbf{u}^i = \mathbf{u}^{i-1} = \left[u_0^{i-1^T} \; ... \; u_{T_f-1}^{i-1^T} \right]^T = \left[u_{0|0}^{i-1^{opt T}} \; ... \; u_{T_f-1|T_f-1}^{i-1^{opt T}} \right]^T$$

and the run cost corresponding to this (fictitious) input sequence will be given by

$$\tilde{J}_i = \sum_{k=1}^{T_f} \tilde{V}_k^i,$$

where

$$\tilde{V}_k^i := \sum_{j=0}^{N_s-1} \ell \left(e_{k+j|k+j}^{i-1^{opt}}, \overbrace{\bar{u}_{k+j|k+j}^i}^{=0} \right) + \sum_{j=N_s}^{H-1} \ell \left(e_{k+j|k+j}^{i-1^{opt}}, 0 \right) + F \left(x_{k+H|k+H-1}^{i-1^{opt}} \right)$$

$$= \sum_{j=0}^{H-1} l \left(e_{k+j}^{i-1} \right) + F \left(x_{k+H}^{i-1} \right) \tag{26}$$

Since the input reference, $u_{k+j}^{i^r}$ that uses each optimization problems is given by $u_{k+j}^{i^r} = u_{k+j|k+j}^{i-1^{opt}}$, then the resulting output error will be given by $e_{k+j|k+j}^{i^{opt}} = e_{k+j}^{i-1}$ for $j = 0, ..., H$. In other words, the open loop output error predictions made by the MPC optimization at each time k, for a given run i, will be the actual (implemented) output error of the past run $i-1$. Here it must be noticed that e_{k+j}^{i-1} refers to the actual error of the system, that is, the error produced by the implemented input $u_{k+j-1}^{i-1} = u_{k+j-1|k+j-1}^{i-1^{opt}}$. Moreover, because of the proposed inter run convergence constraint, the implemented input will be $u_{T_f-1}^{i-V}$, for $j \geq H$.

Let now consider the optimal MPC costs corresponding to $k = 0, ..., T_f - 1$, of a given run $i-1$. From the recursive use of (12) we have

$$V_1^{i-1^{opt}} + \ell \left(e_0^{i-1}, \bar{u}_0^{i-1} \right) \leq V_0^{i-1^{opt}}$$

$$\vdots$$

$$V_{T_f-1}^{i-1^{opt}} + \ell\left(e_{T_f-2}^{i-1}, \overline{u}_{T_f-2}^{i-1}\right) \le V_{T_f-2}^{i-1^{opt}}$$

Then, adding the second term of the left hand side of each inequality to both sides of the next one, and rearranging the terms, we can write

$$V_{T_f-1}^{i-1^{opt}} + \ell\left(e_{T_f-2}^{i-1}, \overline{u}_{T_f-2}^{i-1}\right) + \cdots + \ell\left(e_0^{i-1}, \overline{u}_0^{i-1}\right) \le V_0^{i-1^{opt}} \qquad (27)$$

From (14), the cost $V_{T_f-1}^{i-1^{opt}}$, which is the cost at the end of the run $i-1$, will be given by,

$$V_{T_f-1}^{i-1^{opt}} = \ell\left(e_{T_f-1}^{i-1}, \overline{u}_{T_f-1}^{i-1}\right) + F\left(x_{T_f}^{i-1}\right). \qquad (28)$$

Therefore, by substituting (28) in (27), we have

$$F\left(x_{T_f}^{i-1}\right) + \ell\left(e_{T_f-1}^{i-1}, \overline{u}_{T_f-1}^{i-1}\right) + \cdots + l\left(e_0^{i-1}, \overline{u}_0^{i-1}\right) \le V_1^{i-1^{opt}} \qquad (29)$$

Now, the pseudo cost (26) at time $k=0$, $\tilde{V}_0^i$, can be written as

$$\tilde{V}_0^i = \sum_{j=0}^{T_f-1} l\left(e_j^{i-1}, 0\right) + F\left(x_{T_f}^{i-1}\right)$$

$$= \sum_{j=0}^{T_f-1} l\left(e_j^{i-1}, \overline{u}_j^{i-1}\right) + F\left(x_{T_f}^{i-1}\right) - \sum_{j=0}^{T_f-1} \left\|\overline{u}_j^{i-1}\right\| \qquad (30)$$

and from the comparison of the left hand side of inequality (29) with (30), it follows that

$$\tilde{V}_0^i = V_0^{i-1^{opt}} - \sum_{j=0}^{T_f-1} \left\|\overline{u}_j^{i-1}\right\|$$

Repeating now this reasoning for $k = 1, ..., T_f - 1$ we conclude that

$$\tilde{V}_k^i = V_k^{i-1^{opt}} - \sum_{j=k}^{T_f-1} \left\|\overline{u}_j^{i-1}\right\|, \qquad k = 0, \ldots, T_f - 1$$

Therefore, from the definition of the run cost J i we have[4]

$$\tilde{J}_i \le J_{i-1} - \sum_{k=0}^{T_f-1}\sum_{j=k}^{T_f-1} \left\|\overline{u}_j^{i-1}\right\|. \qquad (31)$$

---

[4] Notice that, if the run i implements the manipulated variable $u_j^i = u_j^{i-1} + \overline{u}_j^{i-1}, j = 0, 1, \ldots, T_f - 1$ and $\overline{u}_j^{i-1} \ne 0$ for some j; then, according to 31 $\tilde{J}_i < J_{i-1}$. Unnaturally, to have found a non null optimal solution in the run $i-1$ is sufficient to have a strictly smaller cost for the run i.

The MPC costs V_k^i is such that $V_k^{i\,opt} \leq \tilde{V}_k^i$, since the solution $\bar{u}_{k+j|k}^i = 0$, for $j = 0, \ldots, H$ is a feasible solution for problem P2 at each time k. This implies that

$$J_i \leq \tilde{J}^i. \tag{32}$$

From (31) and (32) we have

$$J_i \leq \tilde{J}_i \leq J_{i-1} - \sum_{k=0}^{T_f-1} \sum_{j=k}^{T_f-1} \left\| \bar{u}_j^{i-1} \right\|. \tag{33}$$

which means that the run costs are strictly decreasing if at least one of the optimization problems corresponding to the run $i - 1$ find a solution $\bar{u}_{k+j|k}^{i-1} \neq 0$. As a result, two options arise:

I) Let us assume that $\mathbf{u}^{i^r} \neq \mathbf{u}^{perf}$. Then, by property 0.3, $J^i \neq 0$ and following the reasoning used in the proof of Property 0.2, $\bar{u}_j^i \neq 0$, for some $1 \leq j \leq T_f$. Then, according to 33, $J_{i+1} \leq \tilde{J}_{i+1} \leq J_i - \sum_{k=0}^{T_f-1} \sum_{j=k}^{T_f-1} \left\| \bar{u}_j^{i-1} \right\|$ with $\left\| \bar{u}_j^i \right\| > 0$ for some $1 \leq j \leq T_f - 1$.

The sequence J^i will stop decreasing only if $\sum_{j=0}^{T_f-1} \left\| \bar{u}_j^i \right\| = 0$. In addiction, if $\sum_{j=0}^{T_f-1} \left\| \bar{u}_j^i \right\| = 0$, then $\mathbf{u}^{i^r} = \mathbf{u}^{perf}$, which implies that $J_i = 0$. Therefore: $\lim_{i \to \infty} J_i = 0$, which, by Property 0.2 implies that $\lim_{i \to \infty} \mathbf{e}_i = 0$.

Notice that the last limit implies that $\lim_{i \to \infty} \delta^i = 0$ and consequently, $\lim_{i \to \infty} \mathbf{u}^{i^r} = \mathbf{u}^{perf}$.

II) Let us assume that $\mathbf{u}^{i^r} = \mathbf{u}^{perf}$. Then, by Corollary 0.3, $J_i = 0$, and according to (33), $J_{i+1} = \tilde{J}_{i+1} = J_i = 0$. Consequently, by Property 0.2, $\mathbf{e}^i = 0$. $\square$

8. References

Adam, E. J. (2007). Adaptive iterative learning control applied to nonlinear batch reactor, *XII Reunión de Trabajo en Procesamiento de la Información y Control (RPIC 2007)*, Río Gallegos, Santa Cruz, Argentina, (CD version).

Arimoto, S., Kawamura, S. & Miyazaki, F. (1984). Bettering operation of robots by learning, *Journal of Robotic System* 1(2): 123–140.

Bien, Z. & Xu, J.-X. (1998). *Iterative Learning Control: Analysis, Design, Iteration and Application*, Kluwer Academic Publishers.

Bonvin, D., Srinivasan, B. & Hunkeler, D. (2006). Control and optimization of bath processes - improvement of process operation in the production of specialty chemicals, *IEEE Trans. Control Syst. Mag.* 26(2): 34–45.

Camacho E. F. & Bordons C. (2009). *Model Predictive Control*, 2nd Edition, Springer-Verlag.

Chen, Y. & Wen, C. (1999). *Iterative Learning Control*, Springer Verlag.

Cueli, J. R. & Bordons, C. (2008). Iterative nonlinear model predictive control. stability, robustness and applications, *Control Engineering Practice* 16: 1023–1034.

Doh, T. Y., J. H. moon, K. B. J. & Chung, M. J. (1999). Robust ilc with current feedback for uncertain linear system, *Int. J. Syst. Sci* 30(1): 39–47.

González, H. A., Adam, E. J. & Marchetti, J. L. (2008). Conditions for offset elimination in state space receding horizon controllers: A tutorial analysis, *Chemical Engineering and Processing* 47: 2184–2194.

González, H. A., Adam, E. J., Odloak, D. O. & Marchetti, J. L. (2009). Infinite horizon mpc applied to batch processes. Part I, *XII Reunión de Trabajo en Procesamiento de la Información y Control (RPIC 2009)*, Rosario, Santa Fe, Argentina, (CD version).

González, H. A., Adam, E. J., Odloak, D. O. & Marchetti, J. L. (2009). Infinite horizon mpc applied to batch processes. Part II, *XII Reunión de Trabajo en Procesamiento de la Información y Control (RPIC 2009)*, Rosario, Santa Fe, Argentina, (CD version).

González, H. A., Odloak, D. O. & Sotomayor, O. A. (2008). Stable ihmpc for unstable systems, *IFAC*.

Horowitz, R. (1993). Learning control of robot manipulators, *Journal of Dynamic Systems, Measurement and Control* 115: 402–411.

Kurek, J. E. & Zaremba, M. B. (1993). Iterative learning control synthesis based on 2-d system theory, *IEEE Trans. Automat. Contr.* 38(1): 121–125.

Lee, J. H., Lee, K. S. & Kim, W. C. (2000). Model-based iterative learning control with a quadratic criterion for time-varying linear systems, *Automatica* 36: 641–657.

Lee, K. S. & Lee, J. H. (1997). Model predictive control for nonlinear batch processes with asymptotically perfect tracking, *Computer chemical Engng* 24: 873–879.

Lee, K. S. & Lee, J. H. (1999). Model predictive control technique combined with iterative learning for batch processes, *AIChE Journal* 45: 2175–2187.

Lee, K. S. & Lee, J. H. (2003). Iterative learning control-based batch process control technique for integrated control of end product properties and transient profiles of process variables, *Computer chemical Engng* 24: 873–879.

Moon, J. H., Doh, T. Y. & Chung, M. J. (1998). A robust approach to iterative learning control design for uncertain system, *Automatica* 34(8): 1001–1004.

Moyne, J. E., Castillo E. & Hurwitz A.M. (2001). *Run to run control in semiconductor manufacturing*, CRC Press.

Norrlöf, M. (2000). *Iterative Learning Control. Analysis, design, and experiments*, Ph. D. Thesis, LinkÃűpings Universtet, Sweden.

Rawlings J. B. and Mayne D. Q. (2009). *Model Predictive Control: Theory and Design*, Nob Hill Publishing.

Roover, D. D. (1996). Synthesis of a robust iterative learning control using an h approach, *Proc. 35th Conf. Decision Control*, Kobe, Japan, pp. 3044–3049.

Rossiter, J. A. (2003). *Model-Based Predictive Control*, CRC Press.

Scholten, P. (2000). *Iterative Learning Control: Analysis. A design for a linear motor motion system*, M. Sc. Thesis University of Twente.

Scokaert, P. O. M., Rawlings, J. B. & Meadows, E. S. (1997). Discrete-time stability with perturbations: application to moel predictive control.

Srinivasan, B. & Bonvin, D. (2007) Controllability and stability of repetitive batch processes, *Journal of Process Control* 17: 285–295.

Tan, K.K., Huang S.N., Lee T.H. & Tay, A. (2007). Disturbance compensation incorporated in predictive control system using a repetitive learning approach, *System and Control Letters* 56: 75–82.

Tayebi, A. & Zaremba, M. B. (2003). Robust iterative learning control design is straightforward for uncertain lti system satisfying the robust performance condition, *IEEE Trans. Automat. Contr.* 48(1): 101–106.

Yao, K., Gao F., & Allgöwer, F. (2007). Barrel temperature control during operation transition in injection molding, *Control Engineering Practice* 16(11): 1259-1264.

10

Model Predictive Control Relevant Identification

Rodrigo Alvite Romano[1], Alain Segundo Potts[2] and Claudio Garcia[2]
[1]*Instituto Mauá de Tecnologia - IMT*
[2]*Escola Politécnica da Universidade de São Paulo - EPUSP*
Brazil

1. Introduction

Model predictive control (MPC) is a multivariable feedback control technique used in a wide range of practical settings, such as industrial process control, stochastic control in economics, automotive and aerospace applications. As they are able to handle hard input and output constraints, a system can be controlled near its physical limits, which frequently results in performance superior to linear controllers (Maciejowski, 2002), specially for multivariable systems. At each sampling instant, predictive controllers solve an optimization problem to compute the control action over a finite time horizon. Then, the first of the control actions from that horizon is applied to the system. In the next sample time, this policy is repeated, with the time horizon shifted one sample forward. The optimization problem takes into account estimates of the system output, which are computed with the input-output data up to that instant, through a mathematical model. Hence, in MPC applications, a suitable model to generate accurate output predictions in a specific horizon is crucial, so that high performance closed-loop control is achieved. Actually, model development is considered to be, by far, the most expensive and time-consuming task in implementing a model predictive controller (Zhu & Butoyi, 2002).

This chapter aims at discussing parameter estimation techniques to generate suitable models for predictive controllers. Such a discussion is based on the most noticeable approaches in MPC relevant identification literature. The first contribution to be emphasized is that these methods are described in a multivariable context. Furthermore, the comparisons performed between the presented techniques are pointed as another main contribution, since they provide insights into numerical issues and the exactness of each parameter estimation approach for predictive control.

2. System identification for model predictive control

The dominating approach of the system identification techniques is based on the classical prediction error method (PEM) (Ljung, 1999), which is based on one-step ahead predictors. Predictive control applications demand models that generate reliable predictions over an entire prediction horizon. Therefore, parameters estimated from objective functions based on multi-step ahead predictors, generally result in better models for MPC applications (see Shook et al. (1991) and Gopaluni et al. (2004) for rigorous arguments). Since the last decade, an intense research has been done in order to develop system identification methods focused on providing appropriate models for model predictive control. Such methods are denoted as model relevant identification (MRI) in the literature. Strictly speaking, MRI algorithms deal with the problem of estimating model parameters by minimizing multi-step objective functions.

Theoretically, if the model structure exactly matches the structure of the actual system, then the model estimated from a one-step ahead predictor is equivalent to the maximum likelihood estimate, which also provides optimal multi-step ahead predictions. However, in practice, even around an operating point, it is not possible to propose a linear model structure that exactly matches the system to be identified. Consequently, any estimated model has modeling errors associated with the identification algorithm. In these circumstances, models tuned for multi-step ahead predictions are more adequate for high closed-loop performance when using predictive controllers (Huang & Wang, 1999). In other words, when there is a certain amount of bias due to under-modeling (which is the more typical case), the MRI may be considered a way of distributing this bias in a frequency range that is less important for control purposes (Gopaluni et al., 2003).

Before formulating the parameter estimation problem in the MRI context, the discrete-time linear model structures to be used are specified.

2.1 Model parameterization

Consider a linear discrete-time system S with m inputs and p outputs

$$y(t) = G_0(q)u(t) + H_0(q)e(t),\tag{1}$$

where $y(t)$ is the p-dimensional output column vector at sampling instant t, $u(t)$ is the m-dimensional input column vector and $e(t)$ is a p-dimensional zero-mean white noise column vector with a $p \times p$ diagonal covariance matrix R. The system S is characterized by the filter matrices $G_0(q)$ and $H_0(q)$. The process[1] and the noise models of S are denoted by $G(q, \theta)$ and $H(q, \theta)$, respectively. In this work, the system model is represented using matrix fraction descriptions (MFD) of the form

$$G(q, \theta) = F^{-1}(q)B(q)\tag{2}$$
$$H(q, \theta) = D^{-1}(q)C(q).\tag{3}$$

where $B(q)$, $C(q)$, $D(q)$ and $F(q)$ are matrices of polynomials in the shift operator q with dimensions $p \times m$, $p \times p$, $p \times p$ and $p \times p$, respectively. The parameter vector θ is composed of the coefficients of the polynomials in such matrices. Thus, in order to determine θ, one needs to further specify the polynomial matrices in (2) and (3). The matrix $B(q)$ takes the form

$$B(q) = \begin{bmatrix} B_{11}(q) & \cdots & B_{1m}(q) \\ \vdots & \ddots & \vdots \\ B_{p1}(q) & \cdots & B_{pm}(q) \end{bmatrix},\tag{4}$$

whose entries are $\mu_{ij} - 1$ degree polynomials

$$B_{ij}(q) = b_{ij}^{(1)}q^{-1} + \ldots + b_{ij}^{(\mu_{ij})}q^{-\mu_{ij}},$$

---

[1] Sometimes (Ljung, 1999; Zhu, 2001, e.g.), the process model is referred to as transfer function.

for $i \in \{1, \ldots, p\}$ and $j \in \{1, \ldots, m\}$. One of the simplest choice to parameterize the other matrices is through the diagonal form MFD, in which $C(q)$, $D(q)$ and $F(q)$ are diagonal polynomial matrices and their nonzero polynomials are all monic, e.g.,

$$F(q) = \begin{bmatrix} F_{11}(q) & 0 & \cdots & 0 \\ 0 & F_{22}(q) & & \vdots \\ \vdots & & \ddots & 0 \\ 0 & \cdots & 0 & F_{pp}(q) \end{bmatrix}, \tag{5}$$

where the entries of $F(q)$ are ν_i degree polynomials of the form

$$F_{ii}(q) = 1 + f_{ii}^{(1)} q^{-1} + \ldots + f_{ii}^{(\nu_i)} q^{-\nu_i},$$

for each $i \in \{1, 2, \ldots, p\}$. The diagonal matrices $C(q)$ and $D(q)$, as well as their respective entries, are defined analogously.

When the diagonal form is adopted, it is possible to decouple the multi-input multi-output model into a set of p multi-input single-output (MISO) models in the form

$$y_1(t) = F_{11}^{-1}(q) \sum_{j=1}^{m} B_{1j}(q) u_j(t) + \frac{C_{11}(q)}{D_{11}(q)} e_1(t)$$

$$\vdots = \vdots \tag{6}$$

$$y_p(t) = F_{pp}^{-1}(q) \sum_{j=1}^{m} B_{pj}(q) u_j(t) + \frac{C_{pp}(q)}{D_{pp}(q)} e_p(t),$$

in which y_i and u_j denote the ith output and the jth input, respectively.

Unless otherwise stated, it is assumed that all the nonzero polynomials of the matrices have the same degree n, that is to say $\mu_{ij} = \nu_i = n$, for $i \in \{1, \ldots, p\}$ and $j \in \{1, \ldots, m\}$. Although this degree is in general not the same as the McMillan degree, this choice considerably simplifies the order selection problem and, consequently, makes the model structure more suitable for applications in large scale processes.

Besides being simple to understand, the diagonal form has some relevant properties for applications in system identification (Zhu, 2001). The main of them is that algorithms developed for the SISO (single-input single-output) processes can be directly generalized for the multivariable case. Nevertheless, if there are dynamic iterations between different outputs, the estimated model based on the diagonal form can present a larger bias error (Laurí et al., 2010). Alternatively, one can add elements outside the diagonal of $F(q)$, not necessarily monic polynomials, with the purpose of incorporating the dynamic iteration between the process outputs. This approach gives rise to another MFD named "full polynomial form" (Ljung, 1999), in which any $F(q)$ entry may be nonzero. This parameterization is also employed in one of the identification methods described in Section 3.

Next, the multi-step objective function used as the basis for the development of the MRI algorithms is presented.

2.2 The model relevant identification cost function

Firstly, let us define the $p \times p$ filter matrix

$$W_k(q,\theta) \triangleq \left(\sum_{l=0}^{k-1} h(l)q^{-l} \right) H^{-1}(q,\theta) , \tag{7}$$

where $h(l)$ is the lth impulse response coefficient of $H(q,\theta)$.

Thus, the k-step ahead predictor of the output vector (i.e., the output prediction equation at $t+k$ with data available up to instant t) may be expressed as (Ljung, 1999)

$$\hat{y}(t+k|t,\theta) = W_k(q,\theta)G(q,\theta)u(t+k) + (I - W_k(q,\theta))y(t+k) . \tag{8}$$

According to (8), the k-step ahead prediction error is

$$\begin{aligned} \varepsilon(t+k|t,\theta) &= y(t+k) - \hat{y}(t+k|t,\theta) \\ &= W_k(q,\theta)\left(y(t+k) - G(q,\theta)u(t+k)\right) . \end{aligned} \tag{9}$$

From (7)-(9), note that the k-step prediction error is related to the one-step through the filter matrix

$$L_k(q,\theta) \triangleq \sum_{i=0}^{k-1} h(i)q^{-i} , \tag{10}$$

such that

$$\varepsilon(t+k|t) = L_k(q,\theta)\varepsilon(t+k|t+k-1) . \tag{11}$$

As argued previously, the main objective of the MRI methods is to provide models that are optimized for the generation of predictions over an entire prediction horizon. So, a natural choice for the criterion of the parameter estimation problem is the cost function

$$J_{\text{multi}}(P,\theta) = \sum_{k=1}^{P} \sum_{t=0}^{N-k} \|\varepsilon(t+k|t,\theta)\|_2^2 , \tag{12}$$

where $\|\cdot\|_2$ denotes the ℓ_2 norm. Hence, $J_{\text{multi}}(P,\theta)$ quantifies the mean-square error, based on predictions ranging from 1 to P steps ahead in a dataset of length N.

The challenge in estimating the model parameters by minimizing (12) is that such a criterion is highly nonlinear in the model parameters. Therefore, suitable optimization algorithms are necessary, so that local minima or convergence problems are avoided. Strictly speaking, the identification methods to be presented aims at estimating the model parameters based on J_{multi}.

3. Model parameter estimation methods

In recent years, distinct MRI techniques were proposed based on different principles. One of them, conceived by Rossiter & Kouvaritakis (2001), differs from the others since it proposes the use of multiple models to generate the predictions. Thus, an optimized model is estimated for each k-step ahead prediction. In spite of providing "optimal" predictions for the entire horizon, the number of parameters involved can be quite large, specially for multi-input and multi-output processes. It is known that the variance of the parameter estimates is proportional to the ratio between the number of parameters and the dataset length (Ljung, 1999). Hence, the main drawback of the multi-model approach is the amount of data required to estimate a reasonable model set. Such amount of data may be prohibitive in practical situations (Gopaluni et al., 2003). Moreover, most of the MPC algorithms are based on a single model. For these reasons, the multi-model method is not considered in further analysis.

In the pioneering work by Shook et al. (1991), the MRI is performed in the context of data prefiltering using SISO ARX (Auto Regressive with eXternal input) type models. Huang & Wang (1999) extended the previous method, so that a general model structure (e.g., Box-Jenkins) could be employed. Some authors, such as (Gopaluni et al., 2003; Laurí et al., 2010), deal with the parameter estimation problem directly minimizing the MRI cost function, using nonlinear optimization techniques. In another approach, proposed by Gopaluni et al. (2004), the focus is given to the noise model parameter estimation. In this approach, a non-parsimonious process model is estimated, in order to eliminate bias errors (which are caused by under-modeling). Then, with a fixed process model, the parameters of the noise model are obtained by minimizing the cost function (12).

In the following subsections, the main MRI techniques are described in more details.

3.1 The prefiltering approach

3.1.1 The basic idea

For the sake of simplicity, the basic idea behind the prefiltering approach is shown using the SISO case ($m = p = 1$). Nevertheless, its worth mentioning that the conclusions directly apply to MIMO models represented in the diagonal form MFD.

In this case, based on predictor (9), the MRI cost function (12) can be rewritten as

$$J_{\text{multi}}(P,\theta) = \sum_{k=1}^{P} \sum_{t=0}^{N-k} \left(\frac{L_k(q,\theta)}{H(q,\theta)} \left(y(t+k) - G(q,\theta)u(t+k) \right) \right)^2 . \tag{13}$$

If we introduce an auxiliary variable $\tilde{G}(q,\theta)$ that takes into account the deterministic model mismatch, that is

$$\tilde{G}(q,\theta) \triangleq G_0(q) - G(q,\theta) ,$$

then, substituting (1) into (13) gives

$$J_{\text{multi}}(P,\theta) = \sum_{k=1}^{P} \sum_{t=0}^{N-k} \left(\frac{L_k(q,\theta)}{H(q,\theta)} \left(\tilde{G}(q,\theta)u(t+k) + H_0(q)e(t+k) \right) \right)^2$$

$$= \sum_{k=1}^{P} \sum_{t=0}^{N-k} \left(\frac{L_k(q,\theta)}{H(q,\theta)} \left(\begin{bmatrix} \tilde{G}(q,\theta) & H_0(q) \end{bmatrix} \begin{bmatrix} u(t+k) \\ e(t+k) \end{bmatrix} \right) \right)^2 . \tag{14}$$

Supposing $N \to \infty$ and applying Parseval's relationship to (14) yields

$$J_{\text{multi}}(P,\theta) = \sum_{k=1}^{P} \frac{1}{2\pi} \int_{-\pi}^{\pi} \left| \frac{L_k(e^{j\omega},\theta)}{H(e^{j\omega},\theta)} \right|^2 \begin{bmatrix} \tilde{G}(e^{j\omega},\theta) & H_0(e^{j\omega}) \end{bmatrix} \times$$

$$\begin{bmatrix} \Phi_u(\omega) & \Phi_{eu}(\omega) \\ \Phi_{ue}(\omega) & R \end{bmatrix} \begin{bmatrix} \tilde{G}(e^{-j\omega},\theta) \\ H_0(e^{-j\omega}) \end{bmatrix} d\omega ,$$

where $\Phi_u(\omega)$ is the power spectrum of $u(t)$ and $\Phi_{eu}(\omega)$ is the cross-spectrum between $e(t)$ and $u(t)$. Now, moving the summation to the inside of the integral, it follows that

$$J_{\text{multi}}(P,\theta) = \frac{1}{2\pi} \int_{-\pi}^{\pi} \frac{\sum_{k=1}^{P} \left| L_k(e^{j\omega},\theta) \right|^2}{|H(e^{j\omega},\theta)|^2} \left[\tilde{G}(e^{j\omega},\theta) \quad H_0(e^{j\omega}) \right] \times$$
$$\begin{bmatrix} \Phi_u(\omega) & \Phi_{eu}(\omega) \\ \Phi_{ue}(\omega) & R \end{bmatrix} \begin{bmatrix} \tilde{G}(e^{-j\omega},\theta) \\ H_0(e^{-j\omega}) \end{bmatrix} d\omega . \tag{15}$$

From (15) one can see that the deterministic model mismatch is weighted by the input spectrum, while the filter

$$W_{\text{multi}}(e^{j\omega},\theta) = \sum_{k=1}^{P} \left| W_k(e^{j\omega},\theta) \right|^2 = \frac{\sum_{k=1}^{P} \left| L_k(e^{j\omega},\theta) \right|^2}{|H(e^{j\omega},\theta)|^2} \tag{16}$$

weights the whole expression. But, if P is limited to 1, which implies considering only one-step ahead predictions, we obtain

$$J_{\text{multi}}(P,\theta)\Big|_{P=1} = \frac{1}{2\pi} \int_{-\pi}^{\pi} \frac{1}{|H(e^{j\omega},\theta)|^2} \left[\tilde{G}(e^{j\omega},\theta) \quad H_0(e^{j\omega}) \right] \times$$
$$\begin{bmatrix} \Phi_u(\omega) & \Phi_{eu}(\omega) \\ \Phi_{ue}(\omega) & R \end{bmatrix} \begin{bmatrix} \tilde{G}(e^{-j\omega},\theta) \\ H_0(e^{-j\omega}) \end{bmatrix} d\omega . \tag{17}$$

Comparing (17) with (15), it is observed that the latter is identical to the first weighted by the frequency function

$$L_{\text{multi}}(e^{j\omega},\theta) = \sum_{k=1}^{P} \left| L_k(e^{j\omega},\theta) \right|^2 . \tag{18}$$

Hence, the estimation of the model parameters by minimizing the MRI cost function (15) is equivalent to using standard one-step ahead prediction error estimation algorithms (available in software packages, such as Ljung (2007)) after prefiltering the data with (18). As the prefiltering affects the model bias distribution and may also remove disturbances of frequency ranges that one does not want to include in the modeling, the role of the prefilter may be interpreted as a frequency weighting optimized for providing models suitable for multi-step ahead predictions.

3.1.2 Algorithms and implementation issues

Although the prefiltering artifice is an alternative to solve the problem of parameter estimation in the context of MRI, there is a point to be emphasized: the prefilter $L_{\text{multi}}(q,\theta)$ in (18) depends on the noise model $H(q,\theta)$, which is obviously unknown.

An iterative procedure called LRPI (Long Range Predictive Identification) to deal with the unknown noise model was proposed by (Shook et al., 1991). As mentioned previously, in the original formulation only the SISO case based on the ARX structure was concerned. Next, the LRPI algorithm is extended to the multivariable case. To this end, the following is adopted

$$G(q,\theta) = A^{-1}(q)B(q) \tag{19}$$
$$H(q,\theta) = A^{-1}(q) , \tag{20}$$

where the polynomial matrix $A(q)$, as well as its entries, are defined analogously to (5). According to (19)-(20), the ith output equation may be expressed by

$$A_{ii}(q)y_i(t) = \sum_{j=1}^{m} B_{1j}(q)u_j(t) + e_i(t) . \tag{21}$$

Consider the regression $\varphi_i(t) \in \mathbb{R}^{n(m+1)}$ and the parameter $\theta_i \in \mathbb{R}^{n(m+1)}$, relative to the ith system output

$$\varphi_i(t) = [-y_i(t-1), \cdots, -y_i(t-n), u_1(t-1), \cdots, u_m(t-1),$$

$$\cdots, u_1(t-n), \cdots, u_m(t-n)]^T \tag{22}$$

$$\theta_i = \left[a_{ii}^{(1)}, \cdots, a_{ii}^{(n)}, b_{i1}^{(1)}, \cdots, b_{im}^{(1)}, \cdots, b_{i1}^{(n)}, \cdots, b_{im}^{(n)} \right]^T . \tag{23}$$

From (22) and (23), the one-step ahead prediction of $y_i(t)$ may be expressed as

$$\hat{y}_i(t+1|t, \theta_i) = \varphi_i^T(t)\theta_i . \tag{24}$$

Algorithm 1: Extension of the LRPI algorithm to the multivariable case

Step 1. Set $i = 1$ (that is, only the first output is considered).

Step 2. Initialize $L_{\text{multi},i}(q)$ to 1.

Step 3. Filter $y_i(t)$ and each input $u_j(t)$ for $j \in \{1, \ldots, m\}$ with $L_{\text{multi},i}(q)$, i.e.

$$y_i^f(t) \triangleq L_{\text{multi},i}(q)y_i(t) \tag{25}$$

$$u^f(t) \triangleq \begin{bmatrix} L_{\text{multi},i}(q) & 0 & \cdots & 0 \\ 0 & L_{\text{multi},i}(q) & & \vdots \\ \vdots & & \ddots & 0 \\ 0 & \cdots & 0 & L_{\text{multi},i}(q) \end{bmatrix} u(t) . \tag{26}$$

Step 4. Based on (25)-(26), construct the regression vector analogously to (22), so that

$$\varphi_i^f(t) = \left[-y_i^f(t-1), \cdots, -y_i^f(t-n), u^{f^T}(t-1), \cdots, u^{f^T}(t-1) \right]^T . \tag{27}$$

Step 5. Estimate the parameter vector θ_i by solving the linear least-squares problem

$$\hat{\theta}_i = \arg\min_{\theta_i} \sum_t \left(y_i(t) - \varphi_i^{f^T}(t)\theta_i \right)^2 . \tag{28}$$

Step 6. Update $L_{\text{multi},i}(q)$ through (10) and (18), based on the noise model $A_{ii}^{-1}(q)$ estimated in the previous step.

Step 7. Continue if convergence of θ_i occurs, otherwise go back to Step 3.

Step 8. If $i \neq p$, go back to Step 2, with $i = i + 1$. Otherwise, concatenate the estimated models into a MIMO representation.

Remarks:

- For the multi-output case, there are p different filters $L_{\text{multi}}(q)$, each one associated with the ith output and denoted by $L_{\text{multi},i}(q)$.

- With respect to Step 6, as $L_{\text{multi}}(q)$ is a spectral factor of $L_{\text{multi}}(e^{j\omega})$, spectral factorization routines, such as the one proposed in Ježek & Kučera (1985), can be used for solving (18).

- A natural choice to determine the convergence of the algorithm is to check if the ℓ_2 norm of the difference between the parameter estimates in two consecutive iterations is less than δ. Experience has shown that a reasonable value for δ is 10^{-5}.

Alternatively, instead of using an iterative procedure as previously, in the method proposed by Huang & Wang (1999) named MPEM (Multi-step Prediction Error Method), a fixed noise model estimate is employed in order to get $L_{\text{multi}}(q)$. In what follows, the multi-step prediction error algorithm is described, based on the MFD parameterized by (2)-(5).

Algorithm 2: MPEM algorithm based on the diagonal form matrix fraction description

Step 1. Set $i = 1$.

Step 2. Get initial estimates of $C_{ii}(q)$, $D_{ii}(q)$, $F_{ii}(q)$ and, for $j \in \{1, \ldots, m\}$, $B_{ij}(q)$, using standard prediction error methods, namely, based on a one-step ahead cost function (17).

Step 3. Use a spectral factorization routine to solve (18), in which the filters defined in (10) are calculated through the impulse response of the estimated noise model $\hat{D}_{ii}^{-1}(q)\hat{C}_{ii}(q)$.

Step 4. Filter $y_i(t)$ and each input $u_j(t)$, $j \in \{1, \ldots, m\}$, with $\hat{C}_{ii}^{-1}(q)\hat{D}_{ii}(q)L_{\text{multi},i}(q)$.

Step 5. With the fixed noise model $\hat{D}_{ii}^{-1}(q)\hat{C}_{ii}(q)$, calculated in Step 2, estimate $B_{i1}(q), \ldots,$ $B_{im}(q)$, $F_{ii}(q)$ by minimizing the output-error cost function

$$V_{\text{oe},i}\left(B_{i1}(q), \ldots, B_{im}(q), F_{ii}(q)\right) = \sum_t \left(y_i^f(t) - F_{pp}^{-1}(q) \sum_{j=1}^m B_{1j}(q)u_j^f(t) \right)^2. \quad (29)$$

Step 6. If $i \neq p$, go back to Step 2, with $i = i + 1$. Otherwise, concatenate the estimated models into a multi-output representation.

Remarks:

- Once more the diagonal form MFD property, which allows the independent treatment of each model output, is applied to extend the parameter estimation algorithm to the multivariable framework.

- The prefilters of Step 2 differ from the ones used in the LRPI algorithm by the additional terms $\hat{C}_{ii}^{-1}(q)\hat{D}_{ii}(q)$, each one for $i \in \{1, \ldots, p\}$, which represents the inverse of the ith output noise model. Hence, while the filters $L_{\text{multi},i}(q)$ aim at providing optimal weighting for multi-step predictions, the additional terms intend to remove the noise influence for models represented as (6).

- The minimization of (12) is replaced by two nonlinear optimization problems in the MPEM algorithm. At first, it might seem that there is no relevant advantage in such an approach. Nevertheless, it is important to say that the MISO Box-Jenkins identification from Step 2, as well as the minimization of the output-error cost function in (29), can be performed using available software packages (Ljung, 2007, e.g.). Moreover, for models parameterized as (2)-(5), the numerical complexity of these problems are considered to be lower than the one of minimizing J_{multi} directly.

The LRPI algorithm involves only linear least-squares problems, which have many advantages. The most important one being that (28) can be solved efficiently and unambiguously (Ljung, 1999). The price paid for a simple parameter estimation algorithm is the adoption of a limited noise model structure. Consequently, the estimate of the $H(q, \theta)$ entries may be inaccurate, which affects the calculation of each filter $L_{\text{multi},i}(q)$. In turn, MPEM considers a more flexible noise model structure. However, local minima or convergence issues due to nonlinear optimization methods in Steps 2 and 5 may degrade the quality of the estimates. Therefore, the MPEM should outperform the LRPI algorithm, provided that the global minimum is achieved in the estimation steps. Anyway, it is suggested that models are estimated using more than one method and select the one which yields the best multi-step ahead predictions.

3.2 Direct optimization of the cost function

In the prefilter approach described previously, the filters $L_{\text{multi},i}(q)$ are calculated using any spectral factorization routine. Hence, as these filters are approximations of (18), the identified model ability to generate multi-step ahead predictions depends on the degree of the approximation and on the accuracy of the disturbance model estimate. But there is no need to worry about these aspects if the MRI cost function (12) is minimized directly. On the other hand, the model parameterization should be chosen carefully, to minimize numerical problems in the nonlinear optimization algorithm. In Laurí et al. (2010) a "full-polynomial[2] form" ARX model

$$A(q)y(t) = B(q)u(t) + e(t) , \tag{30}$$

with

$$A(q) = \begin{bmatrix} A_{11}(q) & \cdots & A_{1p}(q) \\ \vdots & \ddots & \vdots \\ A_{p1}(q) & \cdots & A_{pp}(q) \end{bmatrix} = I + A^{(1)}q^{-1} + \ldots + A^{(n)}q^{-n} , \tag{31}$$

whose entries are

$$A_{ij}(q) = \begin{cases} 1 + a_{ij}^{(1)}q^{-1} + \ldots + a_{ij}^{(n)}q^{-n} , \text{ for } i = j \\ a_{ij}^{(1)}q^{-1} + \ldots + a_{ij}^{(n)}q^{-n} \quad , \text{ otherwise} \end{cases}$$

and the polynomial matrix $B(q)$ is defined as in (4).

---

[2] Note that, in order to consider output interaction, the polynomial matrix $A(q)$ is not restricted to being diagonal, as in the LRPI algorithm.

For this model structure, let us introduce the parameter matrix

$$
\Theta = \begin{bmatrix} a_{11}^{(1)}, \cdots, a_{1p}^{(1)}, \cdots, a_{11}^{(n)}, \cdots, a_{1p}^{(n)}, b_{11}^{(1)}, \cdots, b_{1m}^{(1)}, \cdots, b_{11}^{(n)}, \cdots, b_{1m}^{(n)} \\ \vdots \\ a_{p1}^{(1)}, \cdots, a_{pp}^{(1)}, \cdots, a_{p1}^{(n)}, \cdots, a_{pp}^{(n)}, b_{p1}^{(1)}, \cdots, b_{pm}^{(1)}, \cdots, b_{p1}^{(n)}, \cdots, b_{pm}^{(n)} \end{bmatrix}^T \in \mathbb{R}^{n(m+p) \times p}
$$

$$(32)$$

and a particular regression vector denoted by $\breve{\varphi}(t+k|t,\Theta) \in \mathbb{R}^{n(m+p)}$, which is composed of inputs up to instant $t+k$, output data up to t and output estimates from $t+1$ to $t+k-1$, for instance

$$
\breve{\varphi}(t+2|t,\Theta) = \left[-\hat{y}^T(t+1|t,\Theta), -y^T(t), \cdots, -y^T(t-n+2), u^T(t+1), \cdots, u^T(t-n+2) \right]^T
$$

and for an arbitrary k

$$
\breve{\varphi}(t+k|t,\Theta) = \left[-\breve{y}^T(t+k-1|t), \cdots, -\breve{y}^T(t+k-n|t), u^T(t+k-1), \cdots, u^T(t+k-n) \right]^T,
$$

$$(33)$$

where

$$
\breve{y}^T(s|t) \triangleq \begin{cases} \hat{y}(s|t,\Theta) , & \text{for } s > t \\ y(s) , & \text{otherwise.} \end{cases}
$$

From (32) and (33), the k-step ahead prediction of $y(t)$ is given by

$$
\hat{y}(t+k|t,\Theta) = \Theta^T \breve{\varphi}(t+k|t,\Theta) .
$$

$$(34)$$

Although the predictor $\hat{y}(t+k|t,\Theta)$ is nonlinear in the parameters, it is important to notice that it can be calculated recursively, from $\hat{y}(t+1|t)$ for $k \in \{2,\ldots,P\}$ using (34). This is the main reason why the ARX structure was adopted. For another thing, if a more flexible model structure is adopted, the k-step ahead predictor equation would be much more complex.

Thus, based on the MRI cost function (12), the parameter estimation can be stated as a nonlinear least-squares problem

$$
\hat{\Theta} = \arg\min_{\Theta} \sum_{k=1}^{P} \sum_{t=0}^{N-k} \|y(t) - \Theta^T \breve{\varphi}(t+k|t,\Theta)\|_2^2 ,
$$

$$(35)$$

which must be solved numerically. The Levenberg-Marquart algorithm is used in Laurí et al. (2010) in order to minimize (35).

3.3 Optimization of the noise model

In Gopaluni et al. (2004) it is shown that, in the absence of a noise model, there is no significant difference between MRI and one-step ahead prediction error methods. On the other hand, when the signal to noise ratio is small, the one-step ahead predictors yield worse results for P-step ahead predictions than MRI methods. Thus, in these circumstances, a suitable disturbance model is crucial to generate accurate multi-step ahead predictions.

Any identified model has bias and variance errors associated with the identification algorithm. While the former is typically associated to model mismatch (such a mismatch can be either in the process model or in the noise model) and the second is due to the effect of unmeasured disturbances. If that there is no significant cross-correlation between the noise and system input (in open-loop, e.g.), the bias errors in the process model may be eliminated by using high order FIR (Finite Impulse Response) models (Zhu, 2001). Under that assumption, the modeling errors are restricted to the noise model.

With this in mind, in Gopaluni et al. (2004) the authors propose a two-step MRI algorithm in which the process is represented by a FIR structure, with sufficiently high order so that bias errors due to the process model can be disregarded. Then, the noise model parameters are estimated using a multi-step cost function.

Consider the multivariable FIR model

$$G_{\text{FIR}}(q, \theta) = B(q) \tag{36}$$

where the polynomial matrix $B(q)$ is defined as in (4). The noise model $H(q, \eta)$ is parameterized using diagonal MFD. These choices are equivalent to (6) with $F(q) = I$. As the estimation of $G(q)$ and $H(q)$ are performed separately, in this subsection, the parameter vector is split into two parts, such that the noise model parameter vector is explicitly referred to as η. So, the ith output noise model structure is

$$H_{ii}(q, \eta_i) \triangleq \frac{C_{ii}(q)}{D_{ii}(q)} = \frac{1 + c_{ii}^{(1)}q^{-1} + \ldots + c_{ii}^{(\alpha_i)}q^{-\alpha_i}}{1 + d_{ii}^{(1)}q^{-1} + \ldots + d_{ii}^{(\beta_i)}q^{-\beta_i}} . \tag{37}$$

Let us introduce the residual of the process model relative to the ith output

$$v_i(t) \triangleq y_i(t) - \sum_{j=1}^{m} B_{ij}(q)u_j(t) . \tag{38}$$

Then, based on (7)-(9), the k-step ahead prediction error of the ith output[3] can be written as

$$\varepsilon_i(t + k|t, \theta) = y_i(t + k) - \hat{y}_i(t + k|t, \theta)$$

$$= \left(\sum_{l=0}^{k-1} h_i(l)q^{-l} \right) \frac{D_{ii}(q)}{C_{ii}(q)} v_i(t) . \tag{39}$$

As $C_{ii}(q)$ and $D_{ii}(q)$ are monic polynomials, the impulse response leading coefficient $h_i(0)$ is always 1. With this, expanding (39) yields

$$\varepsilon_i(t + k|t, \theta) = v_i(t + k) + h_i(1)v_i(t + k - 1) + \ldots + h_i(k - 1)v_i(t + 1)$$

$$-c_{ii}^{(1)}\varepsilon_i(t + k - 1|t, \theta) - \ldots - c_{ii}^{(\alpha_i)}\varepsilon_i(t + k - \alpha_i|t, \theta)$$

$$+d_{ii}^{(1)}L_{k,i}(q)v_i(t + k - 1) + \ldots + d_{ii}^{(\beta_i)}L_{k,i}(q)v_i(t + k - \beta_i) . \tag{40}$$

---

[3] Part of the notation introduced in Section 2.2 is particularized here to the single-output context. For instance, $h_i(l)$ and $L_{k,i}(q)$ are the equivalent to the ones defined in (10), but related to the ith output.

For a compact notation, we define

$$\eta_{k,i} \triangleq \left[h_i(1), \ldots, h_i(k-1), 1, c_{ii}^{(1)}, \ldots, c_{ii}^{(\alpha_i)}, d_{ii}^{(1)}, \ldots, d_{ii}^{(\beta_i)} \right]^T \tag{41}$$

$$\varphi_{k,i}(t, \eta_{k,i}) \triangleq \Big[-v_i(t+k-1), \ldots, -v_i(t+1), y_i(t+k) - v_i(t+k), \varepsilon_i(t+k-1|t, \theta),$$

$$\ldots, \varepsilon_i(t+k-\alpha_i|t, \theta), -\xi_i(t+k-1), \cdots, -\xi_i(t+k-\beta_i) \Big]^T \tag{42}$$

where

$$\xi_i(t+k) \triangleq F_{k,i}(q) v_i(t+k) .$$

Then, we can rewrite (39) as

$$\varepsilon_i(t+k|t) = y_i(t+k|t) - \varphi_{k,i}^T(t, \eta_{k,i}) \eta_{k,i} . \tag{43}$$

In light of the aforementioned paragraphs, the MRI algorithm that optimizes the noise model is summarized as follows.

Algorithm 3: MRI with optimized noise model

Step 1. Set $i = 1$.

Step 2. Fix an a priori noise model to the ith output, for instance

$$\frac{C_{ii}(q)}{D_{ii}(q)} = 1$$

and estimate a multi-input single-output high order FIR model using standard PEM.

Step 3. With the estimate $\hat{G}_{\text{FIR}}(q)$, from the previous step, solve the optimization problem

$$\hat{\eta}_{P,i} = \arg\min_{\eta_{P,i}} \sum_{t=1}^{N-P} \sum_{k=1}^{P} \left(y_i(t+k|t) - \varphi_{k,i}^T(t, \eta_{k,i}) \eta_{k,i} \right)^2 \tag{44}$$

subject to

$$h_i(l) = \check{h}_i(l)(\eta_i) , \text{ for any } l = \{1, 2, \ldots, P-1\} \tag{45}$$

where $\check{h}_i(l)(\eta_i)$ indicates the lth impulse response coefficient of (37), which is obtained by polynomial long division of $C_{ii}(q)$ by $D_{ii}(q)$.

Step 4. If $i \neq p$, go back to Step 2, with $i = i+1$. Otherwise concatenate the estimated models into a single MIMO representation.

Remarks:

- Besides providing unbiased estimates under open-loop conditions, FIR models are suitable in this case because the parameters of G_{FIR} can be efficiently estimated using linear least-squares.

- A numerical optimization method is required to solve the parameter estimation problem. Nevertheless, the Levenberg-Marquart algorithm mentioned in the previous subsection can not deal with constraints. One of the nonlinear optimization algorithm possibilities is the Sequential Quadratic Programming (SQP), which can handle nonlinear constraints such as (45). In Gopaluni et al. (2004) it is shown that if a noise model of the form

$$H_{ii}(q, \eta_i) = \frac{1 + c_{ii}^{(1)} q^{-1} + \ldots + c_{ii}^{(\alpha_i)} q^{-\alpha_i}}{1 - q^{-1}}$$

is adopted, then the constraint (45) may be expressed through a linear in the parameters equation. In this case, (44) can be solved using the standard Quadratic Programming (QP) method.

4. Simulations

The main features of the aforementioned MRI techniques are analyzed using two simulated examples. At first, a SISO process is considered in order to illustrate the influence of the prediction horizon length P in the modeling errors presented by the identified models. Moreover, the performance of each technique is evaluated based on datasets with distinct signal-to-noise ratios (SNR). After that, the closed-loop performance provided by the estimated models is assessed. To this end, the Quadratic Dynamic Matrix Controller (QDMC) (Camacho & Bordons, 2004) and a multivariable distillation column benchmark (Cott, 1995a;b) are employed.

4.1 SISO process example

Consider the third-order overdamped system proposed in Clarke et al. (1987)

$$G_0(q) = \frac{0.00768q^{-1} + 0.02123q^{-2} + 0.00357q^{-3}}{1 - 1.9031q^{-1} + 1.1514q^{-2} - 0.2158q^{-3}}, \tag{46}$$

with a random-walk disturbance, that is

$$H_0(q) = \frac{1}{1 - q^{-1}}. \tag{47}$$

The process is excited in open-loop by a Pseudo Random Binary Sequence (PRBS) switching between $[-0.1, 0.1]$ with a clock period of 5 times the sampling interval. The noise variance is adjusted such that the signal-to-noise ratio (SNR) is 3 (in variance). A record of 1200 samples is collected, which is shown in Fig. 1. The dataset is split into two halves: the first is used for estimation and the second one for validation purposes.

The following reduced-complexity model structure is assumed[4]

$$G(q, \theta) = \frac{b_1 q^{-1} + b_2 q^{-2}}{1 + a_1 q^{-1}} \tag{48}$$

$$H(q, \theta) = \frac{1 + c_{11}^{(1)} q^{-1}}{1 + d_{11}^{(1)} q^{-1}}. \tag{49}$$

---

[4] Except for the noise model optimization method (Subsection 3.3), in which $d_{11}^{(1)}$ is fixed to -1, so that parameter estimation can be handled using standard quadratic programming.

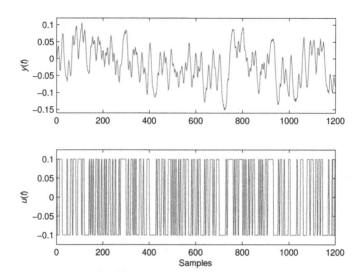

Fig. 1. The dataset from the third-order process (46)-(47).

Before analyzing the capacity of the estimated models to generate accurate multi-step ahead predictions, it is worth noting the influence of the prediction horizon length. The magnitudes of $L_{\text{multi}}(e^{j\omega})$, for $P = \{1, 2, 5, 10, 15\}$, are shown in Fig. 2. As can be seen, $L_{\text{multi}}(q)$ is a low-pass filter, whose cut-off frequency decreases as P increases. Such behavior occurs whenever the disturbance spectrum is concentrated on low frequencies (Gopaluni et al., 2003). Hence, according to (15), the higher the prediction horizon length, the narrower the error weighting.

As a consequence, an increase in P leads to lower modeling errors in low frequencies, but the frequency response of the estimated models are away from the actual one at high frequencies. This behavior is depicted in Fig. 3, which presents the absolute value of the difference between the actual and the estimated (from models obtained using the MPEM algorithm) frequency responses. One can also notice that the effect of increasing P is more prominent in the range $[1, 5]$ than between $[5, 15]$. Furthermore, as shown in Farina & Piroddi (2011), for sufficiently high values of the prediction horizon length, models estimated based on multi-step prediction errors converge to the output (simulation) error estimate.

The cost function J_{multi}, defined in (12), is applied to quantify the model accuracy in terms of multi-step ahead predictions. It is emphasized that such accuracy is quantified using fresh data, that is to say, a distinct dataset from the one used for estimation purposes. In what follows, the performance of the MRI techniques are investigated using two sets of Monte Carlo simulations, each one with 100 distinct white-noise realizations. In order to visualize the SNR effect on different parameter estimation methods, in the first simulation set, the SNR is maintained in 3 and in the other one it is increased to 10. The histograms of J_{multi} for the methods described in Section 3 are depicted in the rows of Fig. 4, for $P = 8$. The left and the right columns present the results for the signal-to-noise ratios of 3 and 10, respectively. The main Monte Carlo simulation results are summarized in Table 1, which reports the mean and the standard deviation of J_{multi}. For comparison, the last column of Table 1 also presents the results produced by the standard (one-step ahead) PEM, based on a Box-Jenkins structure.

The histograms in Fig. 4, as well as Table 1, show that the MPEM and the noise model optimization algorithms presented the smallest J_{multi} (that is, the most accurate multi-step

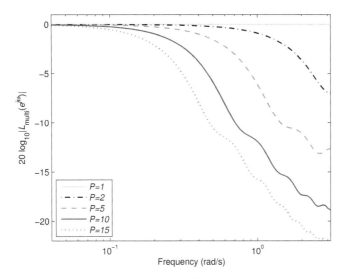

Fig. 2. Magnitude frequency response of $L_{\text{multi}}(e^{j\omega})$ for increasing prediction horizon length.

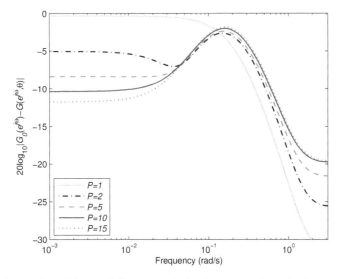

Fig. 3. Absolute value of the modeling error in the frequency domain, for models estimated with $P = \{1, 2, 5, 10, 15\}$.

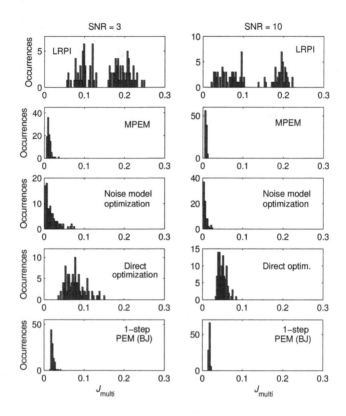

Fig. 4. Histograms of J_{multi} for each MRI method.

predictions) with the lowest variance, which means that these methods are less sensitive to a particular realization. On the other hand, LRPI and direct optimization showed worse performances because these methods are based on the ARX model structure, which is quite different from the process (46)-(47). Another aspect that may be noticed is that, as expected, a higher SNR leads to a smaller J_{multi} mean (more accurate models are expected) and lower deviations of the estimates.

Actually, the performances of the methods based on ARX structure may be interpreted in a broader sense. Although in MRI the effect of bias due to model mismatch is reduced in the

	SNR	LRPI	MPEM	Direct optim.	Noise model optimization	Standard PEM (Box-Jenkins)
mean(J_{multi})	3	0.1526	0.0111	0.0786	0.0178	0.0209
	10	0.1218	0.0074	0.0496	0.0056	0.0172
std(J_{multi})	3	0.0536	0.0045	0.0239	0.0163	0.0042
	10	0.0668	0.0015	0.0104	0.0049	0.0014

Table 1. Mean and standard deviation of the cost function.

parameter estimation step, the task of selecting a suitable model structure is still crucial to the success of a system identification procedure. This statement is also supported by the fact that, according to Table 1, considering a more favorable structure and the one-step ahead PEM is more effective than an inadequate structure whose parameters are estimated based on a multi-step cost function.

4.2 Multivariable system example

The Shell benchmark process is a model of a two-input two-output distillation column (Cott, 1995a;b). The inputs are overhead vapour flow and reboiler duty, denoted here as u_1 and u_2, respectively. The outputs are the column pressure

$$\Delta y_1(t) = \frac{-0.6096 + 0.4022q^{-1}}{1 - 1.5298q^{-1} + 0.574q^{-2}} \Delta u_1(t) + \frac{0.1055 - 0.0918q^{-1}}{1 - 1.5298q^{-1} + 0.574q^{-2}} \Delta u_2(t)$$

$$+ \frac{\lambda}{1 - 1.5945q^{-1} + 0.5945q^{-2}} e_1(t) \tag{50}$$

and the product impurity

$$y_2(t) = 0.0765 \frac{5 \times 10^5}{u_2(t-7) - 1500} + 0.9235 y_2(t-1) + \frac{\lambda}{1 - 1.6595q^{-1} + 0.6595q^{-2}} e_2(t) \tag{51}$$

where Δy_1, Δu_1 and Δu_2 are deviation variables around the nominal operating point[5] (specified in Table 2), that is

$$\Delta y_1(t) = y_1(t) - \bar{y}_1$$
$$\Delta u_1(t) = u_1(t) - \bar{u}_1$$
$$\Delta u_2(t) = u_2(t) - \bar{u}_2 .$$

Variable	Nominal setpoints	Normal operation
Pressure (y_1)	2800	$2700 < y_1 < 2900$
Composition (y_2)	500	$250 < y_2 < 1000$
Overhead vapour flow (u_1)	20	$10 < u_1 < 30$
Reboiler flow (u_2)	2500	$2000 < u_2 < 3000$

Table 2. Summary of distillation column operating conditions.

The disturbances are generated using uncorrelated zero-mean white noises e_1 and e_2, such that $\text{std}(e_1) = 1.231$ and $\text{std}(e_2) = 0.667$. The parameter λ is set to 0.2. The Shell benchmark is widely used to evaluate multivariable system identification or model predictive control strategies (Amjad & Al-Duwaish, 2003; Cott, 1995b; Zhu, 1998, e.g.). Besides being multivariable, the model (50)-(51) offers additional complications: as the disturbances are nonstationary, one of the outputs (product impurity) is slightly nonlinear and the overhead flow (u_1) does not affect the impurity level (y_2).

---

[5] For more details about the simulator operating conditions, the reader is referred to (Cott, 1995b).

The process is excited in open-loop using two uncorrelated random binary sequences (RBS), with u_1 varying from $[15, 25]$ and u_2 from $[2400, 2600]$. The minimum switching time of u_1 and u_2 is 12 and 6, respectively. The dataset is comprised of 1600 samples, where the first half is used for estimation (see Fig. 5) and the rest for validation.

The elements of the transfer function matrix $G(q, \theta)$ and of the noise models are first order. Initially, the input delay matrix

$$n_k = \begin{bmatrix} 0 & 0 \\ 37 & 7 \end{bmatrix} \qquad (52)$$

was estimated applying the function delayest of the Matlab™System Identification toolbox (Ljung, 2007). Notice that except for the entry in which there is no coupling ($u_1 \rightarrow y_2$), the values in n_k coincide with the actual input delays. Thus, before proceeding to the parameter estimation, the input sequences are shifted according to n_k.

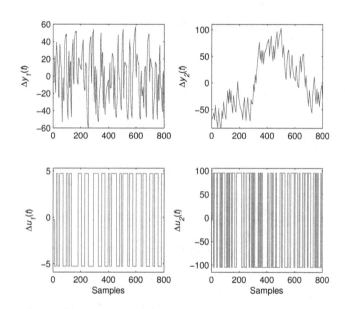

Fig. 5. Estimation dataset from the Shell benchmark simulation.

The models estimated with $P = 40$ are evaluated based on the multi-step prediction errors (12) using the validation dataset, which are presented in Table 3. The most accurate multi-step predictions are generated by the MPEM and the 1-step ahead PEM. This is because, as in the SISO example, the Box-Jenkins structure employed by both methods best suits the process dynamic behavior. Another relevant point is that the noise model optimization yields unstable

	Output	LRPI	MPEM	Direct optim.	Noise model optimization	1-step PEM (Box-Jenkins)
$J_{multi} \times 10^4$	1	0.1154	0.0328	0.1475	∞	0.0322
	2	3.2887	2.5072	3.6831	∞	2.5294

Table 3. Multi-step prediction error.

predictors (due to zeros outside the unitary circle). Consequently, the sum of the prediction errors tends to infinity.

The standard PEM provided multi-step predictions as accurate as the MPEM, even for a sub-parameterized model, which is the case of this example. This result suggests that the under-modeling issue is not the most prominent for this situation. In addition, the fact that the disturbance intensity is very high, besides being concentrated on the low frequencies (the same frequency range that should be weighted to attain improved multi-step ahead predictions) disfavors the MRI approach.

In order to test the robustness of the methods to input-output delay estimation errors, a new estimation is carried out with a modified delay matrix n_k^*, in which the dead-time from u_2 to y_2 is changed from 7 to 8 samples. As shown in Table 4, the MRI methods are less sensitive to this parameter than the 1-step ahead PEM.

	Output	LRPI	MPEM	Direct optim.	Noise model optimization	1-step PEM (Box-Jenkins)
$J_{multi} \times 10^4$	2	3.1669	2.4854	3.8126	∞	3.0794

Table 4. Multi-step prediction error of the 2nd output when there is a slight mismatch in one of the input delay matrix element.

At this point, the performance of the estimated models is investigated when they are employed in a QDMC controller (Camacho & Bordons, 2004) with a prediction and control horizons of 40 and 5, respectively. The output Q and the manipulated R weighting matrices are (Amjad & Al-Duwaish, 2003)

$$Q = \begin{bmatrix} 1 & 0 \\ 0 & 2 \end{bmatrix} \text{ and } R = \begin{bmatrix} 2 & 0 \\ 0 & 2 \end{bmatrix}.$$

The closed-loop responses using the QDMC controller when each set-point is excited with a step of amplitude 1% of the nominal output values are presented in Fig. 6 and 7, where the first one is related to the input delay matrix n_k in (52) and the other refers to n_k^*. The results of the closed-loop validation are also summarized in Table 5, which shows the integrated square error (ISE) for each controlled variable: y_1 and y_2.

In a general way, the first output is closer to the set-point than y_2. This may be explained by the intensity of the disturbance introduced in each output, by the fact that the plant is non-linear whereas the identified models are linear and, finally, due to the presence of a zero in the transfer matrix which consequently affects the quality of the estimated model.

From Fig. 6, one can notice that all the controllers achieved similar responses for the column pressure (y_1). Concerning the other output (product purity), the closed-loop behavior provided by the standard PEM and the MPEM are very close (accordingly to multi-step prediction errors depicted in Table 3). Analogously, the LRPI method yielded a better performance than the direct optimization. Besides, as these two methods showed a worse multi-step prediction accuracy, it reflected in the MPC performance.

As shown in Fig. 7 and according to Table 5, the prediction capacity deterioration of the one-step ahead PEM, due to the delay matrix modification from n_k to n_k^* also leads to a worse closed-loop response. On the other hand, the closed-loop performances provided by the models estimated through MRI algorithms are less sensitive to errors in the time delay determination.

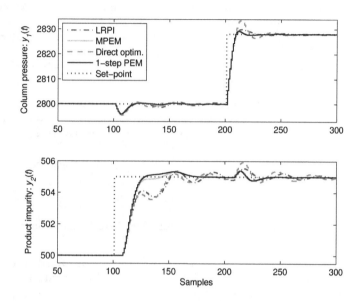

Fig. 6. Closed-loop response based on an accurate input delay estimation.

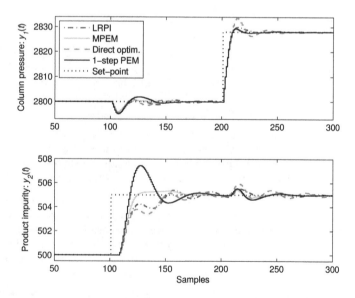

Fig. 7. Closed-loop response for a mismatch in one of the input-output delay matrix entry.

	Output	LRPI	MPEM	Direct optim.	1-step PEM (Box-Jenkins)
ISE $\times10^3$: n_k	1	2.1921	2.0777	2.4091	2.0711
	2	0.3618	0.3182	0.3747	0.3171
ISE $\times10^3$: n_k^*	1	2.1689	2.0883	2.4081	2.2016
	2	0.3519	0.3119	0.3802	0.3667

Table 5. Integrated square error (ISE) of the controlled variables.

5. Conclusions

This chapter focused on parameter estimation algorithms to generate suitable models for predictive controllers. The branch of identification known as MRI was studied and several different ways to obtain models were presented. They must be estimated having in mind that they must be accurate to predict multi-step ahead. Some of these techniques were published considering just the single-input single-output case and in this work they were extended to the multivariable framework. In order to compare the different algorithms, they were implemented and tested, employing a SISO and a MIMO plant. In the comparisons, the standard PEM (built to provide optimal one-step ahead predictions) was also included.

In the analysis with the SISO process, the long range prediction capacity of some of the MRI methods (MPEM and noise model optimization) was superior to the results generated by the standard PEM, based on a Box-Jenkins structure. In addition, the influence of the model structure was also highlighted in a model relevant identification context, since the standard PEM (with a Box-Jenkins) produced more accurate multi-step ahead predictions than the LRPI and the direct optimization algorithms, which are based on a less flexible model structure.

The tests performed with the multivariable plant were more concerned about the use of the MRI and PEM models, when applied to a predictive controller. The results obtained were not so convincing about the advantages of using multi-step prediction based methods in the predictive controller design, since the one-step PEM (with a Box-Jenkins model), even with structure mismatch, provided results that were comparable to the best ones obtained with the model relevant identification methods. However, it was also shown that when there was a slight error in the evaluation of the time delay of one of the input-output pairs, the advantage of the MRI approach became evident.

Although the excitation signal design and the model structure selection are beyond the scope of this work, the examples presented the complete system identification procedure, from the input signal generation, going through the use of different algorithms to estimate the model parameters up to the validation of the models through the verification of their prediction capacity. Besides, the obtained models were applied to a predictive controller to evaluate their performance in controlling a multivariable process.

The system identification for MPC is a subject prone to further research. The effect of multi-step prediction error methods on the closed-loop performance needs to be further investigated. Another important theme to be studied is in which situations the use of MRI methods for developing models for predictive controllers is in fact advantageous as compared to classical prediction error methods.

6. Acknowledgments

The authors gratefully acknowledge the support of Instituto Mauá de Tecnologia (IMT). They also thank the support provided by Petrobras.

7. References

Amjad, S. & Al-Duwaish, H. N. (2003). Model predictive controle of Shell benchmark process, *Proceedings of the 10th IEEE International Conference on Electronics, Circuits and Systems*, Vol. 2, pp. 655–658.

Camacho, E. & Bordons, C. (2004). *Model predictive control*, 2nd edn, Springer-Verlag, London.

Clarke, D. W., Mohtadi, C. & Tuffs, P. S. (1987). Generalized predictive control - part 1 and 2, *Automatica* 23(2): 137–160.

Cott, B. J. (1995a). Introduction to the process identification workshop at the 1992 canadian chemical engineering conference, *Journal of Process Control* 5(2): 67–69.

Cott, B. J. (1995b). Summary of the process identification workshop at the 1992 canadian chemical engineering conference, *Journal of Process Control* 5(2): 109–113.

Farina, M. & Piroddi, L. (2011). Simulation error minimization identification based on multi-stage prediction, *International Journal of Adaptive Control and Signal Processing* 25: 389–406.

Gopaluni, R. B., Patwardhan, R. S. & Shah, S. L. (2003). The nature of data pre-filters in MPC relevant identification – open- and closed-loop issues, *Automatica* 39: 1617–1626.

Gopaluni, R. B., Patwardhan, R. S. & Shah, S. L. (2004). MPC relevant identification – tuning the noise model, *Journal of Process Control* 14: 699–714.

Huang, B. & Wang, Z. (1999). The role of data prefiltering for integrated identification and model predictive control, *Proceedings of the 14th IFAC World Congress*, Beijing, China, pp. 151–156.

Ježek, J. & Kučera, V. (1985). Efficient algorithm for matrix spectral factorization, *Automatica* 21(6): 663–669.

Laurí, D., Salcedo, J. V., García-Nieto, S. & Martínez, M. (2010). Model predictive control relevant identification: multiple input multiple output against multiple input single output, *IET Control Theory and Applications* 4(9): 1756–1766.

Ljung, L. (1999). *System Identification: theory for the user*, 2nd edn, Prentice Hall, Upper Saddle River, NJ.

Ljung, L. (2007). *The system identification toolbox: The manual*, 7 edn, The MathWorks, Inc., MA, USA: Natick.

Maciejowski, M. (2002). *Predictive Control with Constraints*, Prentice Hall, Englewood Cliffs, NJ.

Rossiter, J. A. & Kouvaritakis, B. (2001). Modelling and implicit modelling for predictive control, *International Journal of Control* 74(11): 1085–1095.

Shook, D. S., Mohtadi, C. & Shah, S. L. (1991). Identification for long-range predictive control, *IEE Proceedings-D* 138(1): 75–84.

Zhu, Y. (1998). Multivariable process identification for MPC: the asymptotic method and its applications, *Journal of Process Control* 8(2): 101–115.

Zhu, Y. (2001). *Multivariable System Identification for Process Control*, Elsevier Science, Oxford.

Zhu, Y. & Butoyi, F. (2002). Case studies on closed-loop identification for MPC, *Control Engineering Practice* 10: 403–417.

11

FPGA Implementation of PID Controller for the Stabilization of a DC-DC "Buck" Converter

Eric William Zurita-Bustamante[1], Jesús Linares-Flores[2],
Enrique Guzmán-Ramírez[2] and Hebertt Sira-Ramírez[2]
[1]*Universidad del Istmo*
[2]*Universidad Tecnológica de la Mixteca*
México

1. Introduction

Actually the development of control systems in embedded systems presents a great advantage in terms of easy design, immunity to analog variations, possibility and implement complex control laws and design a short time (Mingyao Ma et al., 2010). One of the devices that allows embedded systems arrangements are field-programmable gate array (FPGA). Several advantages of using FPGAs in industrial applications can be seen in (Joost & Salomon, 2005).

The use of FPGAs to implement control laws of various systems can be observed in different articles. In (Hwu, 2010) performance a technique based on a field programmable gate array to design PID controller applied to the forward converter to reduce the effect of input voltage variations on the transient load response of the output converter. The main characteristic of this technique is the on-line tuned parameters of the PID controller. To validate the topology implemented, they designed a forward converter with an input voltage of 12V, and output dc voltage of 5.12V with a rated output DC current of 10A and a switching frequency at rated load of 195 kHz. The results show than the measured transient load response has no oscillation with on-line tuning applied to the controller.

In the work of LI et al. (Bo Li et al., 2011) presents a digital pulse-width-modulator based sliding-mode controller and FPGA for boost converter. The proposed model they used was higher order delta-sigma modulator. The problem with this modulator is the stability problem. To resolve this problem they implemented a Multi-stage-noise shaping delta-sigma DPWM (MASH sigma-delta DPWM). To verify the function of the proposed controller they implemented a boost converter connected to a Virtex-II Pro XC2VP30 FPGA with and Analog to digital converter as interface. The experimental results show than the MASH sigma-delta DPWM has a faster recovery time in load changes, compared with a PID controller.

In (Mingyao Ma et al., 2010) proposed a FPGA-based mixed-signal voltage-mode controller for switching mode converters. The architecture of the scheme consists of a DPWM generation with a PID controller implemented on FPGA, a DAC and a comparator. The switching mode converters state variables are digitalized via an ADC to the PID controller. The control signal goes to the DPWM module to generate the PWM waveforms. They implemented the PID and the DPWM on a Cyclone II series EP2C25, in other hand; they implemented a single phase full-bridge inverter like the switching mode converter to test the architecture of the controller. Their architecture allows integration of a control system in FPGA.

An implementation of PID controller on FPGA for low voltage synchronous buck converter is presented in (Chander et al., 2010). They use MATlab/Simulink for the PID controller design to generate the coefficients of the controller. They did a comparison between different coefficients to obtain a reasonable controller for the converter. The architecture was implemented in FPGA Virtex-5 XC5VLX50T.

In this article, we will focus on the PID average output feedback controller, implemented in an FPGA, to stabilize the output voltage of a "buck" power converter around a desired constant output reference voltage. The average control inputs are used as a duty ratio generator in a PWM control actuator. The architecture control, used for the classical PID control, has the following features:

- The PWM actuator is implemented through a triangular carrier signal and a comparator. The main function of this modulator is the average signal conversion to a pulsing signal that activates and deactivates the converter power transistor, at a switching frequency of 48kHz.
- The processing time control for the PID is 20.54μs. This processing time were achieved thanks to the parallel execution of units modeled within a FPGA Monmasson & Cirstea (2007)-Rogriguez-Andina et al. (2007).
- The output voltage is obtained through an Analog to Digital Converter (ADC), which is the only additional hardware needed to operate to the controllers. The used ADC is the ADC0820, which is an 8 bits converter.

The rest of the document is organized as follows: section 2 presents the mathematical model of the "buck" converter. The design of the PID control is shown in the section 3, while the simulation of the PID control design is presented in section 4. The architecture of the implemented control is found in section 5. The experimental results of the implementation of the FPGA based controller, are found in section 6. Finally, the conclusions of this work are given section 7.

2. The "buck" converter model

Consider the "buck" converter circuit, shown in Fig. 1. The system is described by the following set of differential equations:

$$L\frac{di_L}{dt} = -v_0 + Eu$$
$$C\frac{dv_0}{dt} = i_L - \frac{v_0}{R} \tag{1}$$
$$y = v_0$$

where i_L represents the inductor current and v_0 is the output capacitor voltage. The control input u, representing the switch position function, takes values in the discrete set 0, 1. The system parameters are constituted by: L and C which are, respectively, the input circuit inductance and the capacitance of the output filter, while R is the load resistance. The external voltage source exhibits the constant value E. The average state model of the "buck" converter circuit, extensively used in the literature (a) Linares & Sira, 2004; b) Linares & Sira, 2004;

Linares et al., 2011; Sira & Agrawal, 2004) may be directly obtained from the original switched model, (1), by simply identifying the switch position function, u, with the average control, denoted by u_{av}. Such an average control input is frequently identified with the duty ratio function in a Pulse Width Modulation implementation. The control input u_{av} is restricted to take values in the closed interval [0, 1]. From (1), the "buck" converter system is clearly a second order linear system of the typical form: $\dot{x} = Ax + bu$ and $y = c^T x$.

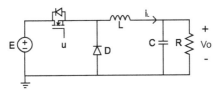

Fig. 1. The electrical circuit of the "buck" converter.

$$A = \begin{bmatrix} 0 & -\frac{1}{L} \\ \frac{1}{C} & -\frac{1}{RC} \end{bmatrix}$$

$$b = \begin{bmatrix} \frac{E}{L} \\ 0 \end{bmatrix} \tag{2}$$

$$c^T = \begin{bmatrix} 0 & 1 \end{bmatrix}$$

Hence, the Kalman controllability matrix of the system $\mathcal{C} = [b, Ab]$, is given by:

$$\mathcal{C} = \begin{bmatrix} \frac{E}{L} & 0 \\ 0 & \frac{E}{LC} \end{bmatrix} \tag{3}$$

The determinant of the controllability matrix is ($\frac{E^2}{L^2C} \neq 0$). Therefore, the system is controllable (Dorf & Bishop, 2011), now we design a classic PID control in the following section.

3. PID controller design

The FPGA implementation of a classical Proportional Integral Derivative (PID) controller was designed based on the corresponding transfer function of the converter (Ogata, 2010), obtained from the average model given in (1), is

$$\frac{V_0(s)}{U_{av}(s)} = \frac{\frac{E}{LC}}{s^2 + \frac{1}{RC}s + \frac{1}{LC}} \tag{4}$$

While the transfer function of the PID controller, is:

$$F_{PID}(s) = K_p \left(1 + \frac{1}{T_i s} + T_d s \right) \tag{5}$$

The block diagram of the PID controlled system is shown in Fig. 2.

The closed loop transfer function is readily found to be

$$H(s) = \frac{(K_p T_d T_i s^2 + K_p T_i s + K_p)(\frac{E}{LC})}{s^3 + \left(\frac{1}{RC} + \frac{EK_p T_d}{LC} \right)s^2 + \frac{(1+EK_p)}{LC}s + \frac{EK_p}{LCT_i}} \tag{6}$$

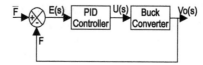

Fig. 2. PID control in closed loop.

The closed loop characteristic polynomial of the PID controlled system is then given by

$$s^3 + (\frac{1}{RC} + \frac{EK_pT_d}{LC})s^2 + \frac{(1 + EK_p)}{LC}s + \frac{EK_p}{LCT_i} = 0 \qquad (7)$$

The coefficients K_p, T_i and T_d are chosen so that (7) becomes a third order Hurwitz polynomial of the form (Dorf & Bishop, 2011; Ogata, 2010):

$$p(s) = (s^2 + 2\zeta\omega_n s + \omega_n^2)(s + \alpha) \qquad (8)$$

Equating the characteristic polynomial coefficients (7) with those of the desired Hurwitz polynomial (8), we obtain the following values of the parameters for the PID controller,

$$K_p = \frac{2\zeta\omega_n\alpha LC + \omega_n^2 LC - 1}{E}$$

$$T_i = \frac{EK_p}{LC\alpha\omega_n^2} \qquad (9)$$

$$T_d = \frac{LC}{EK_p}(\alpha + 2\zeta\omega_n - \frac{1}{RC})$$

4. PID controller cosimulation

In this section, we develop the simulation of the PID controller. This simulation is performed using Matlab/Simulink, ModelSim and PSim Software.

The cosimulation in Matlab/Simulink creates an interface between Matlab and Matlab external program, i.e., the cosimulation allows the interaction of an external simulator with Matlab tools. The cosimulation provides a fast bidirectional link between the hardware description language (HDL) simulator, and Matlab/Simulink for direct hardware design verification Matlab (2008).

Figure 3 shows the scenario between ModelSim and Matlab to obtain the cosimulation. The block that allows interaction with the HDL simulator is called "EDA Simulator Link MQ".

The PSIM software includes an application called *SimCoupler* that presents an interface between PSIM and Matlab Simulink for cosimulation. With the module *SimCoupler* part of the system can be implemented and simulated in PSIM, and the rest of the system in Simulink. With this tool we can access to the broad features of Psim simulation, and the capabilities of Simulink simulation in a complementary way Psim (2006).

The module *SimCoupler* consist of two parts: the link nodes in PSim, and the SimCoupler model block in Simulink. In this work we use the module *SimCoupler* to simulate the buck converter in Psim, while in matlab and through another cosimulation part of the PID control. Figure 4 shows the buck converter circuit in Psim, in this figure one can observe that the circuit

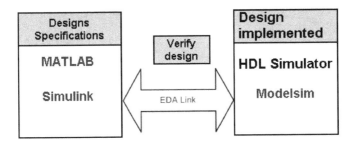

Fig. 3. Cosimulation between Matlab and ModelSim.

input is the signal coming from the PWM control in Simulink, this input signal is connected using the *In link node*. Because the system is feedback, and the feedback signal is the output voltage of the buck converter (V_0), we use the *Out link node* to send the output voltage to Simulink.

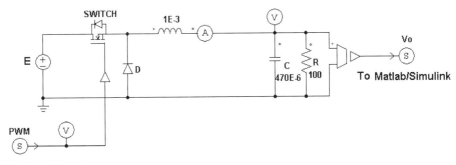

PWM Control Signal
of Matlab/Simulink

Fig. 4. Buck converter circuit for cosimulation with Simulink.

In Simulink we choose the S-function SimCoupler library and the SimCoupler block are added to the design. After adding the block, in its properties is chosen the file path for cosimulation in Psim, a window will automatically appear as shown in Fig. 5 showing all inputs and outputs of the circuit in PSIM, in this case, according to the diagram there are one input and one output on the circuit, the output voltage of the buck converter V_0. While that the input is the PWM signal.

Once set the block are automatically displayed input and output signals in the block for subsequent cosimulation, as shown in Fig. 6.

Before simulating the final system, we proceed to simulate the performance of open-loop buck converter, for this, we just simulate the circuit in Psim. Figure 7 shows the output voltage in simulation for open-loop. The response presents a overshoot of 100% however are able to stabilize around 45ms.

On the other hand, we simulate the PID control with the tools of differentiation and integration of Simulink. Figure 8 shows the performance of the PID controller with

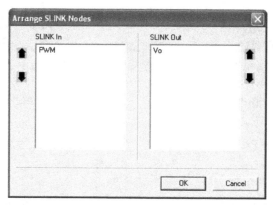

Fig. 5. Inputs and Outputs of the PSIM circuit.

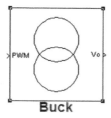

Fig. 6. Simulink block with the inputs and outputs of the PSIM circuit.

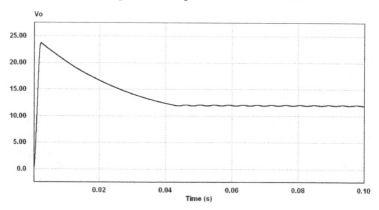

Fig. 7. Output voltage of buck converter in open-loop.

cosimulation between Matlab/Simulink and Psim. The PID control stabilizes the output voltage signal in a time of approximately 18ms, greatly decreasing the overshoot presented at the open-loop response.

Figure 9 shows a cosimulation for the final system with a desired output voltage of 4V, and shows that in the transient response has not overshoot, however, the settling time is about 23 ms, what is intended to improve with the experimental results. Also the Fig. 10 shows the output voltage for a desired voltage of 18 V, which shows that it has a maximum overshoot

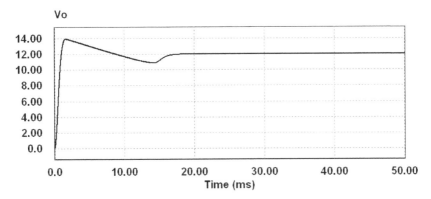

Fig. 8. Output voltage of buck converter with PID controller in cosimulation.

of 7.6 %, and a maximum error of 0.15 V. According to these simulations, we proceed to implement the system on a FPGA NEXYS2 board.

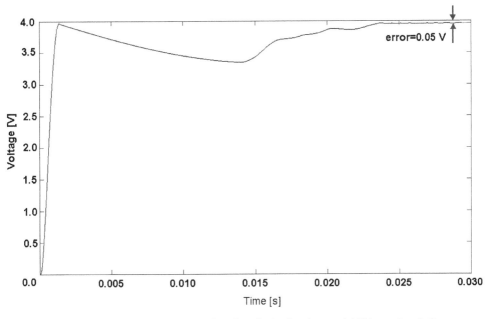

Fig. 9. Output voltage of buck converted with a desired voltage of 4 V in cosimulation.

5. Discrete PID controller implemented on the FPGA

In this section, we explain the hardware implementation of the discrete PID controller. For this purpose, we used the Xilinx ISE Design Suite 12.2 EDA (electronic design automation) -software tool and the Spartan 3E board EDA-hardware tool, it includes a Xilinx Spartan-3E1600 FPGA.

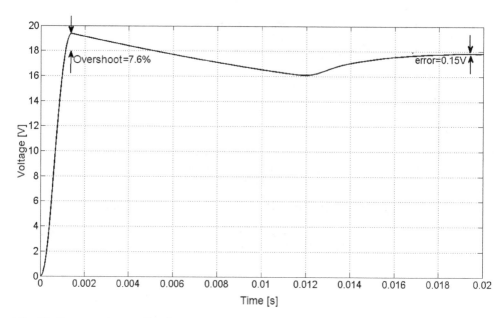

Fig. 10. Output voltage of buck converted with a desired voltage of 4 V in cosimulation.

Now, we must define an efficient design methodology and the abstraction level to model the system, and choose an appropriate sampling period and the suitable format for coefficients and variables.

The PID controller design is based on a hierarchical and modular approach using Top-Down methodology (Palnitkar, 2003), where the modules can be defined with diverse levels of abstraction. Thus, for this design the schematic description was chosen as top level and the controller components were modeled with the VHDL hardware description language (using a behavior level modeling). Previous analysis and simulations showed that due to the range of results generated by the operations involved in the discrete controller is necessary to use a floating point format; for this intention, the IEEE Standard for Binary Floating-Point Arithmetic, IEEE Std 754-1985 (IEEE, 1985) was chosen. Now, based on top-down methodology, an initial modular partitioning step is applied on the FPGA-based PID controller, this process generate four components, Clock manager, ADC control, Control law and PWM generator (see Fig. 11).

The PID controller work with a frequency of 50 MHz (Clk_PID). The Clk_main signal is generated from Clk_main signal by the Clock manager component. The principal element of this component is the Digital Clock Manager (DCM). The DCM is embedded on the Spartan3E FPGA's families and it provides flexible complete control over clock frequency, maintaining its characteristics with a high degree of precision despite normal variations in operating temperature and voltage. The DCM provides a correction clock feature, ensuring a clean Clk_PID output clock with a 50% duty cycle.

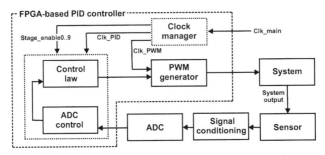

Fig. 11. Block diagram of FPGA-based PID controller.

In order to increase the performance of the control system, we proposed pipeline architecture for the PID controller. Therefore, enable signals of pipeline registers (Stage_enable0..9) are required. These signals are also generated by the Clock manager component.

In addition, the clock manager component generates the frequency required by the PWM for its operation (Clk_PWM). The Clk_PWM signal is derived from the Clk_main by a Digital Frequency Synthesizer (DFS) included in the DCM. The frequency of the Clk_PWM is 25 MHz and has a 50% duty cycle correction too.

The Information from the sensor is analog source, so it must be discretized for that the FPGA can process. For this purpose we have chosen the Analog-Digital Converter (ADC) ADC0820. The ADC0820 is an 8 bits resolution converter, it offers a $2\mu s$ conversion time and it has a 0 to 5 Volts analog input voltage range. The element responsible for this task is the ADC control component.

The ADC control component is composed of two modules, the ADC interface module, which is a simple finite-state machine (FSM) that implements the communications protocol to acquire data of the ADC0820, and the float-point encoder module, which converts the integer value into single-precision floating-point format. A block diagram of ADC interface module is shown in Fig. 12.

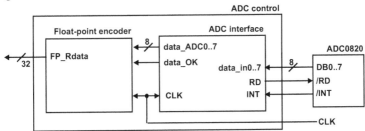

Fig. 12. Block Diagram of the ADC control component.

Now, the information generated by the ADC control component should be processed by the corresponding control law.

The discrete PID controller was synthesized on a FPGA based on equations for the continuous PID controller (Ogata, 2010), defined as

$$u_{av} = K_p(\overline{F}(t) - F(t)) + K_i \int_0^t (\overline{F}(t) - F(t))dt + K_d \frac{d(\overline{F}(t) - F(t))}{dt} \qquad (10)$$

where $K_i = \frac{K_p}{T_i}$ and $K_d = K_p T_d$.

An important aspect in the discretization of (10) is the obtaining of a discrete approximation of the continuous integral and a discrete approximation of the continuous derivative.

For discrete approximation of the continuous integral we have used the Adams-Bashforth method of the second order (Ascher & Petzold, 1998). This method is defined as

$$y[n+1] = y[n] + \frac{1}{2}\Delta t(3\dot{y}[n] - \dot{y}[n-1]) \tag{11}$$

Then, if the continuous integral is defined as $\int_0^t (\overline{F}(t) - F(t))dt$, using the Adams-Bashforth method, its discrete approximation is defined as

$$F[n+1] = F[n] + \frac{1}{2}\Delta t(3(\overline{F}[n] - F[n]) - (\overline{F}[n-1] - F[n-1])) \tag{12}$$

The Fig. 13 shows the proposed architecture for discrete approximation of a continuous integral given by (12).

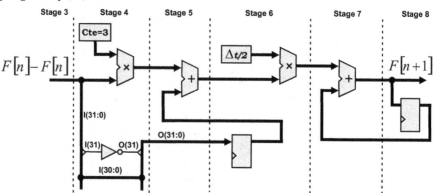

Fig. 13. Block diagram of the discrete approximation of a continuous integral.

On the other hand, the discrete approximation of the continuous derivative is obtained based on finite differences method of the first order (Burden & Douglas, 2000), using the backward difference. This method is defined as

$$\left(\frac{\partial y}{\partial t}\right)_n \approx \frac{y[n] - y[n-1]}{\Delta t} \tag{13}$$

Then, if the continuous derivative is defined as $\frac{\partial(\overline{F}(t)-F(t))}{\partial t}$, using the finite differences method, its discrete approximation is defined as

$$F'[n] = \frac{(\overline{F}[n] - F[n]) - (\overline{F}[n-1] - F[n-1])}{\Delta t} \tag{14}$$

The Fig. 14 shows the proposed architecture for discrete approximation of a continuous derivative given by (14).

The architecture consists of six multipliers and six adders. Then, it is necessary to implement single-precision floating point custom-adder and custom-multiplier.

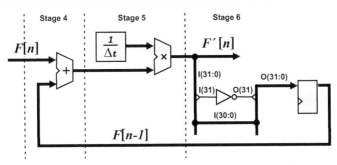

Fig. 14. Block diagram of the discrete approximation of a continuous derivative.

The Xilinx ISE Design Suite 12.2 includes the CORE Generator tool, which allows generating pre-optimized elements for Xilinx's FPGA. Our controller architecture uses multipliers and adders of single-precision floating-point, standard Std-754, generated by this tool. The symbols of the multiplier and adder generated by the CORE Generator tool are showed in the Fig. 15.

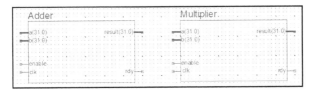

Fig. 15. Adder and Multiplier modules generated for Xilinx CORE Generator tool.

The proposed PID controller architecture is composed of 10 pipeline stages (see Fig. 16) and each of them needs 100 cycles to fulfill its function (2 μs), this indicates that the processing time of one data is 20 μs (time between 2 consecutive data delivered by the controller to the next component, the PWM). The enable signals (Stage_enable0..9) have the control each one of the pipeline registers that composed the proposed architecture.

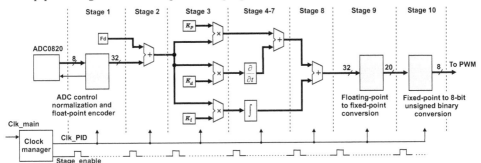

Fig. 16. Architecture proposed for the discrete PID controller implemented into the Spartan-3E1600 FPGA.

In the last stage the PID controller output must be adequacy for the PWM module. This adequation consists of Float-point to 8 bit unsigned binary conversion.

The last component of the proposed architecture is the PWM. The PWM component consists of single up-down counter unit and one magnitude comparator unit (see Fig. 17(a)).

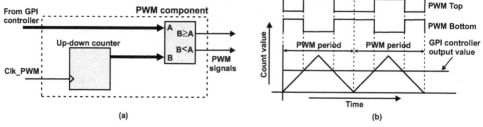

(a)   (b)

Fig. 17. (a) PWM component; (b) PWM outputs.

The PWM output frequency depends of the maximum count value of the counter and the Clk_PWM frequency (see figure 17(b)). Then, the PWM output frequency is defined as

$$PWM frequency = \frac{CLK\_PWM}{2(maximuncount + 1)} = \frac{25MHz}{512} = 48.828KHz \qquad (15)$$

The implementation result of the complete architecture for discrete PID controller are reported in Table 1.

Mod.	Slices	Flip-Flops	4-input's -LUT's	Pre-opt -elem.	Max. Freq (MHz)
1	5668 (38%)	8722 (29%)	8737 (29%)	1 BRAM (2 %) 1 DCM (12 %)	60.37

Table 1. Discrete PID controller implementation results.

6. Experimental results

The PID control and the Pulse Width Modulator (PWM) actuator for the regulation of output voltage of the buck converter were implemented in a Spartan 3E board. The only external hardware connected to the FPGA for measuring the "buck" converter output voltage was the analog digital converter ADC0820. Figure 18 illustrates the block diagram of the FPGA-based control system based on PID controller.

6.1 Requirements of the PID controller

Figure 19 shows the open-loop response of the "buck" converter with the following specifications: $L = 1mH$, $C = 100\mu F$, $R = 100\Omega$, $E = 24V$, $f = 48.828KHz$, $\Delta v_0/v_0 = 0.013\%$, $\Delta i_L = 0.092$ and a duty cycle $D = 0.75$. The output voltage response is a steady-state error of 5.56% and has a settling time of 15ms. On the other hand, we get that the diagram bode of the transfer function given by (4) with the same parameters, has a gain margin $Gm = Inf$ (at Inf rad/sec) and a phase margin $Pm = 0.377$deg (at 1.58×10^4 rad/sec). Given that the buck converter system has infinite gain margin, it can withstand greater changes in system parameters before becoming unstable in closed loop. Since the system has this characteristic, we will design our controllers in closed loop with the following requirements: Overshoot

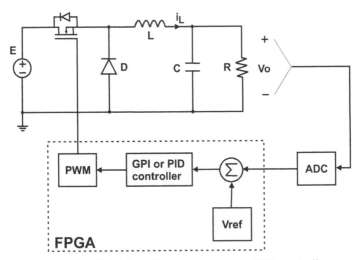

Fig. 18. Block diagram of the FPGA-based control system for PID controller.

less than 4.32%, Setting time less than 5 milliseconds, Steady-state error less than 1%, and Maximum sampling time $40\mu s$.

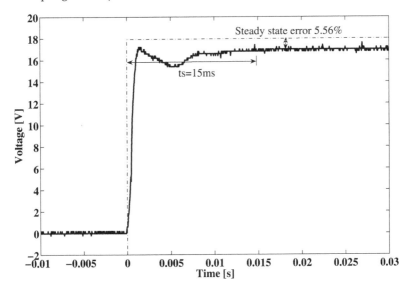

Fig. 19. Output voltage transient response of the "buck" converter with the PID control scheme.

The PID controller gains obtained by the design requirements were:

$$K_p = 0.15; \quad T_i = 1.2 \times 10^{-3}; \quad T_d = 5.9 \times 10^{-4} \tag{16}$$

6.2 PID controller into the FPGA

Figure 20 shows the performance of the PID control law, in the stabilization task for the "buck" converter output voltage. As before, we used a constant reference of 18 V. The continuous line corresponds to the PID controlled response. The settling time of the response of the "buck" converter output voltage through the PID controller, is 13.64 ms. The PID controller tuning was done through a third order Hurwitz polynomial.

Table 2 exhibits the performance of the synthesized controller. The main specifications of the transient response, the bandwidth of the PID controller (see Table 2), these frequencies are calculated in the closed-loop through the damping ratio and settling time (Messner & Tilbury, 1999). The damping coefficient value is 0.707, while the value of settling time is: 13.64 ms.

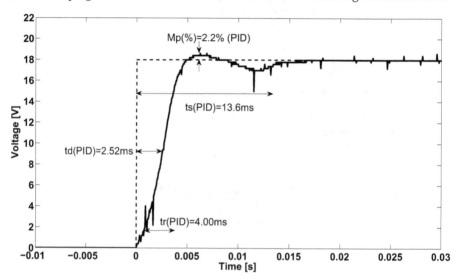

Fig. 20. Output voltage transient response of the "buck" converter with the PID control scheme.

Delay time	t_d	2.52 ms
Rise time	t_r	4 ms
Time of peak	t_p	6.24 ms
Percentage of overshoot	M_p	2.2 %
Settling time	t_s	13.64 ms
Bandwith	B_ω	414.85 Hz

Table 2. Specifications of the Controller transient response PID.

To illustrate the robustness of the PID controller, we made a test with the "buck" converter system by suddenly connecting a dynamic load (DC motor) at the output of the "buck" converter. Figure 21(a) shows the behavior of the perturbed converter's output voltage and the recovery of the output voltage to the desired reference signal when the converter is controlled with the PID controller scheme. Also, in Figure 21(b) is shown the u_{av} control signal, from the PID scheme implemented in the FPGA.

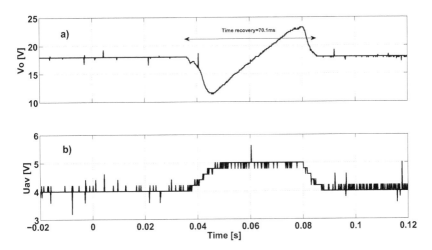

Fig. 21. Output voltage response of the "buck" converter with sudden connection of a DC motor.

7. Conclusions

In this work, we have applied the Proportional Integral Derivative control scheme, synthesized via a Field Programmable Gate Array implementation, for the output voltage regulation in a DC/DC power converter of the "buck" type. The performance of the PID control action was synthesized via a FPGA. The results obtained by cosimulation allowed to study each of the units designed and modeled in VHDL, correcting some errors and, in addition, the cosimulation was a perfect tool allowing faster design process to get a full system simulation before implement the system in the FPGA board. Also we conclude that the PID controller has a good transient response. When we connect a static and a dynamic load to the "buck" converter output, we observed that the PID control results in a significantly faster response, regarding the output voltage recovery time to the desired reference. Finally, the experimental results show the effectiveness of the FPGA realization of both the PID controller, in this case, programmed into the FPGA. This methodology of design can be used to design switched mode power supplies with efficiency greater than 95%.

8. References

Ascher, U. & Petzold, L. (1998). Computer Methods for Ordinary Differential Equations and Differential-Algebraic Equations. SIAM: Society for Industrial and Applied Mathematics.

Bo Li; Shuibao Guo; Xuefang Lin-Shi & Allard, B. (2011) , Design and implementation of the digital controller for boost converter based on FPGA,*IEEE International Symposium on Industrial Electronics (ISIE)*, pp. 1549-1554.

Burden, R. & Douglas, F. (2000). Numerical Analysis. Brooks/Cuelo.

Chander, S.; Agarwal, P. & Gupta, I. (2010) , FPGA-based PID controller for DC-DC converter, *Power Electronics, Drives and Energy Systems (PEDES) & 2010 Power India, 2010 Joint International Conference on*, Dec. 2010, pp.1-6.

Dorf, R. C. & Bishop,R. H., (2011).*Modern Control Systems*.Twelfth Edition: Prentice-Hall.

Hwu, K. I. (2010). Forward Converter with FPGA-Based Selft-Tuning PID Controller. *Tamkang Journal of Science and Engineering*, Vol. 13, No. 2, June 2010, 173-180, ISSN:1560-6686

IEEE Computer Society. (1985). IEEE Standard for Binary Floating-Point Arithmetic, IEEE Std 754-1985.

Joost, R. & Salomon, R., (2005). Advantages of FPGA-based multiprocessor systems in industrial applications, *Industrial Electronics Society, 2005. IECON 2005. 31st Annual Conference of IEEE*, Nov. 2005, pp. 445-450.

Linares-Flores, J. & Sira-Ramírez, H. (2004). DC motor velocity control through a DC-to-DC power converter, *Proc. 43rd IEEE Conf. Dec. Control*, pp. 5297-5302.

Linares-Flores, J. & Sira-Ramírez, H. (2004). Sliding Mode-Delta Modulation GPI Control of a DC Motor through a Buck Converter,*Proc. 2nd Symposium on System, Structure and Control*, México.

Linares-Flores, J.; Antonio-García, A. & Orantes-Molina, A. (2011). Smooth Starter for a DC Machine through a DC-to-DC buck converter, *Revista Ingeniería, Investigación y Tecnología*, Universidad Autónoma de México (UNAM), Vol. XII, No. 2, April 2011, ISSN: 1405-7743.

The MathWorks, Inc. (2008). Release Notes for Matlab R2008a.

Messner, W. C. & Tilbury, D. W., (1999). *Control tutorial for matlab and simulink: A web-Based Approach*, Addison-Wesley.

Mingyao Ma; Wuhua Li & Xiangning He. (2010), An FPGA-based mixed-signal controller for switching mode converters,*IECON 2010 - 36th Annual Conference on IEEE Industrial Electronics Society*, Nov. 2010, pp.1703-1708.

Monmasson, E. & Cirstea, M. N., (2007). FPGA design methodology for industrial control systems-A review, *IEEE Transactions on Industrial Electronics*, vol. 54, no. 4, August 2007, pp. 1824-1842.

Ogata, K. (2010). *Modern Control Engineering*. 5th Edition: Prentice-Hall.

Palnitkar, S. (2003). A guide to digital design and synthesis. 2nd Edition, USA: Printece-Hall.

Phillips, C. L. & Nagle, H. T., (1995). *Digital Control Systems Analysis and Design*, Third Edition: Prentice-Hall.

Powersim, Inc. (2006). *PSIM User's Guide*

Rodriguez-Andina,J. J.; Moure, M. J. & Valdes, M. D., (2007). Features, design tools, and application domains of FPGAs, *IEEE Transactions on Industrial Electronics*, vol. 54, no. 4, August 2007, pp. 1810-1823.

Sira-Ramírez, H. & Agrawal, S. K. (2004).*Differentially Flat Systems*,Marcel Dekker, New York.

12

System Identification Using Orthonormal Basis Filter

Lemma D. Tufa and M. Ramasamy
Chemical Engineering Department,
Universiti Teknologi PETRONAS,
Bandar Seri Iskandar, Perak,
Malaysia

1. Introduction

Models are extensively used in the design and implementation of advanced process control systems. In model predictive control (MPC), model of the plant is used to predict the future output of the plant using the current and future optimal inputs and past outputs. Therefore, the design of MPC, essentially, includes the development of an effective plant model that can be used for predicting the future output of the plant with good accuracy (Camacho & Bordon, 2004; Rawlings, 2000). Models can be developed either from purely theoretical analysis (conservation principles, thermodynamics, etc.) or from experimental data or somewhere in between. The process of model development from experimental data is known as system identification. The identification test can be conducted either in open-loop (open-loop identification) or while the plant is under feedback control (closed-loop identification).

The theory of linear system identification is well developed and there are already numerous literatures. The pioneering work in system identification was done by Ljung (1999) and his book provides detailed theoretical foundation for system identification. The book by Nelles (2001) is also a very practical book and highly recommended for practitioners both on linear and non-liear system identification. Heuberger, *et al.*, (2005) authored a very comprehensive book on modeling and identification using rational orthonormal basis functions, though current developments in application of OBF for MPC, closed-loop identification, etc., were not included.

There are several linear dynamic model structures that are commonly used in control relevant problems. They have two general forms, *i.e.*, the state space and input-output forms. In this chapter, we deal with the latter form also called transfer function. The most common linear input-output model structures can be derived from one general structure (1). The general linear structure consists of a deterministic component, *i.e.*, the plant input , $u(k)$, filtered by a linear filter and a noise component, *i.e.*, a white noise, $e(k)$, filtered by a corresponding linear filter.

$$y(k) = \frac{B(q)}{F(q)A(q)}u(k) + \frac{C(q)}{D(q)A(q)}e(k) \qquad (1)$$

The q in (1) is the forward shift operator defined as $q\,u(t) = u(t + 1)$ and q^{-1} is the delay (backward shift) operator, $q^{-1}u(t) = u(t-1)$.

The various commonly used structures can be easily derived from the general model structure by making some assumptions. The ARX model can be derived from (1) by assuming $F(q) = D(q) = C(q) = 1$. Therefore, the ARX model structure has the form

$$y(k) = \frac{B(q)}{A(q)}u(k) + \frac{1}{A(q)}e(k) \tag{2}$$

The Auto Regressive Moving Average with Exogenous Input (ARMAX) can be derived from (1) by assuming $F(q) = D(q) = 1$.

$$y(k) = \frac{B(q)}{A(q)}u(k) + \frac{C(q)}{A(q)}e(k) \tag{3}$$

Other linear model structures are listed below:

Box Jenkins (BJ):

$$y(k) = \frac{B(q)}{F(q)}u(k) + \frac{C(q)}{D(q)}e(k) \tag{4}$$

Output Error (OE):

$$y(k) = \frac{B(q)}{F(q)}u(k) + e(k) \tag{5}$$

Finite Impulse Response (FIR):

$$y(k) = B(q)u(k) + e(k) \tag{6}$$

It should be noted that in FIR model structures the filters are simple delays. Equation (6) can be expanded into

$$y(k) = (b_1 q^{-1} + b_2 q^{-2} + \ldots + b_m q^{-m})u(k) + e(k) \tag{7}$$

The selection of the appropriate model structure for a specific purpose, among other factors, depends on the consistency of the model parameters, the number of parameters required to describe a system with acceptable degree of accuracy and the computational load in estimating the model parameters. The optimality of model parameters is generally related to the bias and consistency of the model. Bias is the systematic deviation of the model parameters from their optimal value and inconsistency refers to the fact that the bias does not approach zero as the number of data points approach infinity (Nelles, 2001). The most widely used linear models are Step Response, ARX and FIR models (Ljung, 1999; Nelles, 2001). Their popularity is due to the simplicity in estimating the model parameters using the popular linear least square method. However, it is known that all of these three model structures have serious drawbacks in application. The ARX model structure leads to inconsistent parameters for most open-loop identification problems and the FIR and step response model need very large number of parameters to capture the dynamics of a system with acceptable accuracy. The inconsistency in the ARX and also in ARMAX model parameters is caused by the assumption of common denominator dynamics for both the input and noise transfer functions given by $1/A(q)$, which implies that the plant model and the noise model are correlated. In reality, this is rarely the case for open loop identification problems. The output error (OE) and the Box Jenkins (BJ) model structures assume independent transfer function and noise models, and hence they allow consistent parameter estimation. However, determination of the model parameters in both cases involves nonlinear optimization. In addition, in case of BJ, because of the large number of parameters involved in the equation, it is rarely used in practice, especially, in MIMO systems.

Orthonormal Basis Filter (OBF) models have several advantages over the conventional linear models. They are consistent in parameters for most practical open-loop systems and the recently developed ARX-OBF and OBF-ARMAX structures lead to consistent parameters for closed loop identification also. They require relatively a fewer numbers of parameters to capture the dynamics of linear systems (parsimonious in parameters) and the model parameters can be easily estimated using linear least square method (Heuberger, et al., 2005; Heuberger, et al., 1995; Ninness & Gustafsson, 1997; Van den Hof, et al., 1995). MIMO systems can be easily handled using OBF and OBF based structures. In addition, recent works by Lemma and Ramasamy (Lemma & Ramasamy, 2011) prove that OBF based structures show superior performance for multi-step ahead prediction of systems with uncertain time delays compared to most conventional model structures.

Among the earliest works on rational orthonormal bases was contributed by Takenaka (1925) in the 1920's in relation to approximation via interpolation, with the subsequent implications for generalized quadrature formula. In subsequent works, in the 1960s, Walsh (1975) contributed extensively in the applications of orthonormal bases for approximation, both in discrete time and continuous time analysis. In similar periods, Wiener (Wiener, 1949) examined applications of continuous time Laguerre networks for the purpose of building optimal predictor. Van den Hof, *et al.,* (1995) introduced the generalized orthonormal basis filters. They showed that pulse, Laguerre and Kautz filters can be generated from inner functions and their minimal balanced realizations. Ninness and Gustafsson (1997) unified the construction of orthonormal basis filters. Lemma, *et al.,* (2011) proposed an improved method for development of OBF models where the poles and time delays of the system can be estimated and used to develop a parsimonious OBF model. On another work (Lemma, et al., 2010) it was shown that BJ type OBF models can be easily developed by combing structures with AR and ARMA noise model. Some works on closed-loop identification using OBF based structures have also been presented (Badwe, et al., 2011; Gáspár, et al., 1999; Lemma, et al., 2009; Lemma & Ramasamy, 2011).

2. Development of conventional OBF models

Consider a discrete time linear system

$$y(k) = G(q)u(k) \tag{8}$$

where $G(q)$ = transfer function of the system. A stable system, $G(q)$, can be approximately represented by a finite–length generalized Fourier series expansion as:

$$G(q) = \sum_{i=1}^{n} l_i f_i(q) \tag{9}$$

where $\{l_i\}$, $i = 1, 2, ..., n$ are the model parameters, n is the number of parameters, and $f_i(q)$ are the orthonormal basis filters for the system $G(q)$. Orthonormal basis functions can be considered a generalization of the finite length fourier series expansion. Two filters f_1 and f_2 are said to be orthonormal if they satisfy the properties:

$$\langle f_1(q), f_2(q) \rangle = 0 \tag{10}$$

$$\|f_1(q)\| = \|f_2(q)\| = 1 \tag{11}$$

2.1 Common orthonormal basis filters

There are several orthonormal basis filters that can be used for development of linear OBF models. The selection of the appropriate type of filter depends on the dynamic behaviour of the system to be modelled.

Laguerre filter

The Laguerre filters are first-order lag filters with one real pole. They are, therefore, more appropriate for well damped processes. The Laguerre filters are given by

$$f_i = \sqrt{(1-p^2)} \frac{(1-pq)^{i-1}}{(q-p)^i}, \quad |p| < 1 \tag{12}$$

where p is the estimated pole which is related to the time constant, τ, and the sampling interval T_s of the system by

$$p = e^{-(T_s/\tau)} \tag{13}$$

Kautz filter

Kautz filters allow the incorporation of a pair of conjugate complex poles. They are, therefore, effective for modeling weakly damped processes. The Kautz filters are defined by

$$f_{2i-1} = \frac{\sqrt{(1-a^2)(1-b^2)}}{q^2 + a(b-1)q - b} g(a,b,q,i) \tag{14}$$

$$f_{2i} = \frac{\sqrt{(1-b^2)}(q-a)}{q^2 + a(b-1)q - b} g(a,b,q,i) \tag{15}$$

where

$$g(a,b,q,i) = \left(\frac{-bq^2 + a(b-1)q + 1}{q^2 + a(b-1) - b} \right)^{i-1} \tag{16}$$

$-1 < a < 1$ and $-1 < b < 1$ $n = 1, 2, \ldots$

Generalized orthonormal basis filter

Van den Hof, et al., (1995) introduced the generalized orthonormal basis filters and showed the existence of orthogonal functions that, in a natural way, are generated by stable linear dynamic systems and that form an orthonormal basis for the linear signal space l_2^n . Ninness & Gustafsson (1997) unified the construction of orthonormal basis filters. The GOBF filters are formulated as

$$f_i(q,p) = \frac{\sqrt{1 - |p_i|^2}}{(q - p_i)} \prod_{j=1}^{i-1} \frac{(1 - p_j^* q)}{(q - p_j)} \tag{17}$$

where $p \equiv \{p_j : j = 1, 2, 3, \ldots\}$ is an arbitrary sequence of poles inside the unit circle appearing in complex conjugate pairs.

Markov-OBF

When a system involves a time delay and an estimate of the time delay is available, Markov-OBF can be used. The time delay in Markov-OBF is included by placing some of the poles at the origin (Heuberger, et al., 1995). For a SISO system with time delay equal to d samples, the basis function can be selected as:

$$f_i = z^{-i} \text{ for } i = 1, 2, \ldots, d \tag{18}$$

$$f_{i+d}(q,p) = \frac{\sqrt{1 - |p_i|^2}}{(q - p_i)} \prod_{j=1}^{i-1} \frac{(1 - p_j^* q)}{(q - p_j)} z^{-d} \text{ for } i = 1, 2, \ldots, N \tag{19}$$

Patwardhan and Shah (2005) presented a two-step method for estimating time delays from step response of GOBF models. In the first step, the time delays in all input-output channels are assumed zero and the model is identified with GOBF. In GOBF models, the time delay is approximated by a non-minimum phase zero and the corresponding step response is an inverse response. The time delay is then estimated from a tangent drawn at the point of inflection.

2.2 Estimation of GOBF poles

Finding an appropriate estimate of the poles for the filters is an important step in estimating the parameters of the OBF models. Arbitrary choice of poles may lead to a non-parsimonious model. Van den Hof, et al., (2000) showed that for a SISO system with poles $\{a_j : | a_j | < 1$ for $j = 1, 2, \ldots, n\}$, the rate of convergence of the model parameters is determined by the magnitude of the slowest Eigen value.

$$\rho = \max_{j} \prod_{k=1}^{n} \left| \frac{a_j - p_k}{1 - \bar{p}_k a_j} \right| \tag{20}$$

where p_k = arbitrary poles.

Therefore, a good approximation by a small number of parameters can be obtained by choosing a basis for which ρ is small. It is shown that the poles determined by Van den Hof et al. method closely match the dominant poles of the system (Heuberger, et al., 2005; Wahlberg, 1991). Lemma, et al., (2011) proposed a systematic way to estimate the dominant poles and time delays of a system from the input-output identification test data. An OBF model is first developed with randomly chosen real poles and generalized orthonormal basis filters with 10-12 terms. The model is simulated to get a noise free step response of the system. One or two poles of the system are estimated from the resulting noise free step response of the OBF model and it is also observed whether the system is weakly damped or not. This process can be repeated until some convergence criterion is fulfilled. The procedure normally converges after two or three iterations. The procedure is iiterations and is illustrated in Example 1.

2.3 Model parameter estimation

In OBF models, the output can be expressed as a linear combination of the input sequence filtered by the respective filters. For a finite number of parameters, from (9) we get

$$\hat{y}(k) = l_1 f_1(q)u(k) + l_2 f_2(q)u(k) + \dots + l_n f_n(q)u(k) \tag{21}$$

Equation (21) is not linear in its parameters and therefore estimation of parameters using linear least square method is impossible. However, it can be modified such that it is linear in parameters, as

$$\hat{y}(k) = l_1 u_{f1}(k) + l_2 u_{f2}(k) + \dots + l_n u_{fn}(k) \tag{22}$$

where $u_{fi}(k)$ is the filtered input given by

$$u_{fi}(k) = f_i(q)u(k) \tag{23}$$

Once the dominant poles of the system and the types of filters are chosen, the filters $f_1, f_2, \dots, f_n$ are fixed. The filtered inputs, u_{fi}, are determined by filtering the input sequence with the corresponding filter. For an OBF model with n parameters, the prediction can be started from the nth instant in time. Equation (22) can be expanded and written in matrix form as

$$\begin{bmatrix} \hat{y}_{n+1} \\ \hat{y}_{n+2} \\ \cdot \\ \cdot \\ \cdot \\ \hat{y}_N \end{bmatrix} = \begin{bmatrix} u_{f1}(n) & u_{f2}(n-1) & \cdots & u_{fn}(1) \\ u_{f1}(n+1) & u_{f2}(n) & \cdots & u_{fn}(2) \\ \cdot & & & \cdot \\ \cdot & & & \cdot \\ \cdot & & & \cdot \\ u_{f1}(N-1) & u_{f2}(N-2) & \cdots & u_{fn}(N-n) \end{bmatrix} \begin{bmatrix} l_1 \\ l_2 \\ \cdot \\ \cdot \\ \cdot \\ l_n \end{bmatrix} \tag{24}$$

where N is the future time instant.

Equation (24) in vector-matrix notation is given by

$$\hat{y} = X\theta \tag{25}$$

where $\theta = \left[l_1, l_2, ..., l_n\right]^T$ is the parameter vector, $\hat{y}$ is the output vector $\hat{y} = [y_{n+1}, y_{n+2}, ..., y_N]$ and X is the regressor matrix given by

$$X = \begin{bmatrix} u_{f1}(n) & u_{f2}(n-1) & \cdots & u_{fm}(1) \\ u_{f1}(n+1) & u_{f2}(n) & \cdots & u_{fm}(2) \\ \cdot & \cdot & & \cdot \\ \cdot & \cdot & & \cdot \\ \cdot & \cdot & & \cdot \\ u_{f1}(N-1) & u_{f2}(N-2) & \cdots & u_{fm}(N-n) \end{bmatrix} \tag{26}$$

Since (25) is linear in parameters, the model parameters can be estimated using linear least square formula (27).

$$\hat{\theta} = (X^T X)^{-1} X^T y \tag{27}$$

Algorithm 1

1. Use GOBF structure and two randomly selected stable poles and develop (6 to 12) sequence of GOBF filters
2. Develop the regressor matrix (26) using the filters developed at step (1) and the input sequence u(k)
3. Use the linear least square formula (27) to estimate the model parameters
4. Make a better estimate of the poles of the system from the step response of the GOBF model
5. Repeat steps 1 to 4 with the new pole until a convergence criterion is satisfied

The Percentage Prediction Error (PPE) can be a good convergence criterion.

Example 1

An open loop identification test for SISO system is carried out and the input-output data shown in Figure 1 is obtained. A total of 4000 data points are collected at one minute sampling interval with the intention of using 3000 of them for modelling and 1000 for validation. Develop a parsimonious OBF model using the data. No information is available about the pole of the system.

Since there is no information about the poles of the system, two poles: 0.3679 and 0.9672 are arbitrarily chosen for the first iteration of the model. A GOBF model with six terms (you can choose other numbers and compare the accuracy if you need) is first developed with these two poles alternating. Note that, once the poles, type of filter, i.e., GOBF and the number of terms is fixed the filters are fixed and the only remaining value to determine the model is the model parameters.

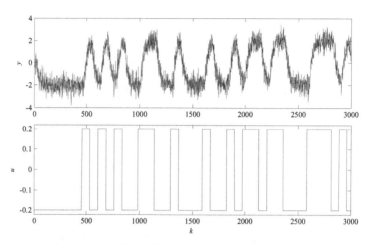

Fig. 1. Input-output data used for identification

To estimate the model parameters the regressor matrix is developed and used together with the plant measured output $y(k)$ in the least square formula (27) to find the model parameters:

[-0.2327   0.8733 -0.2521   0.8854 -0.8767 -0.2357]

The percentage prediction error (PPE) is found to be 9.7357. For the second iteration, the poles of the system are estimated from the noise free step-response of the GOBF model shown in Figure 2.

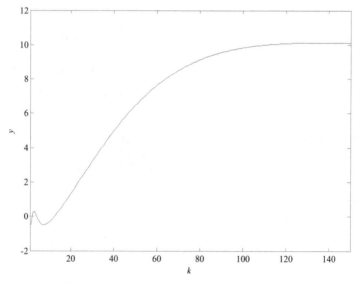

Fig. 2. Step response of the noise free OBF model developed in the first iteration

The OBF model parameters, the PPE and poles estimated at each iteration are presented in Table 1.

Iterations	PPEs	Poles	Model Parameters
1	9.7357	[0.3679 0.9672]	[0.2327 0.8733 -0.2521 0.8854 -0.8767 -0.2357]
2	9.5166	[0.9467 0.9467]	[0.7268 0.7718 0.4069 -0.5214 0.0273 0.0274]
3	9.5149	[0.9499 0.9306]	[0.4992 0.9781 0.4723 -0.3377 0.1387 -0.0305]

Table 1. The results of the OBF iterative identification method

Note that the parameters in the last iteration together with the OBF filters determine the model of the plant. The model accuracy is judged by cross validation. Figure 3 shows the measured output data for sampling instants 3001 to 4000 (this data is not used for modelling) and the result of the OBF simulation for the plant input for the instants 3001 to 4000.

3. BJ- Type models by combining OBF with conventional noise model structures

In Example 1, we developed a GOBF model to capture the deterministic component of the plant. The residual of the model however was just discarded. In reality, this residual may contain useful information about the plant. However, as it is already noted, conventional OBF models do not include noise models. Patwardhan and Shah (2005) showed that the regulatory performance of MPC system improves significantly by including a noise model to the OBF simulation model. In their work, the residual of the OBF model is whitened with

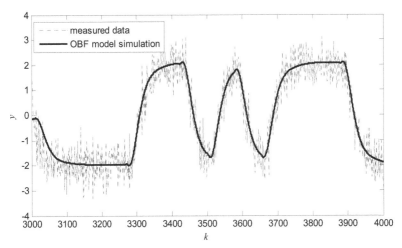

Fig. 3. Validation of the final GOBF model with 6 parameters.

Auto Regressive (AR) noise model. The AR noise model is parameterized in terms of OBF parameters and a minimal order state space model was realized. In this section, an integrated approach for developing BJ models with an OBF plant model and AR or ARMA noise model is presented.

3.1 Model structures

The BJ model structure is known to be the most flexible and comprehensive structure of the conventional linear models(Box & Jenkins, 1970).

$$y(k) = \frac{B(q)}{F(q)}u(k) + \frac{C(q)}{D(q)}e(k) \tag{28}$$

In (28) $B(q)/F(q)$ describes the plant model whereas $C(q)/D(q)$ describes the noise model. The BJ-type model structure proposed by Lemma, et al., (2010) is obtained by replacing the plant model structure with OBF model structure. First, the OBF-AR structure, *i.e.*, with $C(q)=1$ is discussed then the OBF-ARMA structure is discussed.

The OBF-AR model structure assumes an OBF and AR structures for the plant and noise transfer functions, respectively.

$$y(k) = G_{OBF}(q)u(k) + \frac{1}{D(q)}e(k) \tag{29}$$

The OBF-ARMA structure has more flexible noise model than the OBF-AR structure as given by (30).

$$y(k) = G_{OBF}(q)u(k) + \frac{C(q)}{D(q)}e(k) \tag{30}$$

3.2 Estimation of model parameters

The model parameters of both OBF-AR and OBF-ARMA structures are estimated based on the prediction error method as explained below.

Estimation of parameters of OBF-AR model

The prediction error $e(k)$ is defined as

$$e(k) = y(k) - \hat{y}(k \mid k-1) \tag{31}$$

Introducing the prediction error (31) in (29) and rearranging leads to

$$\hat{y}(k \mid k-1) = D(q)G_{OBF}(q)u(k) + \left(1 - D(q)\right)y(k) \tag{32}$$

Assuming that the noise sequence is uncorrelated to the input sequence, the parameters of the OBF model can be estimated separately. These parameters can then be used to calculate the OBF simulation model output using (32).

$$y_{obf}(k) = G_{OBF}(q)u(k) \tag{33}$$

Inserting (33) in (32)

$$\hat{y}(k \mid k-1) = D(q)y_{obf}(k) + \left(1 - D(q)\right)y(k) \tag{34}$$

Equation (34) is linear in parameters since $y_{obf}(k)$ is already known. With $D(q)$ monic, (34) can be expanded and rearranged to yield

$$\hat{y}(k\,|\,k-1) = y_{obf}(k) - d_1 r(k-1) - d_2 r(k-2) - \ldots - d_n r(k-n)$$
(35)

where

n is the order of the polynomial $D(q)$

$$r(i) = y(i) - y_{obf}(i)$$

Note that $r(i)$ represents the residual sequence of the output sequence $y(k)$ of the system from the OBF model output $y_{obf}(k)$. The model parameters in (35) can be calculated by the linear least square formula (27) with the regressor matrix given by (36).

$$X = \begin{bmatrix} y_{obf}(n) & -r(n-1) & -r(n-2) - & \ldots & -r(1) \\ y_{obf}(n+1) & -r(n) & -r(n-1) - & \ldots & -r(2) \\ \cdot & \cdot & \cdot & & \cdot \\ \cdot & \cdot & \cdot & & \cdot \\ \cdot & \cdot & \cdot & & \cdot \\ y_{obf}(N) & -r(N-1) & -r(N-2) - & \ldots & -r(N-n) \end{bmatrix}$$
(36)

where $n = n_D$.

The step-by-step procedure for estimating the OBF-AR model parameters, explained above, is outlined in Algorithm 2.

Algorithm 2

1. Develop a parsimonious OBF model
2. Determine the output sequence of the OBF model $y_{obf}(k)$ for the corresponding input sequence $u(k)$
3. Determine the residuals of the OBF model $r(k) = y(k) - y_{obf}(k)$
4. Develop the regression matrix X given by (36)
5. Determine the parameters of the noise model using (27) enforcing monic condition, i.e., $d_0 = 1$.

Estimation of parameters of OBF-ARMA model

The OBF-ARMA structure is given by (28)

$$y(k) = G_{OBF}(q)u(k) + \frac{C(q)}{D(q)}e(k)$$
(28)

Substituting the prediction error (31) in (28) and rearranging yields

$$C(q)\hat{y}(k\,|\,k-1) = D(q)G_{OBF}(q)u(k) - D(q)y(k) + C(q)y(k)$$
(37)

As in the case of OBF-AR model, if the noise sequence is uncorrelated with the input sequence, the OBF model parameters can be calculated separately and be used to calculate the simulation model output $y_{obf}(k)$ using (33).

Introducing (33) in (37) results in

$$C(q)\hat{y}(k \mid k-1) = D(q)y_{obf}(k) - D(q)y(k) + C(q)y(k) \tag{38}$$

Expanding and rearranging (37) we get

$$\hat{y}(k \mid k-1) = y_{obf}(k) - d_1 r(k-1) - d_2 r(k-2) - \ldots - d_m r(k-m) + \\ c_1 e(k-1) + c_2 e(k-2) + \ldots + c_n e(k-n) \tag{39}$$

The parameter vector and the regressor matrix are derived from (39) and are given by (40) and (41)

$$\theta = [d_1 \, d_2 \ldots d_m \, c_1 \, c_2 \ldots c_n]^T \tag{40}$$

where $n = n_C$, the order of the polynomial $C(q)$

   $m = n_D$, the order of the polynomial $D(q)$

   $mx = max \, (m, n)+1$

$$X = \begin{bmatrix} y_{obf}(mx) & -r(mx-1) & -r(mx-2)-\ldots-r(mx-n) & e(mx-1) & e(mx-2)\ldots e(mx-m) \\ y_{obf}(mx+1) & -r(mx) & -r(mx-1)-\ldots-r(mx-n+1) & e(mx) & e(mx-1)\ldots e(mx-m+1) \\ \cdot & \cdot & \cdot & \cdot & \cdot & \cdot & \cdot \\ \cdot & \cdot & \cdot & \cdot & \cdot & \cdot \\ \cdot & \cdot & \cdot & \cdot & \cdot & \cdot \\ y_{obf}(N) & -r(N-1) & -r(N-2)-\ldots & -r(N-n+1) & e(N-1) & e(N-2)\ldots & e(N-m+1) \end{bmatrix} \tag{41}$$

$$y = [y(mx) \quad y(mx+1) \, \ldots \, y(N)]^T \tag{42}$$

Equation (39) in the form shown above appears a linear regression. However, since the prediction error sequence, $e(k-i)$, itself is a function of the model parameters, it is nonlinear in parameters. To emphasize the significance of these two facts such structures are commonly known as pseudo-linear(Ljung, 1999; Nelles, 2001). The model parameters can be estimated by either a nonlinear optimization method or an extended least square method (Nelles, 2001). The extended least square method is an iterative method where the prediction error sequence is estimated and updated at each iteration using the prediction error of OBF-ARMA model. A good initial estimate of the prediction error sequence is obtained from the OBF-AR model. The parameters for the noise model are estimated using the linear least square method with (40) and (41) as parameters vector and regressor matrix, respectively. From the derivation, it should be remembered that all the poles and zeros of the noise models should be inside the unit circle and both the numerator and denominator polynomials should be monic. If an OBF-AR model with a high-order noise model can be developed, the residuals of the OBF-AR model will generally be close to white noise. In such cases, the noise model parameters of the OBF-ARMA model can be estimated using linear least square method in one step. The step-by-step procedure for estimating OBF-ARMA model parameters is outlined in Algorithm 3.

Algorithm 3

1. Develop a parsimonious OBF model
2. Determine the OBF simulation model output $y_{obf}(k)$ for the corresponding input sequence $u(k)$
3. Determine the residual of the simulation model $r(k) = y(k) - y_{obf}(k)$
4. Develop OBF-AR prediction model
5. Determine the residual of the OBF-AR model, $\hat{e}(k)$
6. Use $y_{obf}(k)$, $r(k)$ and $e(k) \approx \hat{e}(k)$ to develop the regressor matrix (40)
7. Use the linear least square formula (27) to estimate the parameters of the OBF ARMA model
8. Re-estimate the prediction error $e(k) = y(k) - \hat{y}(k)$ from the residual of OBF-ARMA model developed in step 7
9. Repeat steps 6 to 8 until convergence is achieved

Convergence criteria

The percentage prediction error (PPE) can be used as convergence criteria, *i.e.*, stop the iteration when the percentage prediction error improvement is small enough.

$$PPE = \frac{\sum_{k=1}^{n}(y(k) - \hat{y}(k))^2}{\sum_{k=1}^{n}\left((y(k) - \overline{y}(k))^2\right)} \times 100$$

where $\overline{y}$ represents the mean value of measurements $\{y(k)\}$ and $\hat{y}(k)$ predicted value of $y(k)$.

3.3 Multi-step ahead prediction

Multi-step ahead predictions are required in several applications such as model predictive control. In this section multi-step ahead prediction equation and related procedures for both OBF-AR and OBF-ARMA are derived.

Multi-step ahead prediction using OBF-AR model

Using (33) in (29) the OBF-AR equation becomes

$$y(k) = y_{obf}(k) + \frac{1}{D(q)}e(k) \tag{43}$$

i-step ahead prediction is obtained by replacing k with $k + i$

$$y(k+i) = y_{obf}(k+i) + \frac{1}{D(q)}e(k+i) \tag{44}$$

To calculate the i-step ahead prediction, the error term should be divided into current and future parts as shown in (45).

$$y(k+i) = y_{obf}(k+i) + \frac{F_i(q)}{D(q)}e(k) + E_i(q)e(k+i) \tag{45}$$

The last term in (45) contains only the future error sequence which is not known. However, since $e(k)$ is assumed to be a white noise with mean zero, (45) can be simplified to

$$\hat{y}(k+i \,|\, k) = y_{obf}(k+i) + \frac{F_i(q)}{D(q)}e(k) \tag{46}$$

F_i and E_i are determined by solving the Diophantine equation (47) which is obtained by comparing (44) and (45)

$$\frac{1}{D(q)} = E_i(q) + \frac{q^{-i}F_i(q)}{D(q)} \tag{47}$$

Equation (46) could be taken as the final form of the i-step ahead prediction equation. However, in application, since $e(k)$ is not measured the equation cannot be directly used. The next steps are added to solve this problem.

Rearranging (43) to get

$$\frac{1}{D(q)}e(k) = y(k) - y_{obf}(k) \tag{48}$$

Using (48) in (46) to eliminate $e(k)$

$$\hat{y}(k+i \,|\, k) = y_{obf}(k+i) + F_i(q)(y(k) - y_{obf}(k)) \tag{49}$$

Rearranging (49)

$$\hat{y}(k+i \,|\, k) = y_{obf}(k+i)(1 - F_i(q)q^{-i}) + F_i(q)y(k) \tag{50}$$

Rearranging the Diophantine equation (47)

$$\left(1 - q^{-i}F_i(q)\right) = D(q)E_i(q) \tag{51}$$

Using (51) in (50)

$$\hat{y}(k+i \,|\, k) = E_i(q)D(q)y_{obf}(k+i) + F_i(q)y(k) \tag{52}$$

Equation (52) is the usable form of the multi-step ahead prediction equation for the OBF-AR model. Given an OBF-AR model, the solution of the Diophantine equation to get E_i and F_i and the prediction equation (52) forms the procedure for i-step ahead prediction of the OBF-AR model.

Multi-step ahead prediction using OBF-ARMA model

Using (33) in (30) the OBF-ARMA equation becomes

$$y(k) = y_{obf}(k) + \frac{C(q)}{D(q)}e(k) \tag{53}$$

i-step ahead prediction is obtained by replacing k with $k + i$

$$y(k + i) = y_{obf}(k + i) + \frac{C(q)}{D(q)}e(k + i) \tag{54}$$

To calculate the i-step ahead prediction, the error term should be divided into current and future parts.

$$y(k + i) = y_{obf}(k + i) + \frac{F_i(q)}{D(q)}e(k) + E_i(q)e(k + i) \tag{55}$$

Since $e(k)$ is assumed to be a white noise with mean zero, the mean of $E_i(q)$ $e(k+i)$ is equal to zero, and therefore (55) can be simplified to

$$\hat{y}(k + i \mid k) = y_{obf}(k + i) + \frac{F_i(q)}{D(q)}e(k) \tag{56}$$

F_i and E_i are determined by solving the Diophantine equation (57) which is obtained by comparing (54) and (56)

$$\frac{C(q)}{D(q)} = E_i(q) + \frac{q^{-i}F_i(q)}{D(q)} \tag{57}$$

Rearranging (57)

$$\frac{1}{D(q)}e(k) = \frac{1}{C(q)}\left(y(k) - y_{obf}(k)\right) \tag{58}$$

Using (58) in (56) to eliminate $e(k)$

$$\hat{y}(k + i \mid k) = y_{obf}(k + i) + \frac{F_i(q)}{C(q)}\left(y(k) - y_{obf}(k)\right) \tag{59}$$

Rearranging (59)

$$\hat{y}(k + i \mid k) = y_{obf}(k + i)\left(1 - \frac{F_i(q)q^{-i}}{C(q)}\right) + \frac{F_i(q)}{C(q)}y(k) \tag{60}$$

Rearranging the Diophantine equation (60)

$$\left(1 - \frac{q^{-i}F_i(q)}{C(q)}\right) = \frac{D(q)E_i(q)}{C(q)} \tag{61}$$

Using (61) in (60) results in the final usable form of the i-step ahead prediction for OBF-ARMA model.

$$\hat{y}(k+i\,|\,k) = \frac{E_i(q)D(q)}{C(q)}y_{obf}(k+i) + \frac{F_i(q)}{C(q)}y(k) \tag{62}$$

Since y_{obf} $(k+i)$ is the output sequence of the simulation OBF model, if the OBF model parameters are determined its value depends only on the input sequence $u(k+i)$. Therefore, the i-step ahead prediction according to (62) depends on the input sequence up to instant $k+i$ and the output sequence up to instant k.

Multiple-Input Multiple-Output (MIMO) systems

The procedures for estimating the model parameters and i-step ahead prediction can be easily extended to MIMO systems by using multiple-MISO models. First, a MISO OBF model is developed for each output using the input sequences and the corresponding orthonormal basis filters. Then, AR model is developed using $y_{obf}(k)$ and the residual of the OBF simulation model. The OBF-ARMA model is developed in a similar manner, with an OBF model relating each output with all the relevant inputs and one ARMA noise model for each output using Algorithm (Lemma, et al., 2010).

Example 2

In this simulation case study, OBF-AR and OBF-ARMA models are developed for a well damped system that has a Box-Jenkins structure. They are developed with various orders and compared within themselves and with each other. The system is represented by (63). Note that both the numerator and denominator polynomials of the noise model are monic and their roots are located inside the unit circle.

$$y(k) = q^{-6}\frac{1-1.3q^{-1}+0.42q^{-2}}{1-2.55q^{-1}+2.165q^{-2}-0.612q^{-3}}u(k) + \frac{1+0.6q^{-1}}{1-1.15q^{-1}+0.58q^{-2}}e(k) \tag{63}$$

An identification test is simulated on the system using MATLAB and the input–output sequences shown in Figure 4 is obtained.

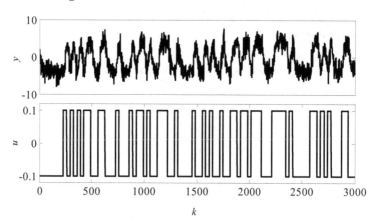

Fig. 4. Input-output data sequence generated by simulation of (63)

The mean and standard deviations of the white noise, $e(k)$, added to the system are 0.0123 and 0.4971, respectively, and the signal to noise ratio (SNR) is 6.6323 . The input signal is a pseudo random binary signal (PRBS) of 4000 data points generated using the '*idinput*' function in MATLAB with band [0 0.03] and levels [-0.1 0.1]. Three thousand of the data points are used for model development and the remaining 1000 for validation. The corresponding output sequence of the system is generated using SIMULINK with a sampling interval of 1 time unit.

OBF-AR model

First a GOBF model with 6 parameters and poles 0.9114 and 0.8465 is developed and the model parameters are estimated to be [3.7273 5.6910 1.0981 -0.9955 0.3692 -0.2252] using Algorithm 1. The AR noise model developed with seven parameters is given by:

$$\frac{1}{D(q)} = \frac{1}{1 - 1.7646q^{-1} + 1.6685q^{-2} - 1.0119q^{-3} + 0.5880q^{-4} - 0.3154q^{-5} + 0.1435q^{-6} - 0.0356q^{-7}} \quad (64)$$

The spectrum of the noise model of the system compared to the spectrum of the model for 3, 5 and 7 parameters is shown in Figure 5. The percentage predication errors of the spectrums of the three noise models compared to spectrum of the noise model in the system is given in Table 2.

n_D	PPE
3	54.3378
5	1.5137
7	0.9104

Table 2. PPE of the three AR noise models of system

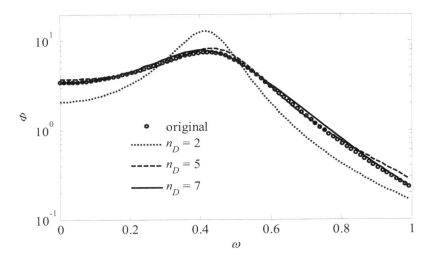

Fig. 5. Spectrums of the AR noise models for n_D = 2, 5 and 7 compared to the noise transfer function of system

It is obvious from both Figure 4 and Table 2 that the noise model with $n_D = 7$ is the closest to the noise transfer function of the system. Therefore, this noise model together with the GOBF model described earlier form the OBF-AR model that represent the system.

4. Closed loop identification using OBF-ARX and OBF-ARMAX structures

When a system identification test is carried out in open loop, in general, the input sequence is not correlated to the noise sequence and OBF model identification is carried out in a straight forward manner. However, when the system identification test is carried out in closed loop the input sequence is correlated to the noise sequence and conventional OBF model development procedures fail to provide consistent model parameters.

The motivation for the structures proposed in this section is the problem of closed-loop identification of open-loop unstable processes. Closed-loop identification of open-loop unstable processes requires that any unstable poles of the plant model should be shared by the noise model $H(q)$ otherwise the predictor will not be stable. It is indicated by both Ljung (1999) and Nelles (2001) that if this requirement is satisfied closed-loop identification of open-loop unstable processes can be handled without problem. In this section, two different linear structures that satisfy these requirements and which are based on OBF structure are proposed. While the proposed models are, specially, effective for developing prediction model for open-loop unstable process that are stabilized by feedback controller, they can be used for open-loop stable process also. These two linear model structures are OBF-ARX and OBF- ARMAX structures.

4.1 Closed–loop identification using OBF-ARX model

Consider an OBF model with ARX structure given by (65)

$$y(k) = \frac{G_{OBF}(q)}{A(q)}u(k) + \frac{1}{A(q)}e(k) \tag{65}$$

Rearranging (65)

$$\hat{y}(k\,|\,k-1) = G_{OBF}(q) - (1 - A(q))y(k) \tag{66}$$

With $A(q)$ monic (66) can be expanded to

$$\hat{y}(k\,|\,k-1) = G_{OBF}(q) - a_1 y(k-1) - a_2 y(k-2) - a_m y(k-m) \tag{67}$$

Note that, (67) can be further expanded to

$$\hat{y}(k\,|\,k-1) = l_1 u_{f1}(k) + l_2 u_{f2}(k) + \ldots + l_m u_{fm}(k) - \\ a_1 y(k-1) - a_2 y(k-2) - \ldots - a_n y(k-n) \tag{68}$$

Therefore, the regressor matrix for the OBF-ARX structure is given by

$$X = \begin{bmatrix} u_{f1}(mx) & u_{f2}(mx-1)...u_{fm}(mx-m) & -y(mx-1)-y(nx-2)...-y(mx-n) \\ . & . & . & . & . & . & . \\ . & . & . & . & . & . \\ . & . & . & . & . \\ u_{f1}(N) & u_{f2}(N-1)...u_{fm}(N-m) & -y(N-1) & -y(N-2)...-y(N-n) \end{bmatrix} \quad (69)$$

where m = order of the OBF model

n = order of $A(q)$

$mx = \max{(n, m)} + 1$

u_{fi} = input u filtered by the corresponding OBF filter f_i

The parameters are estimated using (69) in the least square equation (27). Note that in using (27) the size of y must be from mx to N.

4.2 Multi-step ahead prediction using OBF-ARX model

Consider the OBF-ARX model

$$y(k) = \frac{y_{obf}(k)}{A(q)} + \frac{1}{A(q)} e(k) \quad (70)$$

i-step ahead prediction is obtained by replacing k with $k + i$

$$y(k+i) = \frac{y_{obf}(k+i)}{A(q)} + \frac{1}{A(q)} e(k+i) \quad (71)$$

To calculate the i-step ahead prediction, the noise term can be divided into current and future parts.

$$y(k+i) = \frac{y_{obf}(k+i)}{A(q)} + \frac{F_i(q)}{A(q)} e(k) + E_i(q)e(k+i) \quad (72)$$

Since $e(k)$ is assumed to be a white noise with mean zero, the mean of $E_i(q)$ $e(k+i)$ is equal to zero (72) can be simplified to

$$\hat{y}(k+i\,|\,k) = \frac{y_{obf}(k+i)}{A(q)} + \frac{F_i(q)}{A(q)} e(k) \quad (73)$$

On the other hand rearranging (71)

$$y(k+i) = \frac{y_{obf}(k+i)}{A(q)} + e(k+i)\left(\frac{q^{-i}F_i(q)}{A(q)} + E_i(q) \right) \quad (74)$$

Comparing (70) and (73), F_i and E_i can be calculated by solving the Diophantine equation.

$$\frac{1}{A(q)} = E_i(q) + \frac{q^{-i}F_i(q)}{A(q)} \tag{75}$$

Rearranging (70)

$$\frac{1}{A(q)}e(k) = y(k) - \frac{y_{obf}(k)}{A(q)} \tag{76}$$

Using (76) in (73) to eliminate $e(k)$

$$\hat{y}(k+i\,|\,k) = \frac{y_{obf}(k+i)}{A(q)} + F_i(q)\left(y(k) - \frac{y_{obf}(k)}{A(q)}\right)$$

$$= y_{obf}(k+i)\left(\frac{1}{A(q)} - \frac{q^{-i}F_i(q)}{A(q)}\right) + F_i(q)y(k) \tag{77}$$

Rearranging the Diophantine equation (76)

$$E_i(q) = \frac{1}{A(q)} - \frac{q^{-i}F_i(q)}{A(q)} \tag{78}$$

Finally using (78) in (77), the usable form of the i-step ahead prediction formula, (79), is obtained.

$$\hat{y}(k+i\,|\,k) = E_i(q)y_{obf}(k+i) + F_i(q)y(k) \tag{79}$$

Note that in (79), there is no any denominator polynomial and hence no unstable pole. Therefore, the predictor is stable regardless of the presence of unstable poles in the OBF-ARX model. It should also be noted that, since $y_{obf}(k+i)$ is the output sequence of the simulation OBF model, once the OBF model parameters are determined its value depends only on the input sequence $u(k+i)$. Therefore, the i-step ahead prediction according to (79) depends on the input sequence up to instant $k+i$ and the output sequence up to instant k.

4.3 Closed–loop identification using OBF-ARMAX model

Consider the OBF model with ARMAX structure

$$y(k) = \frac{G_{OBF}(q)}{A(q)}u(k) + \frac{C(q)}{A(q)}e(k) \tag{80}$$

Rearranging (80)

$$\hat{y}(k\,|\,k-1) = G_{OBF}(q) - (1 - A(q))y(k) + (C(q) - 1)e(k) \tag{81}$$

With $A(q)$ and $C(q)$ monic, expanding (74)

$$\hat{y}(k \mid k-1) = l_1 u_{f1}(k) + l_2 u_{f2}(k) + ... + l_m u_{fm}(k) +$$
$$- a_1 y(k-1) - a_2 y(k-2) - ... - a_n y(k-n) + \quad (82)$$
$$c_1 e(k-1) + c_2 e(k-2) + ... + c_n e(k-n)$$

From (83) the regressor matrix is formulated for orders m, n, p

$$X = \begin{bmatrix} u_{f1}(mx) \ u_{f2}(mx-1)...u_{fm}(mx-m) & -y(mx-1) - y(nx-2)... - y(mx-n) \\ \cdot & \cdot & \cdot & \cdot & \cdot & \cdot & \cdot \\ \cdot & \cdot & \cdot & \cdot & \cdot & \cdot \\ \cdot & \cdot & \cdot & \cdot & \cdot & \cdot \\ u_{f1}(N) \ u_{f2}(N-1)... \ u_{fm}(N-m) & -y(N-1) \ -y(N-2)... - y(N-n) \end{bmatrix}$$

$$\begin{bmatrix} -e(mx-1) - e(mx-2)... - e(mx-p) \\ \cdot & \cdot & \cdot \\ \cdot & \cdot & \cdot \\ \cdot & \cdot & \cdot \\ -e(N-1) \ -e(N-2)... - e(N-p) \end{bmatrix} \quad (83)$$

where m = order of the OBF model

$\quad n$ = order of the $A(q)$

$\quad p$ = order of $C(q)$

$\quad mx = \max(n, m, p) + 1$

$\quad u_{fi}$ = input u filtered by the corresponding OBF filter f_i

$\quad e(i)$ = the prediction error

To develop an OBF-ARMAX model, first an OBF-ARX model with high $A(q)$ order is developed. The prediction error is estimated from this OBF-ARX model and used to form the regressor matrix (83). The parameters of the OBF-ARMAX model are, then, estimated using (83) in (27). The prediction error, and consequently the OBF-ARMAX parameters can be improved by estimating the parameters of the OBF-ARMAX model iteratively.

Multi-step ahead prediction using OBF-ARMAX model

A similar analysis to the OBF-ARX case leads to a multi-step ahead prediction relation given by

$$\hat{y}(k+i \mid k) = \frac{E_i(q)}{C(q)} y_{obf}(k+i) + \frac{F_i(q)}{C(q)} y(k) \quad (84)$$

where F_i and E_i are calculated by solving the Diophantine equation

$$\frac{C(q)}{A(q)} = E_i(q) + \frac{q^{-i} F_i(q)}{A(q)} \quad (85)$$

When OBF-ARMAX model is used for modeling open-loop unstable processes that are stabilized by a feedback controller, the common denominator $A(q)$ that contains the unstable pole does not appear in the predictor equation, (84). Therefore, the predictor is stable regardless of the presence of unstable poles in the OBF-ARMAX model, as long as the noise model is invertible. Invertiblity is required because $C(q)$ appears in the denominator. It should also be noted that, since $y_{obf}(k+i)$ is the output sequence of the OBF simulation model, once the OBF model parameters are determined its value depends only on the input sequence $u(k+i)$. Therefore, the i-step ahead prediction according to (84) depends on the input sequence up to instant $k+i$ and the output sequence only up to instant k.

5. Conclusion

OBF models have several characteristics that make them very promising for control relevant system identification compared to most classical linear models. They are parsimonious compared to most conventional linear structures. Their parameters can be easily calculated using linear least square method. They are consistent in their parameters for most practical open-loop identification problems. They can be used both for open-loop and closed-loop identifications. They are effective for modeling system with uncertain time delays. While the theory of linear OBF models seems getting matured, the current research direction is in OBF based non-linear system identification and their application in predictive control scenario.

6. Acknowledgement

We, the authors, would like to express our heartfelt appreciation for the financial and moral support we got from Universiti Teknologi PETRONAS to accomplish this task.

7. References

Akçay, H. (1999). Orthonormal basis functions for modelling continuous-time systems, *Signal Processing*, pp. 261–274.

Badwe, A. S. , Patwardhan, S. C. & Gudi, R. D. (2011). Closed-loop identification using direct approach and high order ARX/GOBF-ARX models, *Journal of Process Control*, Vol. 21, pp. 1056– 1071.

Box, G.E.P. , Jenkins, G.M. (1970). *Time Series Analysis: Forecasting and Control*, Holden-Day, San Francisco.

Camacho, E.F. & Bordon, C. (2004). *Model Predictive Control*, Springer Verlag Limited, London.

Dewilde, P., Vieira, A.C. & Kailath, T. On a generalised Szegö–Levinson realization algorithm for optimal linear predictors based on a network synthesis approach, *IEEE Transactions on Circuits and Systems,Vol.* CAS-25 , No. 9, 663–675.

Finn, C.K., Wahlberg, B. & Ydstie, B.E. (1993). Constrained predictive control using orthogonal expansions, *AIChE Journal*, vol. 39 pp. 1810–1826.

Gáspár, P., Szabó, Z. & Bokor, J. (1999). Closed-loop identification using generalized orthonormal basis functions, *Proceedings of the 38th Conference on Decision & Control*, Phoenix, Arizona USA, December 1999.

Heuberger, P. S. C. , Van den Hof, P. M. J. & Bosgra, O. H. (1995). A Generalized Orthonormal Basis for Linear DynamicalSystems, *IEEE Transactions On Automatic Control,* vol. 40, pp. 451-465.

Heuberger, P. S. C., Van den Hof, P. M. J. & Wahlberg, B. (2005). *Modeling and Identification with Rational Orthogonal Basis Functions,* Springer-Verlag Limited, London.

Lemma, D. T., Ramasamy (2011), M., Closed-loop identification of systems with uncertain time delays using ARX–OBF structure, *Journal of Process control,* Vol. 21, pp. 1148–1154.

Lemma, D.T. , Ramasamy, M. , Shuhaimi, M. (2011). Improved Method for Development of Parsimonious Otrthonormal Basis Filter Models, *Journal of Process Control, Vol* 21, pp. 36-45.

Lemma, D.T. , Ramasamy, M., & Shuhaimi, M. (2009). Closed Loop Identification Using Orthonormal Basis Filter (OBF) and Noise Models, *Proceedings of AIChE Annual Meeting,* Nasville, USA, November 2009.

Lemma, D.T. , Ramasamy, M., Patwardhan, S.C., Shuhaimi, M.(2010). Development of Box-Jenkins type time series models by combining conventional and orthonormal basis filter approaches, *Journal of Process Control, Vol.* 20, pp. 108–120.

Ljung, L. (1999). *System Identification: Theory for the User,* Prentice Hall PTR, New Jersey.

Merched, R. & Sayed, A. H.(2001). RLS-Laguerre lattice adaptive filtering: Error feedback, normalized, and array-based algorithms, *IEEE Transactions on Signal Processing,* vol. 49, pp. 2565–2576, 2001.

Nelles, O. (2001). *Nonlinear System Identification.* Springer-Verlag, Berlin Heidel Berg.

Ninness, B. M., and Gustafsson, F. (1997). A unifying construction of orthonormal bases for system identification, *IEEE Transactions on Automatic Control,* vol. 42, pp. 515– 521.

Patwardhan, S. C. , Manuja , S., Narasimhan , S.S. & Shah, L. (2006). From data to diagnosis and control using generalized orthonormal basis filters. Part II: Model predictive and fault tolerant control, Journal of Process Control, vol. 16, pp. 157-175.

Patwardhan, S. C., & Shah, S. L. (2005). From data to diagnosis and control using generalized orthonormal basis filters, Part I: Development of state observers, *Journal of Process Control,* vol. 15, pp. 819-835.

Rawlings, J.B. (2000). Tutorial overview of model predictive control. *IEEE Contr. Syst. Mag.* 38–52.

Takenaka, S. (1925). On the orthogonal functions and a new formula of interpolation, *Japanese Journal of Mathematics,* pp. 129–145.

Van den Hof, P.M.J. , Heuberger, P.S.C., Bokor, J. (1995). System identification with generalized orthonormal basis functions, *Automatica,* Vol. 31, pp. 1821–1834.

Van den Hof, P. M. J. , Walhberg, B., Heurberger, P. S. C., Ninness, B., Bokor, J., & Oliver e Silva, T. (2000). Modeling and identification with rational orthonormal basis functions, *Proceedings of IFAC SYSID,* Santa Barbara, California, 2000.

Wahlberg, B. (1991). System Identification using Laguerre filters, *IEEE Transactions on Automatic Control,* vol. 36, pp. 551-562.

Walsh, J. L. (1975). Interpolation and Approximation by Rational Functions in the Complex Domain, *American Mathematical Society Colloquium Publications,* vol. XX.

Wiener, N. (1949). *Extrapolation, Interpolation and Smoothing of Stationary Time Series*: M.I.T.-Press, Cambridge, MA 1949.

Permissions

The contributors of this book come from diverse backgrounds, making this book a truly international effort. This book will bring forth new frontiers with its revolutionizing research information and detailed analysis of the nascent developments around the world.

We would like to thank Prof. Ginalber Luiz de Oliveira Serra, for lending his expertise to make the book truly unique. He has played a crucial role in the development of this book. Without his invaluable contribution this book wouldn't have been possible. He has made vital efforts to compile up to date information on the varied aspects of this subject to make this book a valuable addition to the collection of many professionals and students.

This book was conceptualized with the vision of imparting up-to-date information and advanced data in this field. To ensure the same, a matchless editorial board was set up. Every individual on the board went through rigorous rounds of assessment to prove their worth. After which they invested a large part of their time researching and compiling the most relevant data for our readers. Conferences and sessions were held from time to time between the editorial board and the contributing authors to present the data in the most comprehensible form. The editorial team has worked tirelessly to provide valuable and valid information to help people across the globe.

Every chapter published in this book has been scrutinized by our experts. Their significance has been extensively debated. The topics covered herein carry significant findings which will fuel the growth of the discipline. They may even be implemented as practical applications or may be referred to as a beginning point for another development. Chapters in this book were first published by InTech; hereby published with permission under the Creative Commons Attribution License or equivalent.

The editorial board has been involved in producing this book since its inception. They have spent rigorous hours researching and exploring the diverse topics which have resulted in the successful publishing of this book. They have passed on their knowledge of decades through this book. To expedite this challenging task, the publisher supported the team at every step. A small team of assistant editors was also appointed to further simplify the editing procedure and attain best results for the readers.

Our editorial team has been hand-picked from every corner of the world. Their multi-ethnicity adds dynamic inputs to the discussions which result in innovative outcomes. These outcomes are then further discussed with the researchers and contributors who give their valuable feedback and opinion regarding the same. The feedback is then collaborated with the researches and they are edited in a comprehensive manner to aid the understanding of the subject.

Apart from the editorial board, the designing team has also invested a significant amount of their time in understanding the subject and creating the most relevant covers. They scrutinized every image to scout for the most suitable representation of the subject and create an appropriate cover for the book.

The publishing team has been involved in this book since its early stages. They were actively engaged in every process, be it collecting the data, connecting with the contributors or procuring relevant information. The team has been an ardent support to the editorial, designing and production team. Their endless efforts to recruit the best for this project, has resulted in the accomplishment of this book. They are a veteran in the field of academics and their pool of knowledge is as vast as their experience in printing. Their expertise and guidance has proved useful at every step. Their uncompromising quality standards have made this book an exceptional effort. Their encouragement from time to time has been an inspiration for everyone.

The publisher and the editorial board hope that this book will prove to be a valuable piece of knowledge for researchers, students, practitioners and scholars across the globe.

List of Contributors

Ginalber Luiz de Oliveira Serra
Federal Institute of Education, Science and Technology
Laboratory of Computational Intelligence Applied to Technology, São Luis, Maranhão, Brazil

L. F. S. Buzachero, E. Assunção, M. C. M. Teixeira and E. R. P. da Silva
FEIS - School of Electrical Engineering, UNESP, 15385-000, Ilha Solteira, São Paulo, Brazil

E. I. Mainardi Júnior, M. C. M. Teixeira, R. Cardim,M. R. Moreira and E. Assunção
UNESP - Univ Estadual Paulista

Victor L. Yoshimura
IFMT - Federal Institute of Education, Brazil

Javier Fernandez de Canete, Pablo del Saz-Orozco,Alfonso Garcia-Cerezo and Inmaculada Garcia-Moral
University of Malaga, Spain

Kyriakos G. Vamvoudakis
Center for Control, Dynamical-Systems, and Computation (CCDC), University of California, Santa Barbar

Frank L. Lewis
Automation and Robotics Research Institute, The University of Texas at Arlington, USA

Davi Leonardo de Souza
Department of Chemical Engineering and Statistics,Universidade Federal de São João Del-Rei

Fran Sérgio Lobato and Rubens Gedraite
School of Chemical Engineering, Universidade Federal de Uberlândia,Brazil

Víctor H. Andaluz
Universidad Técnica de Ambato, Facultad de Ingeniería en Sistemas, Electrónica e Industrial, Ecuador

Paulo Leica, Flavio Roberti, Marcos Toibero and Ricardo Carelli
Universidad Nacional de San Juan,Instituto de Automática, Argentina

Tomislav B. Šekara and Miroslav R. Mataušek
University of Belgrade/Faculty of Electrical Engineering, Serbia

Eduardo J. Adam
Chemical Engineering Department - Universidad Nacional del Litoral

Alejandro H. González
INTEC (CONICET and Universidad Nacional del Litoral), Argentina

Rodrigo Alvite Romano
Instituto Mauá de Tecnologia - IMT

Alain Segundo Potts and Claudio Garcia
Escola Politécnica da Universidade de São Paulo – EPUSP, Brazil

Eric William Zurita-Bustamante
Universidad del Istmo

Jesús Linares-Flores, Enrique Guzmán-Ramírez and Hebertt Sira-Ramírez
Universidad Tecnológica de la Mixteca, México

Lemma D. Tufa and M. Ramasamy
Chemical Engineering Department, Universiti Teknologi PETRONAS, Bandar Seri Iskandar, Perak, Malaysia

Printed in the USA
CPSIA information can be obtained
at www.ICGtesting.com
JSHW011454221024
72173JS00005B/1072

9 781632 380135